AUDEL®

Complete Roofing Handbook

Installation
Maintenance
Repair

by James E. Brumbaugh
revised by John Leeke

An Audel® Book

SECOND EDITION

Production services by the Walsh Group, Yarmouth, ME.

Library of Congress Cataloging-in-Publication Data

Brumbaugh, James E.
    Complete roofing handbook : installation, maintenance, repair / by James E. Brumbaugh.—2nd ed. / revised by John Leeke.
        p.   cm.
    "An Audel book."
    Includes index.
    ISBN 0–02–517851–2
    1. Roofing.   2. Roofs—Maintenance and repair.   I. Leeke, John.
II. Title.
TH2431.B78   1992
695—dc20   91–43502
                                                                            CIP

10   9   8   7   6   5   4   3   2   1

Contents

Roofs • Wood Shake Roofs • Shingle and Shake Panel Roofs •
Reroofing

Acknowledgments

The knowledge of the following professionals who helped with this revision is matched only by their willingness to share what they have learned through costly research and diligent effort.

Thanks to Gene Leger, Leger Designs, New Boston, New Hampshire for his contribution on attic ventilation; Kay Weeks and Sharon C. Parks of the National Parks Service for help with photos in the historic roofing chapters.

Foreword

The goal of this book is to help you expand your knowledge of roofing. Knowledge is made up of two parts: information and experience. This Audel *Complete Roofing Handbook* gives you a foundation of information. As you work at roofing, you gain the experience that builds your knowledge of the trade.

By studying this book and using the reference section the amateur and do-it-yourselfer will find enough information to get into and back out of a roofing project successfully.

Those working in construction trades other than roofing will find this book useful as an education in a wide range of roofing types that will help them coordinate their own work with roofing operations.

You will find this book useful even if you are not intending to do any roofing yourself. One of the key requirements for the project manager, whether a homeowner or professional, is to judge the quality and completeness of work done by others. This book provides detailed information on materials and methods that will give an informed view of roofing.

Keeping up with changes in roofing is essential as new products are being developed constantly. Any single new product may be a good one, but in combination with another, it may cause problems. You will not necessarily learn about these problems from the manufacturer of either product. You can learn about them through independent sources of late breaking information such as trade journals. See the Reference section which includes a listing of publications, associations, manufacturers, and suppliers.

This book covers single family unit residential roofing. Multifamily and commercial structures require more strict fire and safety details that are not included here.

Two new chapters on roofing for historic buildings cover general considerations in determining historic character and value, historic and alternative materials, with a focus on the replacement, repair, and maintenance of historic wooden shingle roofs.

In addition, the following topics are updated or added:

- Roofing equipment, with additional safety rules for the use of ladders and scaffolding;
- Attic ventilation—a completely new chapter has been added on this important subject, reflecting current trends and thinking.

The editor for this revision, John Leeke, is an experienced writer as well as a practicing tradesman and contractor. His main focus is the preservation of historic architecture which includes solving many roofing problems. Leeke has written many articles on restoration and roofing and is a contributing editor for professional journals and books. This broad experience coupled with that of other professionals, provides a wide range of top-notch resources, many of which are referenced at the end of this book.

A Note from the Revisions Editor:

Too often roofing decisions are based on materials alone. The skills and methods used in their application have equal, if not greater, importance. Neither the best materials installed poorly nor poor materials installed well result in success. Long lasting roofs that perform well throughout their life are the result of both good materials and good installation and maintenance methods.

A knowledgeable approach to the selection of materials and methods is appropriate because of their expected lifespan. An error in judgment now can seriously lower the future value of the major investment made in a building.

There are too many examples of failures in new materials and methods to follow a policy of always using the newest, assuming it is the best.

Of course, it is the introduction of new products and methods that holds the promise for advancements in roofing. If you are compelled to use new materials, research their application and performance through the manufacturer as well as independent sources of information. Use them on small jobs you can monitor closely over the years to determine their performance and effectiveness. Then use the results of your observations to decide where the new materials and methods can best be used, or if they should be used at all.

For critical projects always use materials and methods that have stood the test of time. Is the roof of any building less than critical?

This conservative approach reflects my own traditional bias. Many of the roofing lessons I have learned from historic buildings are applied to modern roofing in this book.

Please feel free to write to me if you have any comment on this current edition. Your ideas and remarks will be used as the next revision is developed.

John Leeke
RR1 Box 2947
Sanford, Maine 04073

CHAPTER I

Types of Roofs

Roofs are constructed in a wide variety of different sizes, shapes, and designs, and there are a number of different ways of classifying them. The three most common methods of classification are based on roof design, roof construction, and the type of roofing material used to cover them. Roofing materials are described in the next chapter.

Roof Design

The easiest way to classify a roof is by its architectural design. The principal types of roof designs are illustrated in Figs. 1–1 through 1–3. A *gable roof* is composed of two sloping surfaces that meet and join together along a common ridge line (Fig. 1–1A). The triangular end wall formed at each end of the roof is called a gable. The degree of slope or pitch will vary on different roofs.

A roof that slopes away in *four* directions from a common ridge line is called a *hip roof* (Fig. 1–1B). The so-called hip is found where a gable would be located on a gable roof. Sometimes a hip roof is used to cover a wing that joins the main portion of the structure covered by a gable roof. This type of roof is frequently called a *gable-and-hip roof.*

Some roofs are designed with a change of slope or double pitch.

1

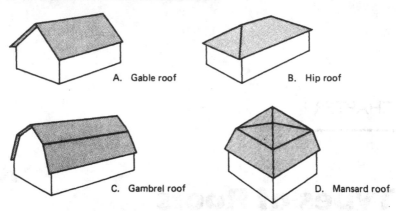

A. Gable roof   B. Hip roof

C. Gambrel roof   D. Mansard roof

Fig. 1–1. Gable, hip, gambrel, and mansard roofs.

The *gambrel roof* roughly resembles the gable roof except that the gable formed at each end has a pentagonal shape (Fig. 1–1C).

The *mansard roof* design was first introduced in France by the architect François Mansart (1598–1666). It enjoyed initial popularity in the United States in areas of French settlement, particularly in and around New Orleans. During the nineteenth century it spread across the country with the French empire and Victorian styles of architecture. During the early- and mid-twentieth century it was ignored by American architects and builders. It is now experiencing a revival in popularity, especially in apartment construction.

On a mansard roof, a change of slope or double pitch occurs on all four sides of the roof (Fig. 1–1D). The lower slope is steeper than the upper one and has a pitch that is sometimes almost vertical. A mansard roof differs from a gambrel roof by having a nearly flat top or deck, which is usually covered by built-up roofing materials.

In a *pyramidal roof* the four sloping surfaces meet at a central point (Fig. 1–2A). Examples of this type of roof design are found in Victorian architecture where they are used to cover towers. Church steeple roofs provide other examples. The *conical roof* is used to cover round towers and silos.

The *flat roof* is as common on commercial and industrial structures as the gable and hip roofs are in residential construction (Fig. 1–3A). Usually flat roofs on residential buildings are built to slope slightly in

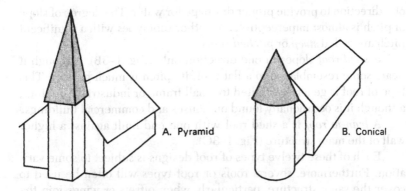

Fig. 1–2. Pyramid and conical roofs.

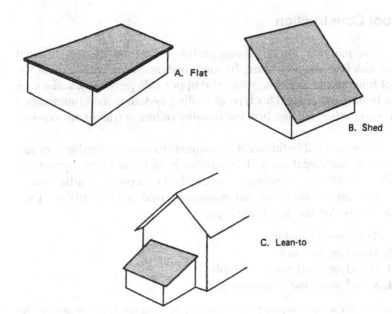

Fig. 1–3. Flat, shed, and lean-to roofs.

one direction to provide proper drainage for water. The degree of slope of pitch is almost imperceptible. All other roof types with a significant pitch are called *steep* or *pitched roofs*.

A *shed roof* slopes in one direction only (Fig. 1–3B). Although it bears some resemblance to a flat roof, its pitch is much greater. This type of roof is generally limited to small frame or industrial buildings, although it is occasionally found on houses and commercial buildings.

A *lean-to roof* is a shed roof with one end built against a higher wall of the main structure (Fig. 1–3C).

Each of these twelve types of roof designs is subject to some variation. Furthermore, several roofs or roof types will often be used to cover the same structure, particularly when offsets or wings join the main part of the structure.

The term *major roof* refers to the main roof covering the largest section of the structure. The term *minor roof*, on the other hand, designates any roof that covers a smaller section of the structure and intersects with the major roof to form a roof valley.

Roof Construction

The roof of a structure forms an integral part of an assembly that may also include the ceiling. In some instances, the underside of the roof functions as a ceiling, while the upper side provides a surface for the roofing materials. This type of roofing system is sometimes called an *open timber roof*, because the framing rafters or trusses are exposed to view.

In residential buildings it is common to join the sloping roof rafters with horizontal joists. The underside of these joists support the ceiling, and the upper side may be finished to serve as an attic floor.

Several different roof and ceiling assemblies are available. They may be divided into four basic types.

1. Truss roof assembly
2. Wood joist assembly
3. Wood joist and rafter assembly
4. Wood plank and beam assembly

Each of these four types of roof and ceiling assemblies enjoys widespread use in residential construction. Both the truss and wood joist and rafter type assemblies provide a flat ceiling surface for the rooms or spaces below. Because this type of construction seals off an area immediately below the roof, adequate ventilation must be provided to prevent the build-up of condensation and excessive heat in these spaces.

One advantage of the wood-joist and rafter-type construction is that is creates an open area or attic that provides additional storage or living space. Truss-type construction limits use of the space because of the bracing between members.

The surface on which the roofing materials are laid is called the *roof deck*. The roof deck can be made of either wood or nonwood materials. Most pitched roof decks found on houses are constructed of wood boards or plywood panel sheathing.

The roof deck is either flat or slopes downward from both sides of a ridge line. A flat or low-pitched roof is level or almost level, and the frame members supporting the roof deck also serve as ceiling joists in most cases. Flat or low-pitched roofs are said to be of "single roof" construction. A *pitched roof*, which consists of rafters tied together by ceiling joists, has an intermediate or steep slope. The area enclosed by the rafters and joists forms an attic or attic crawl space.

Pitched Roof Framing

The deck of a pitched roof is supported primarily by common rafters and jack rafters (Fig. 1–4). The *common rafter* runs square with the wall plate and extends to the *ridge* or *ridge board*. The common rafters are the longest and certainly the most numerous rafters on gable, gambrel, and hip roofs.

A *hip rafter* is used on a hip roof, and extends from the outside angle of the wall plate toward the apex of the roof (Fig. 1–5). A rafter that runs square with the wall plate and intersects the hip rafter is called a *jack rafter*. Jack rafters never extend the entire distance from the wall plate to the ridge or ridgeboard. Jack rafters can be divided into hip jacks, valley jacks, and cripple jacks.

A *hip jack* is a jack rafter that runs from the wall plate to a hip rafter. A jack rafter that extends from the ridgeboard to a valley rafter is called a *valley jack*. A *cripple jack* or *cripple rafter* cuts between a valley and a hip rafter.

A structure with a wing added to it will have a roof system that contains a valley where the minor roof (covering the wing) joins the major or main roof. A *valley* is the internal angle formed by the two

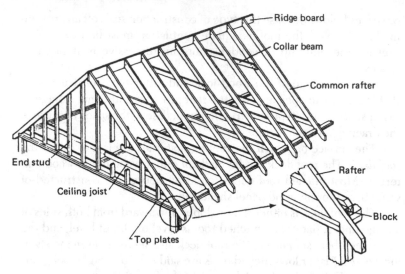

Fig. 1–4. Gable roof framing with rafter and joist assembly.

slopes of the intersecting roofs. A *valley rafter* extends from an inside angle of a wall plate toward the ridge or center line of the structure, and defines the angle of the roof valley (Fig. 1–6).

Rafters are nailed to a *wall plate* or *plate*, which generally consists of two 2 × 4s running horizontally across the top of the wall studs. The roof rafters extend a short distance beyond the wall plate on almost all pitched roofs forming an overhang or "eave" to permit water drainage from the roof surface at a suitable distance from the exterior walls and foundation. The portion of the rafter extending beyond the outer edge of the wall plate is called the *eave* or *tail* of the rafter.

A roof rafter is generally cut near its lower end to fit down on the wall plate. This is referred to as a *seat, bottom,* or *heel cut.* A *ridge, top,* or *plumb cut* is made at the other end of the rafter to allow it to fit against the ridgeboard or, when there is no ridgeboard, against an opposing rafter. Finally, a *side* or *cheek cut* is a bevel cut on the side of a rafter to fit it against another frame member. The various types of rafter cuts are illustrated in Fig. 1–7.

Some roof framing terms are used specifically for making layout

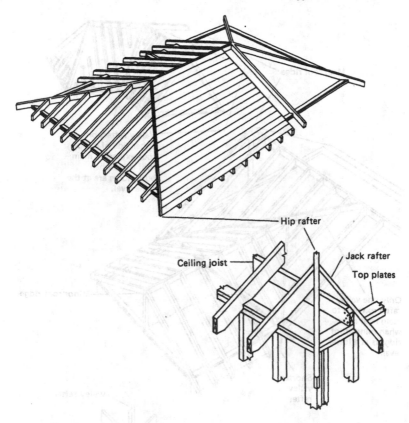

Fig. 1–5. Hip roof framing.

measurements, for example, in determining rafter length or the angle of rafter cut. *Pitch* and *slope* are two of the most commonly used terms belonging to this category of roof-framing terminology. The two terms are often, but incorrectly, used synonymously. Both terms may be used in a general sense to refer to the angle or incline which the roof surface makes with a horizontal plane, but each term more specifically describes a distinctly different mathematical relationship between the total rise and the span of a roof.

The horizontal (level) distance over which the roof rafter passes is

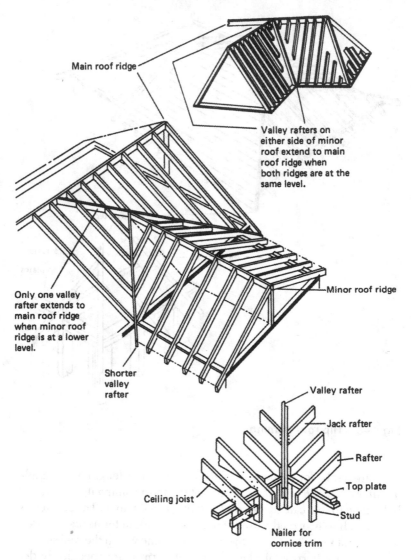

Main roof ridge

Valley rafters on either side of minor roof extend to main roof ridge when both ridges are at the same level.

Only one valley rafter extends to main roof ridge when minor roof ridge is at a lower level.

Minor roof ridge

Shorter valley rafter

Valley rafter

Jack rafter

Rafter

Top plate

Stud

Ceiling joist

Nailer for cornice trim

Fig. 1–6. Valley framing.

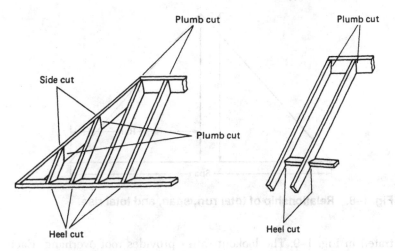

Fig. 1-7. Various types of rafter cuts.

called the *total run* or *horizontal run.* It represents the area covered by the rafter from the outer edge of the wall plate to a vertical line extending down from the exact middle of the ridgeboard, and really has nothing to do with the actual length of the rafter. The total run of a common rafter is one half the width of the structure.

The *span* of the roof is double the total run, and represents the exact width of the structure between the outer edges of opposing wall plates. The *total rise* or *vertical rise,* on the other hand, is the distance extending vertically from the plate line to the top of the ridgeboard. The relationships of the terms run, span, and total rise are illustrated in Fig. 1-8. These relationships determine the "pitch" or "slope" of the roof. Both pitch and slope are explained in Chapter 2 because they are determining factors in the type of roofing materials used on the roof.

Flat Roof Framing

The flat or low-pitched roof is usually constructed with a slight slope or pitch to improve drainage. The frame members that support the roof deck and also serve as ceiling joists are sometimes called *roof joists.*

Construction details of flat or low-pitched roof framing are illus-

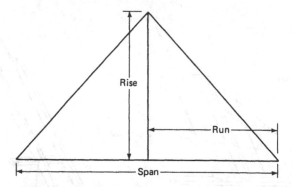

Fig. 1–8. Relationship of total run, span, and total rise.

trated in Fig. 1–9. The lookout rafter provides roof overhang. Each *lookout rafter* is toenailed to the wall plate and nailed at the other end to a double header. The distance from the double header to the wall line is generally twice the overhang measurement. A nailing header for securing the soffit and fascia boards is sometimes nailed to the ends of the lookout rafters.

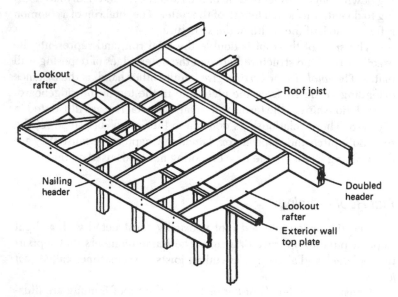

Fig. 1–9. Flat roof framing.

CHAPTER 2

Roofing and Reroofing

A properly constructed roof should last from 5 to over 50 years, depending on weather conditions and the materials used in its construction. As a general rule, the popular asphalt shingle roof has a service life of 15 to 20 years. A roof made of roll roofing lasts about 5 to 10 years. Both slate and tile have a service life of about 50 years or more.

A roof requires a certain amount of regular maintenance and repair. If the roof has deteriorated to a point where simple repairs prove ineffective, it will have to be reroofed or replaced. These two tasks are referred to as *roofing* and *reroofing*.

Roofing is a specialized task in building construction that involves the application of roofing materials to the roof deck. This can occur in new construction or after the existing (but deteriorated) roofing materials have been stripped from the roof deck. Because of its specialized nature, roofing is usually subcontracted out by the building contractor to specialists in this type of work. As a result, professional roofers and roofing companies can be found in most localities.

Reroofing involves the application of new roofing materials on an existing roof and might include removing the existing roofing or leaving it in place. Depending on the type of roofing material used, reroofing is a task that can often be handled by those without special skills or training. Professional roofers are also available for this type of work.

Roofing Terminology

Certain roofing terms need to be explained because they are frequently encountered in building plans, architectural specifications, and literature from roofing material manufacturers. These terms are also used in the descriptions of the roofing and reroofing methods described in this book.

A knowledge of the pitch or slope of a roof is important in determining the type and application method of the shingle or roofing material that is to be used. Although the terms *pitch* and *slope* both refer to the incline of the roof and are sometimes used synonymously in roofing literature, they are calculated and expressed differently.

Pitch may be defined as the ratio of the rise to total span (or twice total run) and is expressed as a fraction (Fig. 2–1). The fraction used to designate pitch becomes larger as the steepness of the roof increases. Thus, a roof with a pitch of ⅓ is steeper than one with a pitch of ⅙.

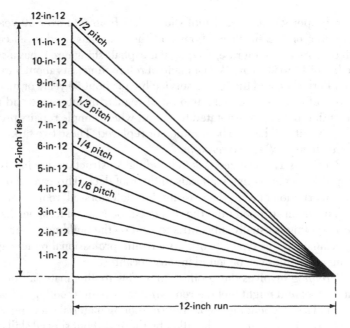

Fig. 2–1. Roof slopes and pitches.

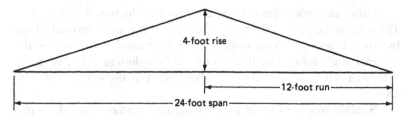

Fig. 2–2. Low-pitched roof.

Roof pitch is calculated by dividing span by rise. For example, a roof with a rise of 4 feet and a span of 24 feet has a pitch of ⁴⁄₂₄ or ⅙ (Fig. 2–2). This is a relatively low-pitched roof an limits the roofing material to roll roofing and various types of hot or cold process built-up roofing. Most other types of roofing materials require a roof with a pitch greater than ⅙. A roof with a relatively steep pitch is shown in Fig. 2–3. Its rise of 8 feet and its span of 24 feet give it a pitch of ⅓ (⁸⁄₂₄ reduced to ⅓).

In contrast to its pitch, the slope of a roof indicates the ratio of the rise *to the run* (instead of the span) and is expressed as *x* number of inches per foot of run. For example, the rise of the roof shown in Fig. 2–3 is 8 feet and the run is 12 feet (one half the span dimension). Slope equals a ratio of 8 feet to 12 feet, or 2 feet of rise for every 3 linear feet. In inches, that equals ²⁴⁄₃₆, which reduces to ⁸⁄₁₂. In other words, the roof shown in Fig. 2–3 rises 8 inches per foot of horizontal run. The slope for this roof is expressed as "8-in-12."

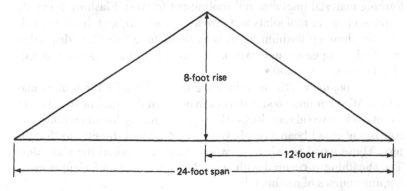

Fig. 2–3. Roof with steep pitch.

Other commonly used roofing terms are illustrated in Fig. 2–4. The *ridge* is the peak of the roof or the apex of the angle formed where two roof slopes meet A *hip ridge,* on the other hand, is the apex of the external angle formed by the meeting of two sloping sides. Both the main roof ridge and hip ridge must be capped with specially cut shingles.

A *gable roof* consists of two sloping roof surfaces joined along a horizontal ridge line. The triangular-shaped endwall formed at each end of the roof is called a *gable*. The roof *rake* is the inclined edge of a gable roof at the endwalls. Each end of a gable roof will have two rakes that incline toward and meet at the ridge. In shingling, it is common to begin at the left rake of the roof and work toward the opposite or right rake.

The roof *eave* is the lower edge of the roof slope that overhangs and extends beyond the face of the exterior wall. Shingle courses are laid to run parallel to the eaves. The ends of the roof rafters at the eaves are covered by a length of trim called the *fascia, fascia board,* or, in this case, an *eave fascia*. The same type of trim attached to a rake is called a *rake fascia*. The underside of the rafters between the eave and the exterior wall may be left open or covered by a *soffit*.

A gable roof rises by inclined planes from two sides of the structure. A *hip roof,* on the other hand, rises by inclined planes from all four sides. As a result, a hip roof has at least four eaves, but no rakes.

The internal angle formed by the junction of two sloping sides of a roof is called a *valley*. Because a roof valley serves as a major channel for water runoff, the valley joint must be made watertight by installing *flashing* material (metal or roll roofing cut to size). Flashing prevents water seepage at roof joints and provides more efficient drainage. Valley and chimney flashing serve both these functions. The drip edge installed along eaves and rakes is an example of flashing used to provide better water runoff.

The *roof deck* is the foundation or nailing base for the roofing materials. Wood or nonwood materials may be used in the construction of a roof deck. A wood roof deck—the type commonly found on houses—consists of wood boards or plywood panels nailed directly to the rafters. These wood boards or plywood panels are called the *sheathing*. The sheathing is covered with an *underlayment* (or *underlay*) of overlapping courses of roofing felt.

Some roofing terms refer to particular aspects of shingling. For example, a shingle *course* is a single row of shingles running the entire

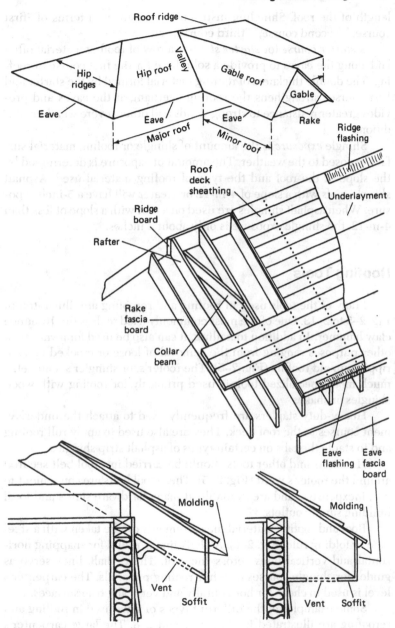

Fig. 2–4. Common roofing and roof construction terminology.

length of the roof. Shingling instructions are given in terms of "first course," "second course," "third course," etc.

A *starter course* (or *starter strip*) is a row of roofing material often laid along the eaves to provide a solid base for the first course of roofing. The double thickness of roofing material formed by the starter and first courses strengthens the roofing overhang at the eaves and provides greater resistance to strong winds and other severe weather conditions.

Shingle *exposure* is the amount of shingle or roofing material surface exposed to the weather. The amount of exposure is determined by the slope of the roof and the type of roofing material used. Asphalt shingle roofs with a slope of 4-in-12 or greater will have a 5-inch exposure. When asphalt shingles are used on a roof with a slope of less than 4-in-12, the shingle exposure is only about 3 inches.

Roofing Tools

Many of the tools used in roofing and reroofing are illustrated in Fig. 2–5 to 2–13. One of the most commonly used tools is the 16-ounce claw hammer. In addition to nailing, it can also be used for a variety of other purposes ranging from the removal of loose or crooked nails to ripping up old roofing (Fig. 2–5). The roofer's, or shingler's, hatchet, a much more specialized tool, is used primarily for roofing with wood shingles or shakes.

Heavy-duty staplers are frequently used to attach the underlayment courses to the roof deck. They are also used to apply roll roofing and to staple the tabs on certain types of asphalt strip shingles.

Hammers and other tools should be carried in a tool belt secured around the roofer's waist (Fig. 2–5). These tool belts are convenient to use, inexpensive, and are available at many hardware stores and local building supply outlets.

Roof and roofing material measurements can be taken with a steel tape or folding rule (Fig. 2–6). A chalk line is useful for snapping horizontal and vertical lines across the roof. These chalk lines serve as guides for shingle courses or other roofing materials. The carpenter's level is used to check for horizontally and vertically true surfaces.

Some examples of the different types of saws used in roofing and reroofing are illustrated in Figs. 2–7 and 2–8. The large carpenter's

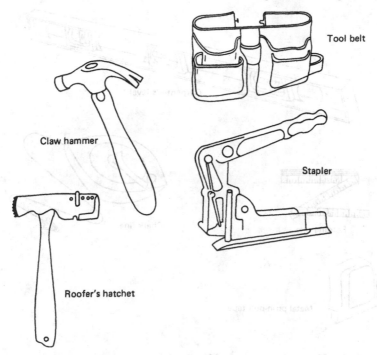

Fig. 2–5. Roofer's hatchet, claw hammer, stapler, and tool belt.

hand saw is used for heavy-duty work such as cutting plywood or wood board sheathing. The keyhole saw cuts on the pull-stroke to prevent the buckling of the blade. It is often used to cut small circular holes for vent pipes or stacks because it can enter slots or holes easily. Both the keyhole saw and back saw can be used for making repairs to wood shingle and shake roofs. The hacksaw is useful for starting cuts in metal roofing, cutting metal flashing, or cutting off nails when removing roofing. The circular power saw shown in Fig. 2–8 is recommended for cutting metal roofing panels, concrete and clay tile, and plywood or wood board sheathing.

A roofing knife is useful for cutting shingles, roll roofing, or other asphalt-based roof covering materials (Fig. 2–9). A utility knife may also be used for this purpose. Tin snips can be used to cut metal flashing and thinner gauges of metal or vinyl roofing. The shingle ripper,

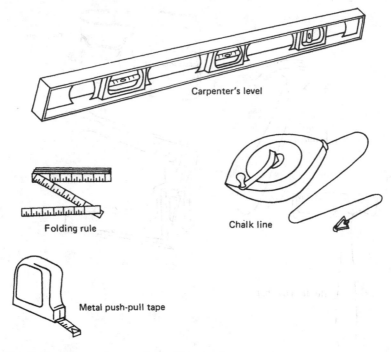

Fig. 2–6. Measuring and marking tools.

another cutting tool common to roofing, is used to "rip" or cut nails holding down individual shingles, tile, or slate. It is a useful tool for roof repairs.

A pry bar is essentially a reroofing tool. It is used to remove nails and pry loose shingles, shakes, tiles, or slates when preparing the roof for new roofing materials. A caulking gun is used to apply beads of caulk along flashing joints and other points where different materials meet.

A putty knife can be used for applying sealant or other materials of a consistency too thick and heavy to brush onto the surface. When it is necessary to insert flashing into the wall or chimney joints, the mortar can be removed with a homemade gouging tool.

Roof coatings, asphalt adhesive cements, masonry primers, and other roofing materials are generally of a thin enough consistency to be applied with a brush, mop, broom, or spray gun. Only inexpensive brushes, mops, or brooms should be purchased for this type of work,

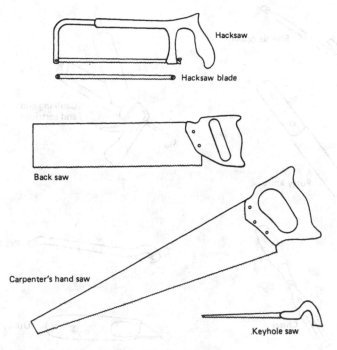

Hacksaw

Hacksaw blade

Back saw

Carpenter's hand saw

Keyhole saw

Fig. 2–7.   Saws commonly used in roofing.

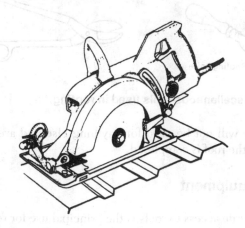

Fig. 2–8.   Power circular saw.

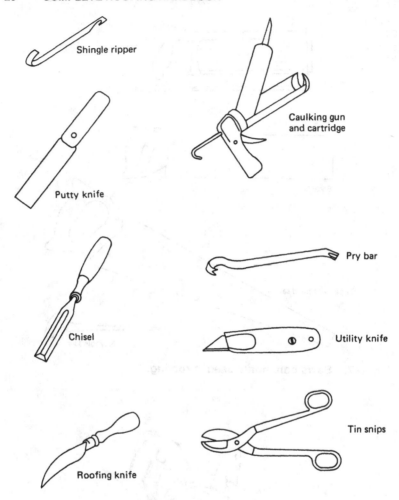

Fig. 2–9. Miscellaneous tools used in roofing.

because they will not be used for anything else and are usually discarded after the roof is completed.

Roofing Equipment

Gaining safe access to roofs is the principal use for roofing equipment. Ladders, roof brackets and scaffolding are commonly used. Accidents related to ladders and scaffolding are a major source of injuries and fatalities among construction workers. Following safety guidelines provides by equipment manufacturers will prevent many accidents.

Equipment can be purchased through local building supply outlets or by mail order from specialty roofing suppliers. If your need for equipment is infrequent, consider renting from a tool and equipment rental store. There are companies that will erect scaffolding for you.

Portable Ladders

Single length, extension, and step ladders can be moved easily around the work site as needed. Because the conditions surrounding their use can be so varied you must use them with great care. Following are guidelines for the setup, use, and maintenance of ladders.

Selection and Location

- Only use a ladder for the purpose for which it was designed. For example, do not use a ladder for a horizontal plank between pump jacks.
- Use a ladder with nonconductive side rails if the ladder could contact exposed energized electrical equipment.
- If ladders are set where they can be displaced by workplace activities or traffic, they must be secured to prevent accidental movement. If they must be placed in passageways, doorways, or driveways, and cannot be secured, erect a barricade to keep traffic or activities away from the ladder.
- Ladders must not be tied or fastened together to create longer sections unless they are specifically designed for such use.
- When portable ladders are used for access to an upper landing surface, the side rails must extend at least 3 feet about the upper landing.
- Always carry a ladder in a horizontal position to avoid contact with overhead electrical wires and equipment.

Setup

- To raise a ladder, place the bottom end of the ladder against the wall and "walk" it up until it is in a vertical position (Fig. 2–10). Lift the ladder and move it away from the wall a distance approximately one-fourth its length.

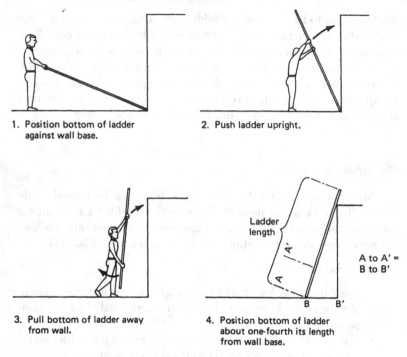

1. Position bottom of ladder against wall base.

2. Push ladder upright.

3. Pull bottom of ladder away from wall.

4. Position bottom of ladder about one-fourth its length from wall base.

Ladder length

A to A' = B to B'

Fig. 2–10. Raising and positioning a ladder.

- Position the ladder so it is absolutely vertical, and not leaning left or right. Adjust the position so the feet rest solidly and evenly on the ground and so the rails both meet the edge of the roof above.
- Climb the ladder part way. If the ladder wobbles, get down immediately and reposition it until it is stable. Then climb the ladder and fasten it at the top if needed. Ladders can be fastened by lashing with ropes or with hardware brackets (Fig. 2–11).
- Set ladders only on a stable and level surface, unless secured, to prevent accidental movement.
- The top of the ladder must be secured to the edge of the landing above.
- If a ladder must sit on a slippery surface, it should have slip-resistant feet to prevent accidental movement. Slip-resistant

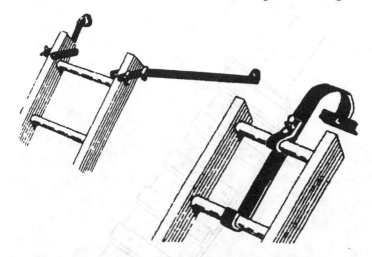

Fig. 2–11. Ladders arms (left) hold the ladder away from the wall. A hook attachment (right) for a ladder top. *(Courtesy Roofmaster Products Co.)*

feet must not be used as a substitute for care in placing, lashing, or holding a ladder on slippery surfaces.

- If the two top rails of a ladder are not supported equally, equip the ladder with a single support attachment.
- When raising a ladder, do not slam it against the gutter. It is easy to damage a gutter in this way.

Use

- When ascending or descending a ladder, face the ladder. Grasp the ladder with at least one hand when moving up or down the ladder. Never lean from a ladder, because unevenly placed weight may cause it to fall. Keep your hips between the two rails as you climb and use the ladder.
- A crawling board or so-called "chicken" ladder (Fig. 2–12) is often used on roof of tile, slate, or other brittle and easily breakable roofing materials, because it spreads your weight more effectively. These job-made ladders typically are built of 1 × 10- or 1 × 12-inch boards to which 1 × 2-inch battens are nailed. A chicken ladder is hooked over the roof ridge and should extend from the ridge to the eave.

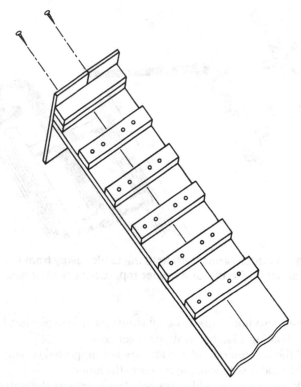

Fig. 2–12. Crawling board or chicken ladder.

- Fasten a lifeline of ¾-inch diameter rope beside each crawling board for a handhold.
- An ordinary ladder can be used in the same way with wood brackets (Fig. 2–13) or a metal hook attachment (Fig. 2–11). The cross-sections of ladder rungs are typically round or "D" shaped. Use a metal hook attachment that matches the cross-section of the rungs.
- Do not load a ladder beyond its rated capacity.
- Be cautious of ladder deflection under a load which could cause the ladder to slip off its support.
- Keep the areas around the tops and bottoms of ladders clear.
- Do not move, shift, or extend a ladder while someone is on it.

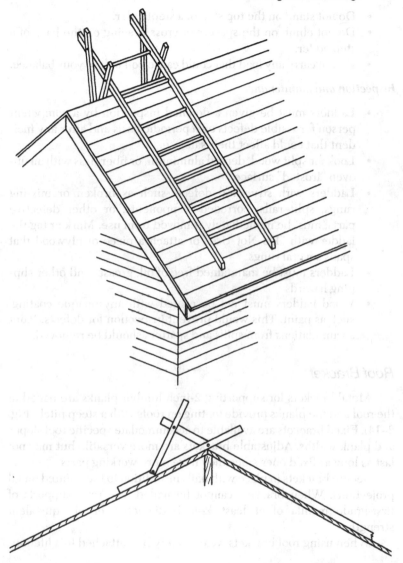

Fig. 2–13. Ladder with braces or supports attached to the rails and set to the angle of the roof.

- Do not stand on the top step of a stepladder.
- Do not climb on the spacers or cross bracing on the back of a stepladder.
- Do not carry any load that could cause you to lose your balance.

Inspection and maintenance

- Ladders must be given a detailed inspection by a competent person for visible defects on a periodic basis and after any incident that could affect their safe use.
- Look for split wood, dented aluminum, or fiberglass with an uneven "frosted" surface.
- Ladders with structural defects such as broken or missing rungs, split rails, corroded components, or other defective parts, must be immediately removed from use. Mark or tag the ladder with "Do Not Use" or attach a piece of plywood that spans several rungs.
- Ladders must be maintained free of oil, grease, and other slipping hazards.
- Wood ladders must not be coated with any opaque coating, such as paint. This would prevent inspection for defects. Paint accumulations from spills and spatters should be removed.

Roof Brackets

Metal brackets for supporting 2-inch lumber planks are nailed to the roof and the planks provide footing on roofs with a steep pitch (Fig. 2–14). Fixed brackets are available to accommodate specific roof slopes and plank widths. Adjustable brackets are more versatile but may not last as long as fixed ones since they have more working parts.

Secure brackets in place with nails in addition to the pointed metal projections. When brackets cannot be nailed, use rope supports of first-grade manila of at least ¾-inch diameter, or of equivalent strength.

When using roof brackets wear a safety belt attached to a lifeline.

Scaffolding

Scaffolding is a temporary structure that provides a safe surface in high places for holding materials and workers. The principal advantage

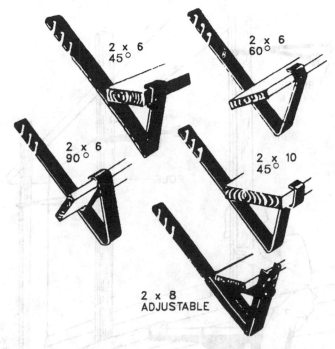

Fig. 2–14. Roof brackets. *(Courtesy Roofmaster Products Co.)*

is that the scaffold does not lean against any part of the building, thus eliminating possible damage to the building.

When using scaffolding, hoist materials up and down with ropes. Do not hand and drop materials.

Attach a "tag line" to control the movement of hoisted loads. Do not allow materials and debris to accumulate and cause a hazard. Do not work on scaffolds during storms or high winds.

Pump jacks—Pump jacks are a combination of special metal brackets and braces with vertical wood poles and horizontal wood or fabricated planks. They set up quickly and provide stable access to the edge of a roof (Fig. 2–15). Once set up, each jack is pumped by foot-power to raise the jack and planks.

The working load, which includes workers and materials, cannot exceed 500 pounds.

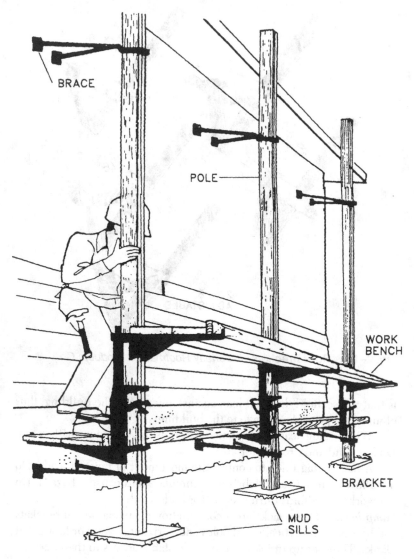

Fig. 2–15. Pump jack scaffolding setup. *(Courtesy Roofmaster Products Co.)*

Fig. 2–16. Pump jack bracket. *(Courtesy Roofmaster) Products Co.*

Poles are made up of 2 × 4 lumber with straight grain, free of defects such as cross grain, shakes, and large knots that could impair strength. Spike the 2 × 4s together with tenpenny (10d) common nails no more than 12 inches on center. Stagger butt joints of the 2 × 4s at least 6 feet. Do not locate the joint near a defect in the neighboring 2 × 4. Poles should not exceed 30 feet in height. Pole bottoms should bear on mud sills made of planks to prevent settling. Do not use small chunks of wood, bricks, or concrete blocks for mud sills.

When assembling the poles and brackets make the seam of the poles parallel to the bracket.

Pump jack brackets should have two positive gripping mechanisms to prevent failure or slipping.

Secure the poles to the building with ridged triangular braces (Fig. 2–17). Place braces at the top, bottom, and other points as necessary to provide a maximum vertical spacing of no more than 10 feet between braces.

When wood scaffold planks are used, space the poles no more than 10 feet center to center. Completely fill the width of the bracket with planks. Wood planks should be rated and stamped for scaffolding use. When fabricated planks with a rated span of more than 10 feet are used, pole spacing may exceed 10 feet. Lap planks over the bracket at least 8 inches but not more than 12 inches. Nail or screw cleats across

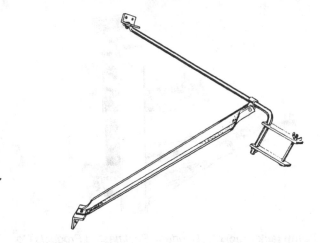

Fig. 2–17. Pump jack brace. *(Courtesy Roofmaster Products Co.)*

the ends of the planks on the bottom surface to prevent the plank from sliding off the bracket.

When raising the jacks and planks, keep the jacks even in height. Install an extra brace within four feet when passing a brace while raising the jacks.

No more than two workers are allowed between any two poles. Guardrails or an integral workbench are needed unless workers wear safety belts with lifelines. Do not stand on the workbench, it is for tools and materials only.

Always follow the bracket and brace manufacturer's recommendations for assembly, use, and load limits.

Roofing Materials

Roofing materials may be grouped in the following categories: (1) surface materials, (2) underlayment, (3) flashing, (4) fasteners, (5) roofing cements, (6) roof coatings, and (7) roofing tape.

Surface Materials

Roof surface materials form the top layer or external cover of the roof. They are also sometimes called *roofing* or *roof covering materials*. These materials do not include the underlayment or flashing. The principal types of roof covering materials are asphalt shingles, asphalt roll roofing, wood shingles and shakes, tile, slate, metal roofing, and mineral fiber shingles.

Slight shade variations will occur in different production runs of all types of factory-colored roofing materials. When applied to the roof, these shade variations produce a roof surface condition called *color patterning, checkerboarding, shading,* or *stair-stepping.*

Color patterning in asphalt roofing is caused primarily by variations in the depth of granular imbedment into the hot asphalt. Patterning in aluminum or other types of metal roofing panels can be caused by variations between batches of paint or other coloring agents.

A color range spread throughout a slate or clay tile roof is generally considered very attractive. With clay tiles, color patterning may occur when tiles manufactured from clay of varying colors are applied to the same roof. The location of the tiles within the kiln during the baking process may also produce color variations. Concrete tiles are subject to shade variations, but to a much lesser degree than clay tiles.

Slight variations in shades are discernible only after application to a roof and when viewed from a relatively long distance. Because of this, the roofer should scrutinize the work carefully from street level several times during the day. By following this procedure, color patterning can be detected and eliminated by blending the roofing materials over the entire roof deck.

Asphalt shingles—Asphalt shingles are the most common type of covering material used on pitched roofs. They are made by saturating an organic or fiberglass base with asphalt and then covering the asphalt surface with mineral granules (Figs. 2–18 and 2–19). Organic-base asphalt shingles have a class A, B, or C fire rating, depending on how they are made. Fiberglass-base shingles have an A fire rating. Fiberglass-base asphalt shingles are the more durable of the two. Their principal disadvantage is their tendency to become brittle when installed at temperatures of 50°F or lower.

An asphalt shingle roof has a service life of approximately 15 to 20 years. Asphalt shingles are lightweight, comparatively inexpensive, easy to install, and available in many different sizes, shapes, and colors. They are also easily repaired and require little maintenance.

Asphalt shingles are usually manufactured in 12 × 36-inch strips, but individual shingles are also available. The three-tab, self-sealing asphalt shingle is the most popular type in current use. It is usually laid with a 5-inch exposure and a 2-inch horizontal headlap.

Fig. 2–18. Asphalt shingle roof. *(Courtesy Johns-Manville Corp.)*

Asphalt Roll Roofing—Roll roofing is recommended for flat or low-pitched roofs with a slope of 4-in-12 or less. Roll roofing is more difficult to install than asphalt shingles. It has a class A or C fire rating, depending on its composition. Roll roofing is less expensive than asphalt shingles, but it results in a less attractive roof and has a shorter service life.

Fig. 2–19. Fiberglass shingle roof. *(Courtesy Johns-Manville Corp.)*

Wood Shingles and Shakes—Individual wood shingles are available in lengths of 16, 18, and 24 inches and come in random widths, (Fig. 2–20), while shakes are machine or hand split into 18- and 24-inch lengths. Wood shingles have a smoother surface than shakes because they are sawn instead of split. Most shingles and shakes are made from the highest (No. 1) grade of western red cedar or redwood. Neither is

Fig. 2–20. Wood shake roof. *(Courtesy Johns-Manville Corp.)*

fire resistant unless treated. In many areas, local building codes require that only treated wood shingles and shakes be used on a roof. Chemically treating wood raises the fire resistance to a class C rating.

Wood shingles are also available in the form of roof panels (Fig. 2–21). Each panel consists of sixteen 18-inch No. 1 grade cedar shingles bonded to an 8-foot long ½-inch thick plywood base. The use of panels instead of individual wood shingles reduces the roofing cost because of reduced installation time.

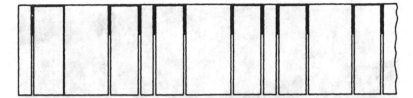

Fig. 2–21. Shingle roof panel.

Tile—Roofing tile was introduced to this country by the Spanish, and it is still a popular roofing material in Florida, California, and the Southwest (Fig. 2–22). Both clay and concrete tiles are available. Kiln-fired red clay tiles are traditional and are available in mission (barrel-shaped), flat, and other styles in a red, brown, or buff color. Concrete extruded tiles are a recent innovation and are available in a wider range of shapes and colors than clay tile. Both clay and concrete tiles are produced integrally colored or with a cured-on glaze. The most widely used tile is integrally colored.

A typical roofing tile is 12 × 7 inches. Most are approximately ½ inch thick. Tiles will not deteriorate, wear, or require painting. Although tiles may crack, split, or become brittle, they are easily replaced. A properly constructed tile roof is fireproof and has a service life of 50 years or more.

Tile is a little less expensive than slate, but about twice as expensive as asphalt shingle. Roofing tiles are heavy, expensive to ship, and difficult to install. Because tiles weigh more than most other roofing materials, a strong roof framework is required to support them. In some cases, it may be necessary to strengthen the framework by installing additional rafters and bracing. Roof reinforcement adds to the cost of a tile roof.

Slate—Slate is a heavy, durable, nonporous roofing material with a service life of 50 years or more. Because roofing slate resists deterioration, a slate roof requires little maintenance. Cracked or broken slate can be easily replaced.

Although a slate roof is attractive and enjoys a long service life, it is expensive and difficult to install. Like tile, it is heavy and requires a strong roof framework and deck to support it. Its weight also makes it expensive to ship. If the structure being roofed is located far from the few centers in the eastern United States where slate is produced, the shipping costs will add significantly to the total cost. Slating is almost

Fig. 2–22. Flat tile roof. *(Courtesy Johns-Manville Corp.)*

always done by professional contractors who specialize in this type of roofing.

Metal Roofing—Metal roofing is used on main roof decks or on the flat decks of dormers, porches, and entryways. The roofing metals include aluminum, galvanized steel, copper, copper-coated galvanized steel or aluminum, terne metal, tin, and lead. Metal roofs are used primarily on commercial, industrial, and farm buildings (Fig. 2–23).

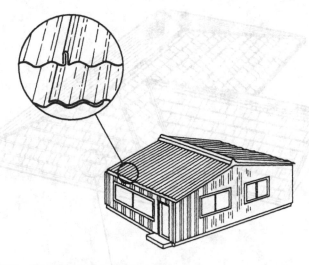

Fig. 2–23. Corrugated metal roofing.

Metal roofing is available in the form of shingles or shakes, tiles, panels, or sheets. The panels or sheets are produced in several different styles including flat, corrugated or ribbed, and standing seam. Metal roofing is fireproof and has good weather resistance. When properly installed and maintained, a metal roof should have a service life of 20 years or more. Metal roofing usually weighs less than most nonmetal roofing materials per roof square.

Aluminum is available in the form of shingles or shakes, which are produced from lightweight aluminum sheets by a stamp-and-die technique (Fig. 2–24).

Metal roofing is noisy when it rains. The panels are subject to a certain amount of expansion and contraction as temperatures change. This may cause the fasteners to pull loose from the nailing base. Aluminum is especially susceptible to denting and scratching. Most metal roofing should be installed by a contractor.

Mineral Fiber Shingles—Mineral fiber (or fiber-cement) shingles are made of cement with a fiberglass reinforcement (Fig. 2–25). They are fireproof, very rigid, durable, and strong, and have a service life of 50 years or more. Mineral fiber shingles are thicker and heavier than asphalt shingles, but are not quite as heavy as slate or tile. Like slate or

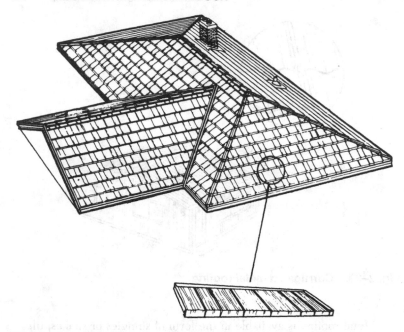

Fig. 2–24. Aluminum roofing shingles *(Courtesy Reynolds Metals Co.)*

tile, however, they require a strong roof framework to support their weight.

Mineral fiber shingles are more expensive than asphalt shingles, but cost considerably less than slate, tile, or wood shingles and shakes. They are easy to install and require little maintenance. Mineral fiber shingles can be damaged by impact, such as by a branch falling on the roof, but the individual shingles are easy to replace.

Miner fiber shingles are manufactured in a variety of colors, shapes, and textures. They are available in standard shingle sizes or smaller units resembling slate or wood shakes.

Underlayment

The *underlayment* (or *underlay*) is a moisture-resistant layer of roofing felt, which is nailed to the roof deck sheathing before laying the covering material. It is used to provide a dry, flat surface for the roof-

Fig. 2–25. Mineral fiber (fiber-cement) shingle roof. *(Courtesy Johns-Manville Corp.)*

ing, to prevent a chemical reaction from occurring between the wood resins in the sheathing and the roof covering material, and to protect the sheathing, rafters, and interior spaces from water damage during roofing.

Underlayment is used under asphalt shingles, mineral fiber shingles, slate, and tile. It can also be used under wood shingles or shakes, but this can cause premature decay.

The roofing felt underlayment is made of dry felt impregnated with an asphalt or coal tar. It is produced in 36-inch wide rolls and in several different weights. The rolls vary in length from 72 feet for a No. 30 weight felt to 144 feet for a No. 15 felt. The most widely used underlayment is No. 15 asphalt-saturated roofing felt; it weighs approximately 15 pounds per roof square (100 square feet of roof surface). No. 15 felt is used as an underlayment under asphalt shingles. Tile and slate roofs use No. 30 asphalt-saturated roofing felt or two layers of No. 15 felt as an underlayment.

Flashing

Flashing is used to protect seams or joints from water seepage. It is installed at the junction formed by the roof and a vertical wall, along roof rakes and eaves, along ridges, in roof valleys, around chimneys, vent pipes and stacks, at intersections of different roof planes, and at other points on the roof where water from rain or melting snow could penetrate the roof and enter the structure.

Flashing is made of metal, plastic, or mineral-surfaced roll roofing. Metal and composition materials are used on both residential and non-residential structures. Plastic flashing is sometimes used on the built-up roofs of commercial and industrial structures.

Galvanized steel is the most commonly used metal flashing, particularly on houses, but metals such as aluminum, copper, and stainless steel are also used. Copper flashing is often used on wood shingle or shake roofs. Aluminum and stainless steel flashing is frequently used on roofs located near saltwater bodies where corrosion is a problem.

Metal flashing may be cut and formed at the site or preformed valleys, drip edges, gravel stops, and other types of metal flashing may be purchased from a supplier. The valleys, vent pipes, stacks, and eaves of asphalt shingle roofs can also be flashed with 90-pound mineral-surfaced roll roofing, which is available in 36-inch wide rolls. The valley flashing on these roofs consists of an 18-inch wide sheet covered by a 36-inch wide one. A roll of 18-inch wide roll roofing suitable for valley flashing is available from roofing manufacturers.

Fasteners

Roofing nails are usually made of galvanized steel, aluminum, or copper. They are sharp pointed nails with heads ranging from ⅜ to ⁷⁄₁₆ inch in diameter, and are available in a variety of different lengths. The length of the nails used for a particular job will depend on the type and thickness of the roofing material, the type of roof-deck sheathing, and whether the roofing material is being applied in new construction or over an existing roof. The nails should penetrate at least ¾ inch into solid wood-deck sheathing boards. If approved plywood is used in the construction of the deck, the nails should completely penetrate the plywood. Nails with barbed or otherwise deformed shanks provide the best holding power. Recommended nail lengths for several different types of roofing applications are listed in Table 2–1.

The number and type of nails required for a specific roofing application will be stated in the instructions provided by the roofing material manufacturer. These instructions will also state the number of nails required per roof square. Additional information about roofing nails and nailing procedures is included in the chapters covering specific types of roofing applications.

Staples may be used instead of nails to fasten the underlayment or roll roofing to the roof deck. They are also sometimes used to hold down asphalt shingle tabs. Before using staples, however, check to determine whether their use is covered by the roofing manufacturer's

Table 2–1. Recommended Nail Lengths for Common Roofing and Reroofing Applications

Application	Nail Length
Roll roofing on new deck	1 inch
Asphalt strip shingles on new deck	1¼ inch
Individual asphalt shingles on new deck	1¼ inch
Asphalt shingles over asphalt shingles or roll roofing	1¼ to 1½ inch
Asphalt shingles over wood shingles	1¾ inch
16-inch and 18-inch wood shingles on new deck	1¼ inch
24-inch wood shingles on new deck	1½ inch
16-inch and 18-inch wood shingles over existing roof	1¾ inch
24-inch wood shingles over existing roof	2 inch

warranty. The staples used in roofing are 1 inch wide and up to 1¼ inches long. They are applied with a stapling machine.

Special types of fasteners are often used to fasten roofing materials to nonwood decks. The type and number of fasteners and the fastening procedure will usually be specified by the manufacturer of the roof-deck materials.

Roofing Cements and Sealers

Various types of cements and sealers are available for waterproofing seams or bonding overlapping layers of roofing material. The three principal types are plastic asphalt cement, lap cement, and asphalt adhesive cement.

Plastic Asphalt Cement—Plastic asphalt cement is an asphalt material used to seal the joint between the roof and a chimney, vent pipe, adjoining wall, or other type of vertical surface. Because plastic asphalt cement forms a part of the flashing assembly, it is sometimes referred to as flashing cement.

A good quality plastic asphalt cement will have enough elasticity to compensate for normal expansion and contraction without cracking. Furthermore, it should not flow when the outdoor temperatures are high, or become brittle when they are low.

Lap Cement—A lap cement is used to create a waterproof bond between overlapping sections of roll roofing. Lap cement will vary in consistency, depending on the manufacturer, but all lap cements are thinner and easier to work with than plastic asphalt cement. Lap cement is applied with a brush and should cover the entire lapped area.

Asphalt Adhesive Cement—Free-tab strip shingles are manufactured without an adhesive on the bottom of each tab. The disadvantage of this is that a strong wind or gust can lift the tab and bend it back. The free tabs on these shingles can be sealed down to the surface by coating the bottom of each tab with an asphalt adhesive cement. These cements are also used for sealing laps of roll roofing when it is applied by the blind nailing method (see Chapter 10).

Asphalt adhesive cement is mixed with a solvent that evaporates rapidly when exposed to the air. Consequently, it is a quick setting adhesive and is sometimes difficult to work with. It may be applied with a brush, trowel, or gun, depending on consistency.

Roof Coatings

A number of different types of asphalt coatings are used in roofing. Most are used to construct, restore, or resurface the roof membrane of a built-up roof. Some may also be used to resurface roll roofing or to cover a metal roof that shows signs of wear. All are of a thin enough consistency to be applied with a spray gun, broom, or mop.

The membrane of a built-up roof consists of alternate layers of waterproofing bitumen (roofing asphalt or pitch) and roofing felt. Roofing asphalts are available in several grades or classes based on their approximate softening point range. The grade selected will be determined by the roof slope, climatic conditions, type of roof assembly, and other factors. A roof coating called *steep asphalt* is generally recommended for roof slopes of more than 1-in-12.

Asphalt water emulsions are a special type of roof coating consisting of asphalt, or asphalt combined with other ingredients, and emulsified with water. A principal disadvantage of this type of coating is that it must not be exposed to rain for at least 24 hours after application. Another disadvantage is that stored emulsions must not be allowed to freeze during cold weather.

Other Roofing Materials

Masonry primer, roofing tape, and caulk are also used in roofing and reroofing.

Masonry Primer—The joint between the roof and a masonry wall or chimney must be sealed with a suitable asphalt coating or cement to prevent leaks; however, the bond will not be satisfactory unless the masonry surface is first coated with a primer. Masonry primers have been developed for this purpose.

A masonry primer (or asphalt primer) is a fluid substance that can be sprayed or brushed onto a masonry surface. If applied properly, it will be absorbed by the pores of the masonry and will provide a suitable bonding surface for an asphalt coating or cement. Incorrect application will be indicated by the formation of a surface film. If this should occur, the primer should be thinned to the proper consistency and reapplied. Always follow the manufacturer's instructions when thinning a masonry primer.

Roofing Tape–Roofing tape is used on built-up roofs to seal and reinforce the joints between rigid insulation boards or panels. The tape also produces a more uniform surface for the application of the roof membrane and prevents the loss of asphalt coating at the insulation joints. Roofing tape is usually made by saturating a porous material, such as cotton or glass fiber, with asphalt. Roofing tape is available in 4- to 36-inch wide rolls. The amount (linear length) of roofing tape contained in a roll will vary depending on the roofing manufacturer.

Acrylic Latex Sealer—Loose flashing or flashing with only minor damage can be sealed with a bead of acrylic latex. The sealer is available in cartridges for application with a caulking gun (Fig. 2–26).

Reroofing

Reroofing may involve the complete removal of the existing roof, in which case the method used to apply the new roofing materials is identical to that employed in new construction, or it may involve the application of new roofing materials over an existing roof after it has been prepared to serve as a suitable nailing base.

There are several obvious advantages to reroofing over an existing roof without having to remove the old roofing materials. Two principal advantages are the savings in time and money. A third advantage is the increased protection and insulation provided by the additional layers

Fig. 2–26. Applying caulking compound along edge of chimney flashing. *(Courtesy Borden Chemical Division/Borden, Inc.)*

of new roofing materials. Unfortunately, not all roofs can be reroofed in this manner.

Built up layers of roofing can affect roof drainage and may cause damage to roof railing or other roof trim.

Asphalt shingles are the most commonly used reroofing material. Existing asphalt shingle roofs and wood shingle or shake roofs are frequently reroofed with asphalt shingles. Asphalt shingles can be laid over almost any kind of roofing material *except* tile, slate, or asbestos-cement shingles. These roofing materials are too hard and brittle for nailing and provide a poor base for the new roofing materials. As a result, tile, slate, and asbestos-cement shingle roofs must be completely removed before new roofing materials are applied to the roof deck.

The roof should be thoroughly inspected before a decision is made to reroof it. This inspection should not only include the condition of the existing roof covering materials, but also the condition of the sheathing, rafters, and the frame members. Bear in mind that the existing roof must provide a rigid, smooth, and uniform base for the new roof covering materials. If portions of the old roofing materials are missing or too extensively damaged to function as a suitable base, then reroofing over the existing roof is out of the question. The roof will have to be stripped and new roofing materials applied as in new construction.

Safety Precautions

Roofing involves working atop buildings and so requires certain minimal safety precautions. This attention to safety is particularly important for the layman, who may not be accustomed to moving about at these elevations. Although professionals sometimes regard their own safety with a certain disdain, it is not an attitude that should be imitated by the nonprofessional, nor does it weaken the argument for a constant observance of safety standards. More than one professional roofer has been seriously injured or killed because familiarity with the working conditions led to carelessness.

Special attention should be paid to the type of clothing worn on the job. A suitable pair of shoes is especially important, because the roofer is usually working on a sloping surface. Tennis shoes or rubber soled shoes provide the best footing. Shoes with leather soles are *not* recommended, because they will not grip the surface. Working barefoot provides good footing, but the roof surfaces are usually too hot to walk on during the summer months. There is also the possibility of stepping on a loose roofing nail or picking up a splinter from the wood sheathing on an exposed roof deck. Stepping on a nail usually results in stumbling or moving quickly to avoid further pain—an instinctive, but possibly fatal reaction when standing at the edge of a roof deck.

Do not wear clothing that is either too loose or too tight. If the clothing is too loose, it may snag on a shingle or nail and cause you to lose your balance. Shirts should always be buttoned and tucked in. Clothing that is too tight will restrict your movements and will be uncomfortable, especially in warm weather.

Safety Lines

A safety line system consists of the line, or rope itself, the anchor, and a body belt or harness (Fig. 2–27). The system is used to provide protection against falling when working on steep slopes or near edges.

- The end of the line should be fastened to an anchor. This could be a bracket at the ridge of the roof (Fig. 2–27) or a tree on the other side of the house.
- The body belt or harness is attached to the line with a special device that can be adjusted up and down the line. In case of a fall, a cam action arrestor instantly locks, providing a positive braking action.
- A retractor can be used at the ridge anchor point which automatically adjusts the length of line during use. This eliminates the need to adjust the rope grab device constantly.
- Always use a safety line system that has been rated and tested to meet Occupational Safety and Health Administration (OSHA) standards.

Most professional roofers seem to prefer working without a safety rope or harness, because either one tends to restrict movement. This practice is *not* recommended.

Weather

Weather conditions should be considered before going up on a roof. This may present a problem, especially if time is limited or only a specific time during the day is free for the work. Nevertheless, do not

RETRACTOR OR SNAP

ANCHOR BRACKET

SAFETY LINE

ROPE GRAB

Fig. 2–27. Safety line system. *(Courtesy Roofmaster)*

attempt to work on a roof during unfavorable weather conditions. Safety should always come first.

Roofing should never be done when the surface is damp. A damp roof, whether it be from the morning dew or a recent rain, provides poor footing. More than one roofer has been seriously or fatally injured by a fall from a damp and slippery roof.

Working on a roof is not recommended when a rainstorm is approaching. There is always danger from lightning even though the clouds may not be directly overhead. Furthermore, there is a tendency to hurry the work when a storm is approaching, which often results in a careless and sloppy job. When you observe a storm approaching cover the roofing materials and get down off the roof. Do not go back up until the roof has had time to dry.

Other sources of danger to roofers are chimneys, roof vents, cables, television antennas, and electrical wires. Never lean against an old chimney or use it for support. Loose bricks can be dislodged or the entire chimney above the roof line may collapse. This could be dangerous not only for the roofer, but also for people standing below. When working within 10 feet of electrical lines always have the local electrical utility company install insulating covers. This service is usually done free of charge.

Sometimes gas- or oil-fired heating equipment is vented to the outdoors through a metal chimney or vent pipe that protrudes through the surface of the roof. Grabbing one of these for support can often result in nasty burns. If these pipes are placed near the edge of the roof, sudden reaction to the pain could result in loss of balance and a fall from the roof. Even when cold, these objects are usually too weakly constructed to support the weight of an adult.

Metal cables are often used to support television antennas, metal chimneys, vent pipes, or advertising sign boards mounted on roofs. These sometimes seem to have been deliberately placed to trip the unwary worker. Mentally mark the location of all cables and other potential causes of accidents before doing any work on the roof.

Television antennas, service entrance wires, and other types of electrical wiring are potential sources of electrical shock. Keep hands off of them and avoid brushing up against them. A severe electrical shock can result in a temporary loss of consciousness, and a roof, particularly a sloping one, is the last place you want this to happen.

If possible, never work on a roof alone. There should be at least one other person on the job, or at least in the house, in case of an accident. Injuries suffered from a fall require immediate attention.

Every type of construction work has certain inherent dangers, some more than others. Awareness of these dangers and observance of safety procedures that specifically apply to the type of work being done will minimize the possibility of injury.

Estimating Roofing Materials

An important initial step in any roofing or reroofing project is to correctly estimate the amount of roofing material required for the job. Overestimating is expensive and wasteful. Underestimating means that additional roofing material must be purchased to complete the job. This often results in slight shade variations between the colors of the new and old materials and results in a condition known as color patterning (see the discussion of color patterning in this chapter).

Roofing is estimated and sold in squares. Each square of roofing is the amount required to cover 100 square feet of roof area. To estimate the amount of roofing required to cover a roof, first compute the total roof area in square feet, add 10 percent for cutting and waste, and divide by 100.

The roof areas of the basic roof types illustrated in Fig. 2–28 are calculated by multiplying the rake line (or the sum of the rakes) by the eave line. For example, the area of the rectangular shed roof (Fig. 2–29) is calculated by multiplying the rake (line A) by the eave (line B). The area of the gable roof in the same illustration is calculated by multiplying the sum of the two rakes (lines A and B) by the eave (line C). Finally, the area of the more complicated gambrel roof is calculated by multiplying the sum of the rakes (lines A, B, C, and D) by the eave (line E). Because these are simple roofs, it is not difficult to climb onto their surfaces and measure them. However, a different method of calculating the total roof surface area is used for roof designs more elaborate than the three basic types illustrated in Fig. 2–28. The method used does not require climbing onto the roof or resorting to complicated formulas and computations. It requires only that the roof pitch and its horizontal area in square feet be known.

Roof pitch is the relation between the rise and the span of the roof (see the discussion of roofing terminology in this chapter) and is expressed as a fraction. Roof pitch can be determined by forming a triangle with a carpenter's folding rule (Fig. 2–29) and holding it at arm's length so that the roof slopes are aligned with the sides of the rule. The

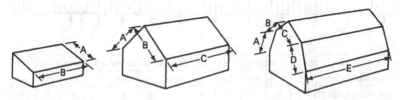

Fig. 2–28. Basic roof types. *(Courtesy Asphalt Roofing Manufacturers Association)*

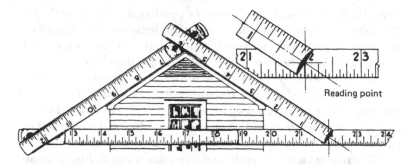

Reading point

Fig. 2–29. Using a folding rule to determine roof pitch. *(Courtesy Asphalt Roofing Manufacturers Association)*

reading is taken on the base section of the rule and the reading point is converted to pitch by using the data in Table 2–2. The numerical value used in the line headed *rule reading* will always be the one closest to the reading point on the base section of the rule. In the example shown in Fig. 2–29, the reading point and rule reading were both 22. However, a reading point of 22⅛ on the rule would also have resulted in a rule reading of 22 because it is closest to that figure. Under the numerical value 22 in Table 2–2, the roof has a ⅓ pitch and a slope of 8-in-12.

After the roof pitch has been determined, the total ground area (horizontal surface) of the roof should be measured and transferred to a roof plan drawing similar to the one illustrated in Fig. 2–30. All mea-

Table 2–2. Converting Rule Readings to Pitch and Rise

RULE READING	20 1/2	20 7/8	21 1/4	21 5/8	22	22 3/8	22 3/4	23 1/16	23 3/8	23 5/8	23 13/16	23 15/16
PITCH FRACTIONS	1/2	11/24	5/12	3/8	1/3	7/24	1/4	5/24	1/6	1/8	1/12	1/24
RISE-INCHES PER FT.	12	11	10	9	8	7	6	5	4	3	2	1

(Courtesy Asphalt Roofing Manufacturers Association)

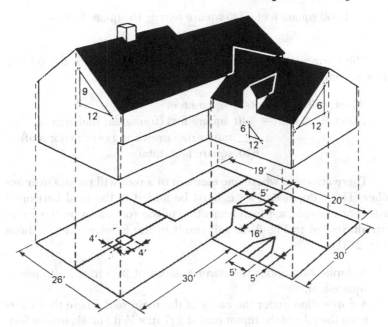

Fig. 2–30. Roof plan. *(Courtesy Asphalt Roofing Manufacturers Association)*

surements can be made from the ground or within the attic. Climbing on the roof is not necessary.

The horizontal areas are calculated after all measurements have been made, a roof plan drawn, and the pitches of the various elements of the roof have been determined with a carpenter's folding rule. *Include only those areas in each calculation that come under elements of the roof having the same pitch.* The horizontal area under the main roof of the structure illustrated in Fig. 2–30 includes

$$
\begin{array}{ll}
26\ \text{feet} \times 30\ \text{feet} = & 780\ \text{square feet} \\
19\ \text{feet} \times 30\ \text{feet} = & \underline{570}\ \text{square feet} \\
& 1{,}350\ \text{square feet (total)}
\end{array}
$$

The overlapping triangular area of the minor roof (8 feet × 5 feet or 40 square feet) and the opening for the chimney (4 feet × 4 feet or 16 square feet) must then be subtracted from the 1,350 total square feet to obtain the horizontal area under the main roof. This will give

$$1{,}350 \text{ square feet} - (40 \text{ square feet} + 16 \text{ square feet}) =$$
$$1{,}294 \text{ square feet}$$

The horizontal area under the minor (ell) roof with the 6-in-12 slope is computed as follows:

20 feet × 30 feet = 600 square feet
8 feet × 5 feet = 40 square feet (triangular roof area of
_____ minor roof projecting over major roof)
640 square feet (total)

There are cases where one element of a roof will project over another. These duplicated areas must be added to the total horizontal area. For example, a 4-inch projection of the roof eaves on the structure illustrated in Fig. 2–30 will result in the following duplications (Fig. 2–31).

1. A duplication under the dormer eaves of 2 (5 ft. × ⅓ ft.) or 3⅓ square feet.
2. A duplication under the eaves of the main roof where they overhang the rake of the minor roof of 2 (7 ft. × ⅓ ft.) or 4⅔ square feet.
3. A duplication where the rake of the wide section of the main roof overhangs the rake of the small section in the rear or 9½ feet × ⅓ feet or 3⅙ square feet.

The total square footage of the duplicated areas is 11⅙ square feet. For these computations, the square footage is rounded off to the next highest number, or 12 square feet, and divided as follows:

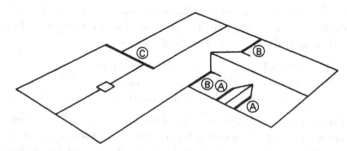

Fig. 2–31. Duplications.

1,294 square feet + 8 square feet = 1,302 square feet (major roof)
 640 square feet + 4 square feet = <u> 644</u> square feet (minor roof)
                                   1,946 square feet

The horizontal areas of 1,302 square feet (major roof) and 644 square feet (minor roof) can be converted to slope areas by using the data in Table 2–3. The horizontal areas are given in column 1 and the corresponding slope areas are given in columns 2–12. To convert horizontal areas to slope areas, begin by finding the slope area in the column under the pitch determined for the roof. The total horizontal area of the main roof of the structure shown in Fig. 2–30 is 1,302 square feet. Referring to the 9-inch rise column in Table 2–3 for the main roof, the following is found:

Horizontal Area	*Slope Area under 9-inch Rise*
1,000	1,250.0
300	375.0
<u>2</u>	<u>2.5</u>
1,302	1,627.5

The total area for the minor roof with a 6-inch rise (6-in-12 slope) or ¼ pitch is 644 square feet.

Horizontal Area	*Slope Area under 6-inch Rise*
600	670.8
40	44.7
<u>4</u>	<u>4.5</u>
644	720

The total area for both the major and minor roofs combined is 2,347.5 square feet (1,627.5 square feet + 720 square feet). Add 10 percent for wastage to bring the figure to 2,582 square feet or roughly 26 squares of roofing.

The drip edges, valley flashing, and other roof accessories required to complete the job will be determined by linear measurements made along eaves, rakes, ridges, and valleys. Eaves and ridges are horizontal measurements and may be taken off the roof plan. Rakes and valleys run on a slope and their true lengths must be taken from conversion tables (see Tables 2–3 and 2–4).

The amount of metal flashing used as drip edges is estimated by

Table 2–3. Conversion of Horizontal Distances or Areas to Slope Distances or Areas

RISE (Inches per ft. of horizontal run)	1″	2″	3″	4″	5″	6″	7″	8″	9″	10″	11″	12″
PITCH (Fractions)	1/24	1/12	1/8	1/6	5/24	1/4	7/24	1/3	3/8	5/12	11/24	1/2
CONVERSION FACTOR	1.004	1.014	1.031	1.054	1.083	1.118	1.157	1.202	1.250	1.302	1.356	1.414
HORIZONTAL (Area in Sq. Ft. or Length in Feet)												
1	1.0	1.0	1.0	1.1	1.1	1.1	1.2	1.2	1.3	1.3	1.4	1.4
2	2.0	2.0	2.1	2.1	2.2	2.2	3.2	2.4	2.5	2.6	2.7	2.8
3	3.0	3.0	3.1	3.2	3.2	3.2	3.5	3.6	3.8	3.9	4.1	4.2
4	4.0	4.1	4.1	4.2	4.3	4.5	4.6	4.8	5.0	5.2	5.4	5.7
5	5.0	5.1	5.2	5.3	5.4	5.6	5.8	6.0	6.3	6.5	6.8	7.1
6	6.0	6.1	6.2	6.3	6.5	6.7	6.9	7.2	7.5	7.8	8.1	8.5
7	7.0	7.1	7.2	7.4	7.6	7.8	8.1	8.4	8.8	9.1	9.5	9.9
8	8.0	8.1	8.3	8.4	8.7	8.9	9.3	9.6	10.0	10.4	10.8	11.0
9	9.0	9.1	9.3	9.5	9.7	10.1	10.4	10.8	11.3	11.7	12.2	12.0
10	10.0	10.1	10.3	10.5	10.8	11.2	11.6	12.0	12.5	13.0	13.6	14.0

RISE (Inches per ft. of horizontal run)	1"	2"	3"	4"	5"	6"	7"	8"	9"	10"	11"	12"
PITCH (Fractions)	1/24	1/12	1/8	1/6	5/24	1/4	7/24	1/3	3/8	5/12	11/24	1/2
CONVERSION FACTOR	1.004	1.014	1.031	1.054	1.083	1.118	1.157	1.202	1.250	1.302	1.356	1.414
20	20.1	20.3	20.6	21.1	21.7	22.4	23.1	24.0	25.0	26.0	27.1	28.3
30	30.1	30.4	31.0	31.6	32.5	33.5	34.7	36.1	37.5	39.1	40.7	42.4
40	40.2	40.6	41.2	42.2	43.3	44.7	46.3	48.1	50.0	52.1	54.2	56.6
50	50.2	50.7	51.6	52.7	54.2	55.9	57.8	60.1	62.5	65.1	67.8	70.7
60	60.2	60.8	61.9	63.2	65.0	67.1	69.4	72.1	75.0	78.1	81.4	84.8
70	70.3	71.0	72.2	73.8	75.8	78.3	81.0	84.1	87.5	91.1	94.9	99.0
80	80.3	81.1	82.5	84.3	86.6	89.4	92.6	96.2	100.0	104.2	108.5	113.1
90	90.4	91.3	92.8	94.9	97.5	100.6	104.1	108.2	112.5	117.2	122.0	127.3
100	100.4	101.4	103.1	105.4	108.3	111.8	115.7	120.2	125.0	130.2	135.6	141.4
200	200.8	202.8	206.2	210.8	216.6	223.6	231.4	240.4	250.0	260.4	271.2	282.8
300	301.2	304.2	309.3	316.2	324.9	335.4	347.1	360.6	375.0	390.6	406.8	424.2
400	401.6	405.6	412.4	421.6	433.2	447.2	462.8	480.8	500.0	520.8	542.4	565.6
500	502.0	507.0	515.5	527.0	541.5	559.0	578.5	601.0	625.0	651.0	678.0	707.0
600	602.4	608.4	618.6	632.4	649.8	670.8	694.2	721.2	750.0	781.2	813.6	848.4
700	702.8	709.8	721.7	737.8	758.1	782.6	809.9	841.4	875.0	911.4	949.2	989.8
800	803.2	811.2	824.8	843.2	864.4	894.4	925.6	961.6	1000.0	1041.6	1084.8	1131.2
900	903.6	912.6	927.9	948.6	974.7	1006.2	1041.3	1081.8	1125.0	1171.8	1220.4	1272.6
1000	1004.0	1014.0	1031.0	1054.0	1083.0	1118.0	1157.0	1202.0	1250.0	1302.0	1356.0	1414.0

(Courtesy Asphalt Roofing Manufacturers Association)

Table 2–4. Determination of Valley and Hip Lengths

RISE (Inches per ft. of horizontal run) PITCH	4"	5"	6"	7"	8"	9"	10"	11"	12"	14"	16"	18"
Degrees	18°26'	22°37'	26°34'	30°16'	33°41'	36°52'	39°48'	42°31'	45°	49°24'	53°8'	56°19'
Fractions	1/6	5/24	1/4	7/24	1/3	3/8	5/12	11/24	1/2	7/12	2/3	3/4
CONVERSION FACTOR	1.452	1.474	1.500	1.524	1.564	1.600	1.642	1.684	1.732	1.814	1.944	2.062
HORIZONTAL (Length in Feet)												
1	1.5	1.5	1.5	1.5	1.6	1.6	1.6	1.7	1.7	1.8	1.9	2.1
2	2.9	2.9	3.0	3.0	3.1	3.2	3.3	3.4	3.5	3.6	3.9	4.1
3	4.4	4.4	4.5	4.6	4.7	4.8	4.9	5.1	5.2	5.4	5.8	6.2
4	5.8	5.9	6.0	6.1	6.3	6.4	6.6	6.7	6.9	7.3	7.8	8.2
5	7.3	7.4	7.5	7.6	7.8	8.0	8.2	8.4	8.7	9.1	9.7	10.3
6	8.7	8.8	9.0	9.1	9.4	9.6	9.9	10.1	10.4	10.9	11.7	12.4
7	10.2	10.3	10.5	10.7	10.9	11.2	11.5	11.8	12.1	12.7	13.6	14.4
8	11.6	11.8	12.0	12.2	12.5	12.8	13.1	13.5	13.9	14.5	15.6	16.5
9	13.1	13.3	13.5	13.7	14.1	14.4	14.8	15.2	15.6	16.3	17.5	18.6
10	14.5	14.7	15.0	15.2	15.6	16.0	16.4	16.8	17.3	18.1	19.4	20.6

RISE (Inches per ft. of horizontal run)	4"	5"	6"	7"	8"	9"	10"	11"	12"	14"	16"	18"
PITCH Degrees	18°26'	22°37'	26°34'	30°16'	33°41'	36°52'	39°48'	42°31'	45°	49°24'	53°8'	56°19'
Fractions	1/6	5/24	1/4	7/24	1/3	3/8	5/12	11/24	1/2	7/12	2/3	3/4
CONVERSION FACTOR	1.452	1.474	1.500	1.524	1.564	1.600	1.642	1.684	1.732	1.814	1.944	2.062
20	29.0	29.5	30.0	30.5	31.3	32.0	32.8	33.7	34.6	36.3	38.9	41.2
30	43.6	44.2	45.0	45.7	46.9	48.0	49.3	50.5	52.0	54.4	58.3	61.9
40	58.1	59.0	60.0	61.0	62.6	64.0	65.7	67.4	69.3	72.6	77.8	82.5
50	72.6	73.7	75.0	76.2	78.2	80.0	82.1	84.2	86.6	90.7	97.2	103.1
60	87.1	88.4	90.0	91.4	93.8	96.0	98.5	101.0	103.9	108.8	116.6	123.7
70	101.6	103.2	105.0	106.7	109.5	112.0	114.9	117.9	121.2	127.0	136.1	144.3
80	116.2	117.9	120.0	121.9	125.1	128.0	131.4	134.7	138.6	145.1	155.5	165.0
90	130.7	132.7	135.0	137.2	140.8	144.0	147.8	151.6	155.9	163.3	175.0	185.6
100	145.2	147.4	150.0	152.4	156.4	160.0	164.2	168.4	173.2	181.4	194.4	205.2

(Courtesy Asphalt Roofing Manufacturers Association)

calculating the total length of all the roof eaves and rakes. The length of the roof rake is determined by first measuring the horizontal distance over which it extends. For the structure illustrated in Fig. 2–30, the rakes on the ends of the main roof span distances of 26 feet and 19 feet respectively. Additional rake footage of 13 feet plus 3½ feet is found at the point where the higher roof section joins the lower one. Thus, the total rake footage for the structure illustrated in Fig. 2–30 is 26 + 19 + 13 + 3½ = 61½ feet. By referring to Table 2–3 under the column for a roof with 9-inch rise, the following is found:

Horizontal Run	Length at Rake
60.0	75.0
1.0	1.3
.5	.6
61.5	76.9 feet (actual length of major roof rake)

The minor roof of the structure illustrated in Fig. 2–30 has a rake with a horizontal distance of 30 feet. The dormer roof rake adds an additional 5 feet for a total of 35 feet. Referring to Table 2–3 under the column for a roof with a 6-inch rise, the following is found:

Horizontal Run	Length at Rake
30	33.5
5	5.6
35	39.1 feet (actual length of minor roof rake)

The horizontal length of the eaves for both the major and minor roof should also be measured and added to the total actual length of the rakes. The resulting figure will serve as an estimate of the total amount of metal flashing required as a drip edge along the roof eaves and rakes.

The point at which the minor roof meets the major roof of the structure illustrated in Fig. 2–30 creates two valleys. To estimate the amount of flashing material required for these valleys, it is first necessary to determine the run of the common rafter and then the length of the valley. The run of the common rafter is always one half the roof span. When used to determine the length of a valley, the run of the common rafter should be taken at the lower end of the valley. For example, the portion of the minor roof that projects over the major roof on the structure in Fig. 2–30 has a span of 16 feet at the lower end of the valley. The run of the common rafter at this point will be 8 feet (run = ½ span). Because there are two valleys formed at this roof intersection, however, the total run of the common rafter to be used in the calculations is 16 feet.

Horizontal length (16 feet) is converted to true valley length by means of the data provided in Table 2–4. Because the intersecting roofs have different rises, however, the length for each rise must be found and the average of the two used. Under the 6-inch rise and 9-inch rise columns in Table 2–4, the following is found:

Horizontal Length	6-inch Rise (Minor Roof)	9-inch Rise (Major Roof)
10	15	16.0
6	9	9.6
	24	25.6 = 49.6 feet

49.6 feet ÷ 2 = 24.8 feet (true length of each valley)

The same calculation is used to find the true length of the dormer valleys. As shown in Fig. 2–30, the run of the common rafter at the dormer is 2.5 feet (½ span of 5 feet). The following is found by using Table 2–4:

Horizontal Length	6-inch Rise Dormer Roof
2.0	3.00
.5	.75
2.5	3.75 feet (true length of each dormer valley) × 2 = 7.5 feet

Total valley length for the structure in Fig. 2–30 is 24.8 feet + 7.5 feet = 32.3 feet.

Hiring a Roofing Contractor

A roofing contractor should be hired to do the work if the job requires skills and experience beyond those of the average homeowner.

A roofing contractor can be found by looking in the Yellow Pages of the local telephone directory under "Roofing Contractor." Other sources helpful in recommending roofing contractors include the bank willing to finance the work, the local chapter of the National Association of Home Builders or Home Builders Association, local government offices for government-funded or nonprofit-operated home improvement assistance centers, relatives, friends, and neighbors. From these sources, obtain a list of four or five roofing contractors from which to make a final selection.

Ask each contractor for a list of past customers and check whether these individuals were satisfied with the work. Call the local Better Business Bureau to determine if complaints have been lodged against any of the contractors on the list.

Give the same specifications for the job to each of the roofing contractors and ask for an estimate. After the estimates have been reviewed and a contractor selected, draw up a written contract for the job or have one drawn up by a lawyer. Check the contract for work content and warranty before signing it. To make it a valid contract, both the building owner and the contractor must sign it. The contract should contain the same specifications contained in the bid or estimate submitted by the contractor. It should also state the cost of both materials and labor, and the payment method.

Check to determine if the roofing contractor has enough insurance to cover injuries to workers or accidental damage to the structure. If not, the building owner would be wise to purchase the necessary insurance.

CHAPTER 3

Roof Deck Preparation

The roof deck is the platform or base over which the roofing materials are laid. It is nailed or otherwise fastened to the rafters, the structural framework that supports the roof. The roof deck must provide a smooth, flat, and rigid surface for the roofing materials. If the deck is not rigid, vibration or movement resulting from structural instability may affect the lay of the roofing. An uneven surface will also cause roofing problems. On a flat roof, an uneven roof deck surface may result in drainage problems, wrinkling or buckling of roll roofing, or cracking of built-up roof membranes. These problems may be avoided by careful roof deck preparation before the roofing materials are laid.

Roof Deck Construction

The roof deck may be constructed of either wood or nonwood materials. Most pitched roof decks on houses are made of wood sheathing covered by an underlayment of asphalt-saturated felt. As a general rule, nonwood materials are used in the construction of roof decks on commercial and other nonresidential structures.

61

Wood Decks

A wood roof deck consists of wood sheathing and, when required, an underlayment or covering of asphalt-saturated felt. The wood sheathing may be boards, plywood panels, or structural panels.

Wood Sheathing Boards—Wood sheathing boards should be made of clear, sound, well-seasoned lumber of not less than 1 inch nominal thickness and not more than 8 inches nominal width. Wider boards are more likely to swell or shrink in width and cause the roofing material to wrinkle or buckle. Badly warped boards or boards containing excessively resinous areas or loose knots should be rejected. Tongue-and-groove, shiplap, or splined boards are recommended with each board flush nailed to the supporting rafters. Tongue-and-groove boards have the greatest resistance to warping or buckling. All boards and joints should be staggered and provided with adequate bearing or support.

Wood sheathing boards are nailed to each supporting rafter with two eightpenny (8d) common nails or two sevenpenny (7d) threaded nails. Each board must be supported by at least two roof rafters. Vertical joints must be centered over rafters unless end-matched tongue-and-groove boards are used. The joints formed between the ends of these boards may occur between rafters. All board end joints should be staggered.

The roof deck may be of closed or open construction, depending on the type of roofing material used to cover it. On a closed roof deck, the boards are applied without any spacing between them. They are usually nailed parallel to the roof eave. In areas of the country where high winds are a problem, greater racking resistance can be obtained by applying the boards diagonally to the eave. At the rake, the boards may extend beyond the end wall to provide an overhang or they may be cut flush with the wall (Fig. 3–1).

Spaced or open sheathing is customarily used as a roof deck for wood shingles or shakes. The spaces between the boards provide enough ventilation to prevent the bottoms of the shingles or shakes from absorbing and retaining moisture. The width of the space between the boards is determined by the shingle or shake weather exposure. Spaced or open sheathing is also used as a roof deck for tile roofing and metal or vinyl roofing panels. As shown in Fig. 3–2, many open roof decks are constructed with closed sheathing along the eaves and sometimes along the eaves as well. Additional information about spaced or open roof decks is provided in Chapters 12 and 14.

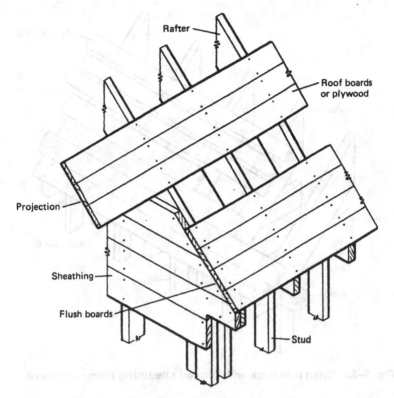

Fig. 3–1. Rake details showing sheathing projection or cut flush with endwall.

Plywood Sheathing— Plywood roof sheathing is available in 4 × 8-foot panels and in different thicknesses. Each panel is nailed to the rafters with its face grain running parallel to the roof eaves, as in Fig. 3–3. The ends of the panels should be positioned over the center of rafters with vertical seams staggered by at least one rafter spacing to eliminate the possibility of a single seam running up the slope of the roof deck. Provide a ⅛-inch edge spacing and a ¹⁄₁₆-inch end spacing between plywood panels to allow for expansion and contraction.

Only structural plywood that meets the American Plywood Association's performance rating for roof deck sheathing should be used. Each panel will be stamped with an APA registered trademark indicating its performance rating (Fig. 3–4).

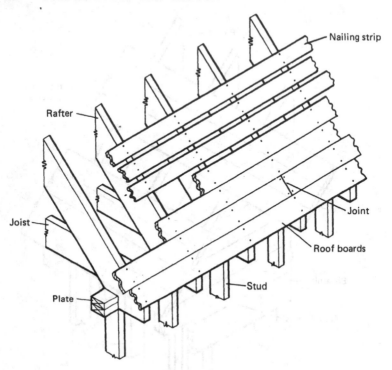

Fig. 3–2. Open roof deck with closed sheathing along roof eave.

When roofing with asphalt shingles, roll roofing, or wood shingles and shakes, plywood panels with a *minimum* thickness of ⅜-inch are recommended for use as sheathing when the rafters are spaced the standard 16 inches apart. A ½-inch to ⅝-inch thick plywood panel is recommended when the rafters are spaced 24 inches apart. Preclips or H clips should be installed on the horizontal joints of adjoining panels when the rafters are spaced 24 inches apart to insure a flat, even deck surface. Better nail penetration and holding power, improved racking resistance, and a smoother roof appearance can be obtained by using ⅝-inch thick plywood for 16-inch spacing of rafters and 1-inch thick plywood for 24-inch rafter spacings. For 16-inch rafter spacings, ⅝-inch thick plywood is considered minimum when slate and other heavy roofing materials are used.

Nail each plywood sheathing panel 6 inches on center along all bearing edges and 12 inches on center along intermediate members. A

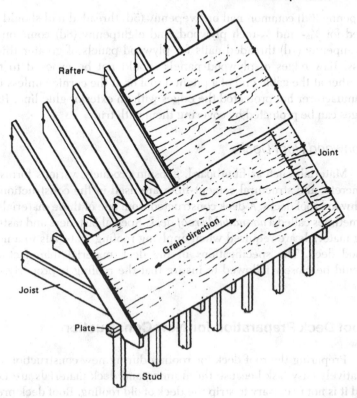

Fig. 3–3. Plywood roof sheathing.

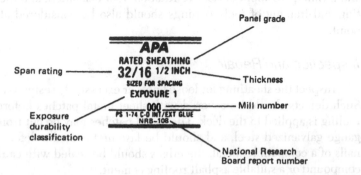

Fig. 3–4. APA plywood sheathing performance rating. *(Courtesy American Plywood Association)*

sixpenny (6d) common nail or fivepenny (5d) threaded nail should be used for ⁵⁄₁₆- and ⅜-inch plywood, and eightpenny (8d) common or sevenpenny (7d) threaded nails for plywood panels of greater thickness. Raw edges of plywood panels should not be exposed to the weather at the gable end of a pitched roof or at the cornice unless the manufacturer has protected the edges with an exterior glue line. Raw edges can be protected by covering them with trim.

Nonwood Roof Decks

Materials such as fiberboard, gypsum products, various forms of concrete, and structural cement-fiber are used in the construction of nonwood roof decks. A deck constructed from one of these materials is sometimes called a *nonnailable roof deck*. Special fasteners and fastening materials are required when applying roofing materials to a nonwood deck. The specifications of the deck material manufacturer should be closely followed to insure that the roofing is properly applied.

Roof Deck Preparation for New Construction

Preparing the roof deck for roofing during new construction is a relatively easy task because the framing and deck materials are new and it is not necessary to strip the deck of old roofing. Roof deck preparation for new construction should begin with a careful inspection for defects in the framing and sheathing before the underlayment (if used) and flashing are applied. Roof ventilation, roof insulation, and the cutting and framing of roof openings should also be considered at this point.

Inspection and Repair

Inspect the sheathing for loose knots or excessively resinous areas. Such defects should be covered with sheet metal patches before the roofing is applied to the deck. The metal patches may be cut from 26-gauge galvanized steel, and should be fastened over the defect with nails of a compatible metal; the edges should be sealed with caulking compound or a suitable asphalt roofing cement.

Make certain that the number and spacing of the rafters is adequate to support the roof deck, the roofing materials, the weight of the workers doing the roofing, and the usual snow and wind loads. Slate, tile, and fiber-cement shingle roofs require greater support than asphalt shingles or roofs covered with relatively lighter roofing materials. Additional rafters and bracing should be installed if the framework is not strong enough to provide adequate support.

Ventilation

Condensation will occur when warm moist air comes in contact with a cold surface. Sources of moisture inside a structure include cooking, bathing, washing clothes, and air conditioning. The moisture rises with warm air and condenses when it reaches the cold roof sheathing. This condensation can damage the sheathing of a wood roof deck, the roof rafters, and the insulation. The problem can be prevented or minimized by permitting air to circulate below the roof deck. This can be accomplished by proper roof ventilation. Roof ventilating methods are described in Chapter 4.

Insulation

Insulation can be used to reduce the flow of heat into the attic or attic crawl space or to provide thermal resistance to heat flow in the roof. A vapor barrier installed on the side of the insulation facing toward the interior heated spaces of the structure will reduce moisture penetration.

Roof Openings

It is usually necessary to cut one or more openings in the roof deck for certain types of construction or for objects that protrude through the roof surface. If the opening is wider than the space between two roof rafters, it will have to be framed. The best time to cut and frame roof openings for chimneys, vent pipes, stacks, skylights, and dormers is during new construction before the underlayment and flashing are applied. An existing roof will require the removal of some of the roofing materials to expose the deck before the opening can be cut and framed. Cutting and framing methods for roof openings are described in Chapter 5.

Underlayment

Some roof decks are covered with a protective layer or underlayment of roofing felt to protect the structure from water penetration both during roofing installation and during service.

Protection During Roofing Application—An underlayment ensures that shingles will be laid over a dry surface. Moist or wet wood board sheathing may cause buckling and distortion of the shingles. Trapped moisture may also cause the boards to warp.

Plywood manufacturers usually recommend applying underlayment as soon as possible to prevent excessive moisture absorption of the sheathing. While plywood rated for roof sheathing can get wet without diminishing its structural strength, the moisture can cause other problems. Moisture on the plywood surface can cause checking and raised grain. If the plywood is laid without expansion spaces along each edge, the plywood will swell, forming ridges along the joints. These raised defects can telegraph through thin fiberglass shingles and affect the visual appearance.

Using an underlayment is especially important if using wafer board or oriented strand board. Both are very susceptible to moisture damage and can lose their structural strength.

Protection During Service—Any moisture that penetrates the roofing will be kept out of the house. At the same time moisture vapor from inside the structure can escape through vapor-permeable roofing felt. This protection from water penetration is considered necessary when shingles may be lifted or removed by high winds, when rain is driven under the roofing or if the roofing is damaged any time during its life.

Underlayment prevents any direct contact between the shingles and highly resinous areas on wood board sheathing such as knots and resin pockets. Some shingle materials are chemically incompatible with resins and prolonged contact could damage the shingles.

Underlayment or Not—There are a few reasons for not using underlayment. You can avoid the cost of underlayment if you select certain materials and install them in certain ways.

- Plywood sheathing should be spaced properly to allow for expansion.
- Complete the roofing before the chance of rain. This is more likely on smaller roofs or if you have a well trained roofing crew with a good production rate.

- Footing is less secure when walking on underlayment. In any case, do not walk on underlayment that has not been nailed down.
- Roofing felt laid in the hot afternoon will buckle if left exposed to dew or rain overnight. Asphalt-treated roofing felt swells as it absorbs moisture, causing buckling between nailing points. This can make application of roofing first thing the next morning difficult. Usually the felt will dry and flatten with half a day of drying weather.

In deciding whether or not to use a roofing felt underlayment, always follow the roofing manufacturer's recommendation. The difficulty comes when the roofing manufacturer's recommendations are at odds with the sheathing manufacturer's recommendations. Consider carefully the effects moisture and movement will have on the various materials that make up the roof.

The method used to apply the underlayment on a roof deck is determined by its pitch or slope. Both pitch and slope, as well as the methods used to calculate them, are described in Chapter 2.

Wood decks on pitched roofs with a slope of 2-in-12 or greater over which asphalt shingles are to be laid are usually covered with an underlayment of No. 15 asphalt-saturated roofing felt. Nos. 20 and 30 weight roofing felts are available for other types of roofing applications. All roofing felt is available in 36-inch wide rolls and lengths of 72, 108, and 144 feet per roll. *Never use a coated sheet or heavy felt as an underlayment.* Both types of materials will function as vapor barriers and trap moisture or frost between the roofing material and the roof deck.

If the roof deck has a slope of 4-in-12 or greater (Fig. 3–5), nail a metal drip edge along the bottom edge (roof eaves) *before* the felt is laid and proceed as follows:

1. Lay one layer or ply of felt horizontally over the entire roof, lapping each course 2 inches over the underlying one with a 4-inch side lap at end (vertical) joints.
2. Lap the felt 6 inches from both sides over all hips and ridges.
3. Secure the felt to the deck with enough fasteners (roofing nails or staples) to hold it in place until the shingles are applied.
4. Nail a metal drip edge to the rakes of a gable roof *after* the underlayment has been laid.

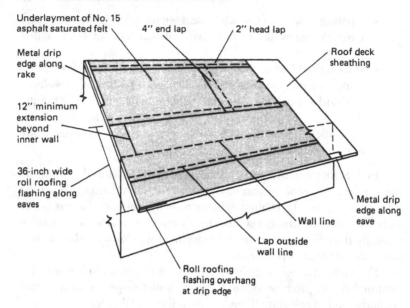

Fig. 3–5. Underlayment applied to roof deck with slope of 4-in-12 or more. *(Courtesy Asphalt Roofing Manufacturers Association)*

Where the January average daily temperature is 25°F or less, or where there is a possibility of ice forming along the eaves and causing roof leaks from a backup of water, apply an eaves flashing strip of 36-inch wide heavy roll roofing (50 pound or heavier) along the eaves. The roll roofing should be laid so that it overhangs the metal drip edge ¼ inch and extends up the roof deck to a minimum point of 12 inches inside the interior wall line of the structure. If a second overlapping sheet of roll roofing must be laid to meet this 12-inch minimum requirement, the horizontal lap must occur *outside* the wall line.

For roof decks with a slope of 2-in-12 to less than 4-in-12, as in Fig. 3–6, apply the underlayment as follows:

1. Coat the roof deck with bituminous plastic cement (at two gallons per 100 square feet of roof) from the roof eaves up the roof to a point approximately 24 inches inside the wall line and nail the first course of roofing felt to the deck.
2. Apply two layers of felt parallel with the eaves, laying the first

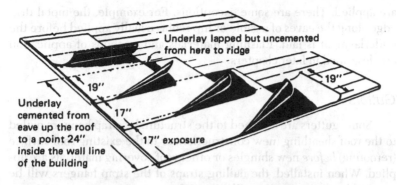

Fig. 3–6. Underlayment applied to roof deck with slope of less than 4-in-12. *(Courtesy Asphalt Roofing Manufacturers Association)*

19-inch wide sheet as a starter course covered by a second layer of 36-inch wide felt.

3. Cover the roof deck with 36-inch wide sheets of felt overlapping the preceding layer by 19 inches to expose 17 inches of the underlying sheet.
4. Secure the underlayment to the roof deck with enough fasteners (roofing nails or staples) to hold the felt in place until the shingles are laid.

A single ply of special ice and water shield material can be used. This type of material is formulated to seal around nails that penetrate it.

Wood decks of pitched roofs with a slope of 2-in-12 or greater over which asphalt shingles are to be laid are usually covered with an underlayment of No. 15 asphalt-saturated roofing felt.

Flashing

Flashing is used to seal the roof against leakage at chimneys, vent pipes, stacks, and other protrusions through the roof surface; at the intersections of different roof planes; and at the abutments of the roof against adjoining vertical walls.

In most cases, the flashing is installed after the underlayment (if used) is laid and before the shingles or other roof covering materials are applied. There are some exceptions. For example, the metal drip edge along the eaves of a gable or hip roof is usually applied before the underlayment is laid. Flashing materials and methods of application are described in later chapters.

Gutters

Some gutters are attached to the structure by strap hangers nailed to the roof sheathing (new construction) or to the existing roof surface (reroofing) *before* new shingles or other roof covering materials are applied. When installed, the nailing straps of the strap hangers will be covered by the roofing (see Chapter 15).

Roof Deck Preparation for Reroofing

Preparation of the roof deck for reroofing will depend on the condition of the existing roof and the types of materials selected for the new roof. The roof deck requirements for new construction also apply to reroofing. The roof deck must be smooth, flat, and rigid, and must provide a solid nailing base for the new roofing materials.

Inspection and Repairs

Inspect the condition of the old roofing materials to determine whether they should be removed or allowed to remain. Check the roof for loose, cracked, broken, or damaged roofing. Old roofing that shows signs of extensive deterioration must be removed. Tile, slate, and fiber-cement shingles must be removed regardless of their condition because they are too brittle for driving nails. The preparation of the roof deck after removing the existing roofing materials will be the same as described for new construction.

Inspect the rafters and the underside of the sheathing from the attic or attic crawl space for damage or defects and make the necessary repairs (see the discussion of roof deck preparation for new construction). Deteriorated plywood panels and warped, broken, or rotting sheathing boards will have to be replaced. In order to replace damaged sections of the sheathing, it will be necessary first to strip the old roofing from the deck. Only the damaged portions of plywood panel or sheathing board have to be cut away and replaced. Make certain that the replacement piece is large enough to be supported by at least two rafters, nail it flush with eightpenny (8d) nails, and caulk the joints to prevent water from entering the structure along the joints. A rafter can be reinforced without removing the old roofing by nailing a second rafter directly parallel to it.

Roofs are exposed to a variety of weather conditions and eventually develop leaks, which must be repaired before reroofing. The most common types of roof leaks can be traced to one or more of the following causes:

1. Warped, corroded or cracked flashing around chimneys, vent pipes, and other structural interruptions in the roof surface
2. Broken, loose, or missing shingles, tiles, or other types of roofing materials
3. Rusty, loose, or missing nails or other types of fasteners
4. Dried out and cracked roofing compounds or sealers used to seal seams in roof-covering materials
5. Blocked or damaged gutters and downspouts
6. Rotted or cracked plywood sheathing panels or boards
7. Ice dams forming along the roof eaves

Most roof leaks are difficult to trace because the water almost never collects directly under the point at which it enters. It will frequently run down a rafter or beneath the sheathing for a considerable distance before dropping to the attic or attic crawl space floor. To trace a leak, find the point at which the water has been collecting and then look for water stains along the joist above it. These stains will usually leave a path to the point at which the water has been entering. Sometimes a small pinhole of light can be seen. If the leak can be located, shove a straight piece of wire through the hole to indicate its position from the outside. If the point at which the water is entering cannot be determined by this method, calculate the approximate location by measuring the distance from the end of the joist to the point at which the stains end. Add to this measurement the length of the roof overhang, and this will be the approximate distance of the leak from the edge of the roof.

If it is raining, trace the flow of water up along the joist to its approximate entry point. Repairs are impossible until the rain stops, but the point at which the water enters can be marked and a pail placed where the puddle is forming. To ensure that the dripping water does not change its position on the joist, run a string or wire from the joist down to the pail. After the leak has been located from the inside (or at least its approximate location determined), go onto the roof and try to find the cause of the leak.

Repairing a leak is *not* recommended during inclement weather. The leak should be stopped with a temporary repair until permanent repairs can be made (Fig. 3–7).

As in new construction, the roof framing must be strong enough to support the deck and the new roofing, especially if heavy roofing ma-

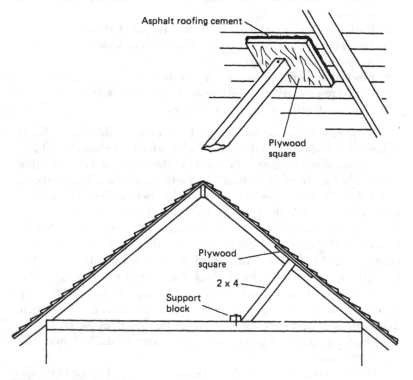

Fig. 3–7. Temporary roof repairs.

terial such as slate, tile, or fiber-cement shingles are used. The framework can be strengthened by installing additional rafters and bracing.

Removing Roofing Materials

Begin by removing the ridge shingles or caps with a claw hammer or shovel. If the structure has a hip roof, the hip ridge shingles or caps should be removed next. Removing the covering over the roof ridge will expose the nails along the top edge of the last course of shingles or roofing. Work down the slope of the roof toward the eaves and remove each course of shingles or roofing. Tile, slate, and fiber-cement shingles are brittle and can be removed by first striking them with a hammer, removing the broken pieces, and then pulling the roofing nails. Built-up roofing membranes can be removed by cutting through the membrane to the roof deck and then prying it up with a shovel. Special care should be taken when removing existing metal flashing, since if kept intact it can be used as a pattern for cutting new flashing.

Deck Preparation

Old Roofing Removed—If the existing roofing is removed, the roof deck should be prepared and the new roofing materials applied as in new construction. After the roofing materials have been removed, the roof deck should be prepared as follows:

1. Repair the roof framework where necessary to provide a level deck surface.
2. Reinforce the roof framework if roofing materials heavier than the original ones are to be used. This will usually be the case when roofing with slate, tile, or fiber-cement shingles.
3. Remove rotted, warped, split, or broken wood sheathing boards, or delaminated plywood sheathing, and replace them with new sheathing of like kind.
4. Pull all protruding and rusted nails and renail the sheathing at new locations with eightpenny (8d) nails. Fill the old nail holes with asphalt cement.
5. Cover all large cracks, slivers, knot holes, loose knots, pitch spots, and excessively resinous areas with a piece of galvanized sheet

metal, securely nail the patch to the sheathing, and coat the edges of the patch with asphalt cement to prevent leakage.

6. Sweep the deck clean of all loose debris, and, if wet, allow it time to dry thoroughly before applying any new roofing material.

Old Roofing Retained—The procedures used to prepare the old roof for reroofing are determined by the type of existing roofing materials, their condition, and the type of materials used for reroofing. These procedures are covered in the chapters that describe specific types of roofing materials and their methods of application. The procedures for inspecting and repairing the roof framework and roof deck are the same for both new construction and reroofing (see the discussion of roof deck preparation for new construction).

CHAPTER 4

Roof and Attic Ventilation

Daytime summer attic temperatures often reach 150°F and higher. This trapped hot air penetrates downward into the living and sleeping areas of the structure, making them uncomfortably warm both day and night. If the structure is air conditioned, the hot attic air will increase the load on the air conditioner, and lower its efficiency. Hot attic air is also potentially dangerous and destructive, sometimes causing spontaneous combustion in overheated attics and attic crawl spaces, and early deterioration of shingles, framing, and insulation.

Most roof decks are covered with an underlayment of waterproof asphalt-saturated felt to prevent rain or melting snow from leaking through the roof deck into the interior of the structure. This waterproof underlayment also tends to block the passage of water vapor from inside the structure through the roof deck to the outdoors. This can be a problem in the winter because moisture from cooking, baths, and a variety of other sources rises with the warm air and condenses when it comes into contact with the cold sheathing boards. The underlayment will block movement of the moisture to the exterior. The wood in the roof framework and roof deck sheathing absorbs large amounts of this moisture. Moisture will eventually cause the wood framework and sheathing to warp or decay. The damage may be minimal, resulting in only a few minor leaks, or it may be so extensive that the entire roof structure is seriously weakened. Moisture may also become trapped in the roof, attic, or attic crawl space insulation, causing it to lose its effectiveness.

The most effective method of removing trapped moisture and reducing air temperatures is by adequately ventilating the spaces beneath the roof. Properly ventilating these spaces will cause the air to circulate freely and will remove warm moist air from the structure before the water vapor condenses and causes any damage.

The best time to inspect and improve roof and attic ventilation is during reroofing or before the roofing materials are applied in new construction.

The increasing number of reports coming in on total roof failures because of little or no attic ventilation indicates the need for better standards and practice. In this chapter we will look at the current standards and practices to see how they can be improved.

Attic ventilation is desirable and necessary for three very good reasons:

1. To prevent moisture problems
2. To prevent heat build-up
3. To reduce damage caused by ice dams

A fourth reason is to prevent shingle manufacturers from claiming your shingles failed by buckling because of no attic ventilation. However, some building scientists challenge this assertion.

If attic ventilation is the answer to these problems, what then must be done to insure proper attic ventilation, and what is the "proper" method of attic ventilation?

Current Standards and Practices

Currently the most widespread form of attic ventilation is the gable end louvered vent ranging in size from a single slit to very large triangular units. For a long time there was little data on how much summer attic ventilation was necessary. Current BOCA Code (1987), which is based on FHA Minimum Property Standards (MPS) states:

709.1.1 Ventilating area: The minimum required net free ventilating area shall be 1/150 of the area of the space ventilated, except that the minimum required area shall be reduced to $\frac{1}{300}$ where at least 50 percent of the required ventilating area is provided by ventilators located in the upper portion of the space to be ventilated at least 3 feet (914 mm) above eave or cornice vents with the balance of the required ventilation provided by eave or cornice vents.

How good are gable end louvers in providing attic ventilation? According to one investigator, "The aesthetic appeal of simple louvers has been a major factor in their continued and popular use." Unfortunately they have limited effect on rain and snow infiltration, especially when they are sized for summer ventilation. More complicated, more expensive louvers are equally limited in stopping rain or snow penetration. Builders are not interested in more expensive units as long as they believe the simple ones work.

Moisture Control

For moisture control, FHA MPS regulations specify that attic vents in houses with a vapor barrier in the ceiling shall have a free area equal to $\frac{1}{300}$ of the attic area. To compensate for the 8×8 mesh screen (64 openings per square inch) which is also required, the screened opening must be 25 percent larger. If screened and louvered openings are used for the vent, the size of the louver must be 125 percent larger than the specified free area. Hence, in a 1,000 square foot house, the attic vents required to satisfy the needs for moisture control must have a free area of 3-$\frac{1}{3}$ square feet, a screened area of 4-1/6 square feet, and a louvered area with screens of 7-$\frac{1}{2}$ square feet. This means that a louver of 3-$\frac{3}{4}$ square feet is required in each gable of such a house.

Temperature Reduction

If any substantial reduction in attic temperatures by gravity air flow is to be achieved, the vent area required by current codes will have to be significantly increased. This $\frac{1}{300}$ ratio is less than one half square inch ($\frac{1}{2}$) per square foot of attic floor. When you remember the code deals with minimums, common sense tells you this is inadequate.

Stack Height

The difference in height between the intake and exhaust vents (stack height) is what determines the amount of vent area required. If one-half of the vents are low in the eave and the other half high in the attic near the ridge, the air flow will be greater and better than if all the vents were high. The greater the difference in height from the low inlet vents to the high outlet vents, the greater the gravity air flow. It is necessary to rely on gravity air flow rather than the greater air flow caused by wind because the prime need for ventilation is NOT when the wind is blowing, but rather when there is no wind to force air into and out of the attic.

In a house with a 6-foot stack height, the free venting area will have to be 1/50 of the attic area. With one-half of the vent a screened intake vent and the other half of the vent a louvered and screened exhaust vent located high in the attic, a 1,000 square foot house will require a screened intake of 12 square feet and two louvered exhausts of 11 square feet each. These "Rambo" sized louvers are impractical and cannot be economically built to control rain and snow infiltration. Attic vents large enough to significantly cool an attic in the summer must be found. If the stack height is less than 6 feet, larger vents will be required. In other words, the lower the stack height, the larger the vent required.

Returning to the rule, if a gable end louver is to be used as the exhaust vent in combination with eave soffit vents, it will be nearly impossible to provide enough exhaust vent to equal the area of intake vent. And even if the exhaust vents are one-half the area of the intake vents, the gable end louver vent will be extremely large. A house 25 feet wide with 4-foot high ridge will have a stack height of 2'-8". To provide the required 13 feet of louver in each gable, the louver must be 12'-6" wide at the base and 2'-0" high. A vent this size would be likely to allow water and snow penetration.

Effective Criteria

Obviously other means of attic ventilation are needed. But first, we should establish criteria all attic ventilation schemes must meet and see how well the traditional methods perform.

1. There must always be an exhaust vent in a negative pressure area.
2. It must be a system in which the separate parts work in unison.
3. The system must work at the most crucial time, i.e., when there is no wind.
4. The system must substantially reduce rain and snow infiltration.
5. Air flow must be along the underside of roof sheathing.

6. The flow of air must be consistent and in the same direction, regardless of wind direction.

7. One-and-one-half (1-½) cubic feet per minute of air per square foot of attic floor area must be moved along the underside of the roof sheathing.

Gable End Vents—When wind is parallel to gable vent louvers, the entering air stream drops down, flows along the attic floor (over the insulation) and exits through the negative pressure vent.

As long as the wind blows as shown in Fig. 4–1 there will always be an exhaust vent in the negative pressure area, and criteria number 1 and 2 seem to be satisfied. But wind does not blow steadily in one direction. A shift in direction can change a positive pressure to negative; intakes become exhausts and exhausts change to intakes, while at the same time each becomes an intake (Fig. 4–2). Louvers work independently of each other, not as a system. There is no ventilation into the upper part of the attic, the underside of the sheathing is not ventilated, and very little moisture is removed. There is no air movement due to thermal effect.

What little effectiveness these vents may have depends on wind. Clearly all six of our criteria cannot be met. Gable end louvered vents do not perform well.

Soffit Vents—Figs. 4–3 and 4–4 show the air flow pattern with soffit vents. Soffit vents provide effective air flow regardless of wind direction and there is a balance between intake and exhaust areas. Air flow is confined to the attic floor. There is no ventilation into the upper part of the attic, the underside of the sheathing is not ventilated, and very little moisture is removed. There is no air movement due to thermal effect.

There is a common but mistaken belief that combining ridge and soffit vents causes them to work together as a system. Although this is a combination of high and low vents, the air flow patterns peculiar to each vent are not changed. As is evident from Figs. 4–5 and 4–6, criteria 2, 3, 4, and 5 are not met. There is no thermal effect, air flow is confined to attic floor and thus over insulation, resulting in the possible degradation of the insulation's effectiveness. That is, its R value can be reduced.

Given a choice between no attic ventilation and gable or soffit vents, one should opt for the latter. But it should be obvious that these ventilation schemes are marginal at best. For those who remain uncon-

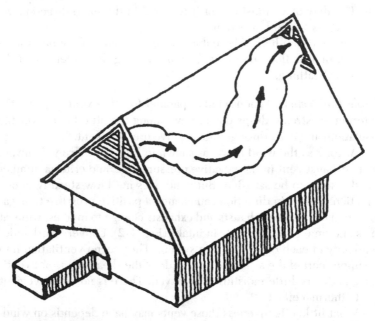

Fig. 4–1. When the wind strikes a building, it creates an area of high positive pressure. The air flows up, around, and over the building, creating areas of negative pressure, or suction, and causing vents located on negative side to become exhaust vents.

vinced, we have a question: If gable louvered vents work as well as many claim, why are 95% of all ice dams found on houses with gable end louvered vents? Of course, you'll find ice dams on houses without attic ventilation, on houses with metal roofs, and on houses with ridge and soffit vents. But this does not alter what is an observable truth. In other words go see for yourself.

Other Ventilation Schemes—There are other combinations of vents such as roof louvers with or without soffit vents, turbine vents, and attic fans. We will not go into an analysis of air flow patterns with these devices. Suffice it to say they are highly ineffective and fail to meet the 7 criteria for effective attic ventilation.

Turbine vents are rare on residential structures in New England, but common in the Midwest. The vanes can rotate in two directions, and as a result rain and snow may enter. In Oklahoma, the insulated

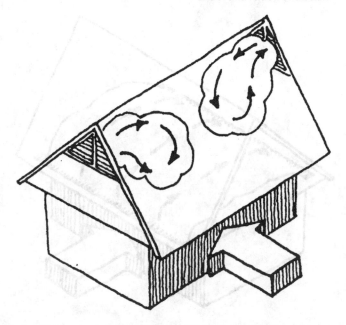

Fig. 4–2.   With wind perpendicular to gable end vents, air moves in and out of the same vent.

attic floors of houses are covered with sand brought in by turbine vents. Alas, they fail at the most crucial time, that is, when there is no wind.

Some manufacturers of modular houses equip the attics with attic fans in the mistaken belief that they "make a difference" by reducing attic temperatures and therefore reducing air conditioning loads. At first blush this seems reasonable, but numerous studies conducted by National Bureau of Standards (NBS), universities, utilities, and others cast serious doubt on these claims. Research has shown that attic ventilation may reduce attic temperatures significantly, but have negligible effect on the total cooling load. They do not and cannot save energy. Simply stated, they increase the electric bill without reducing the air conditioning costs.

Research by NBS and others has shown that significant energy savings are possible using a whole-house fan. These fans draw cooler air into the house through windows and other openings during the night. The air conditioner is operated only during the heat of the day.

Given a choice between no attic ventilation and any of these tradi-

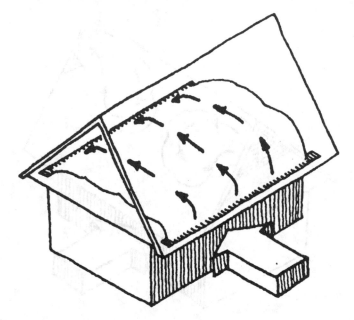

Fig. 4–3. Wind perpendicular to soffit.

tional schemes, you should opt for venting. But it is clear these ventilation schemes are only marginally effective.

Solutions

Soffit-Ridge Vent

Our search has shown the considerable deficiencies in traditional methods of attic ventilation. We come finally, then, to the combination of ridge and soffit vents.

Fig. 4–7 and 4–8 illustrate the effectiveness of this system (criteria 2) in using wind pressure and thermal effect to provide a continuous flow of air under the roof sheathing. Intake at the soffit vents allows air to move up along the underside of the roof sheathing and exit at the ridge. As is obvious from the illustrations, air flow is consistent and in the same direction, *regardless of wind directions.*

Actual testing of ridge vents with and without external baffles in

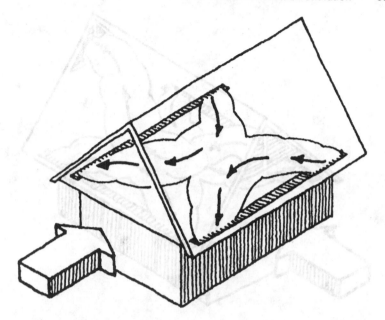

Fig. 4–4. Wind parallel to soffit.

an airstream to which water spray was added, showed water entering the baffleless ridge vent. Approximately 1½ quarts of water entered through the entire length of the ridge vent and fell to the attic floor.

The ridge vent with the baffle, even in a 100 mph wind, took in only one ounce of water.

Photographs of these tests show the baffle preventing the entrance of water and the water jumping over the top of the ridge. The baffle serves both to protect the ridge from water penetration and causes the wind striking it to jump over the top of the ridge. A venturi action is created, which results in negative pressure on both windward and leeward sides of the ridge, resulting in air being sucked out of the attic.

The ridge vent is always in negative pressure and is always an exhaust vent regardless of wind direction. It is the only system that reduces attic floor temperatures when wind speed is at zero, and the only scheme that meets the seven criteria of effective attic ventilation.

Balance—Manufacturers of ridge vents have been debating whether or not balance between ridge and soffit net free venting areas (NFVA)

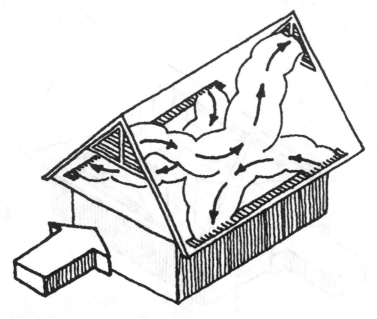

Fig. 4–5. Gable and soffit vent air flow pattern with wind parallel to ridge

is necessary. Some argue that they have found no physical reason why there has to be a balance, but at the same time admit that ridge venting is more efficient with soffit vents. Other manufacturers require the installation of soffit vents with their ridge vents. Of course, with classic capes and gambrels, soffits cannot be used, so balance is impossible.

One manufacturer has developed a filtered ridge vent to prevent the intrusion of rain and snow, as could happen with ridge vents on houses without soffit vents.

While the debate among ridge vent manufacturers continues, the number of ventilation-related roof failures and resulting lawsuits continues at an alarming rate.

Without two openings, ventilation cannot occur.

Ridge vents must be driven or "fed" with a continuous stream of air from the soffits. Otherwise, they can reverse. The windward side becomes an intake vent and the leeward side an exhaust vent. The degree of balance that should exist between the ridge and soffit is not

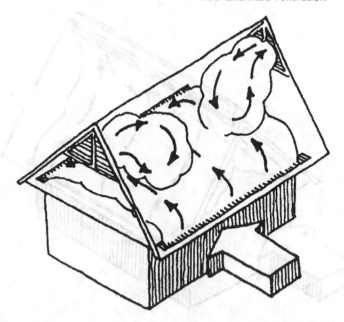

Fig. 4–6. Gable and soffit air flow pattern with wind perpendicular to ridge

known exactly. However, roof failures on houses equipped with ridge vents and blocked soffit vents have occurred. There is some minimum NFVA of soffit vents below which the ventilation system fails.

A good place to start with balance is the one-to-one ratio that exists between the nine square inches per lineal foot of ridge and the nine square inches per lineal foot of double louvered soffit vents. An absolute balance is not necessary. The NFVA of soffit vents is more readily changed, so an imbalance of about 10 percent to 15 percent is acceptable.

Because most houses have only about a 6-inch overhang, aluminum, pressed board, and vinyl soffit vents should not be used. One square foot of most vinyl soffit vents—those with ⅛-inch diameter holes spaced ½-inch on center—have only between 5 and 6 square inches of NFVA. When cut to fit a 6-inch overhang, the NFVA is reduced to about 3 square inches. The experience with ice dams indicates this is about the point at which the system "fails."

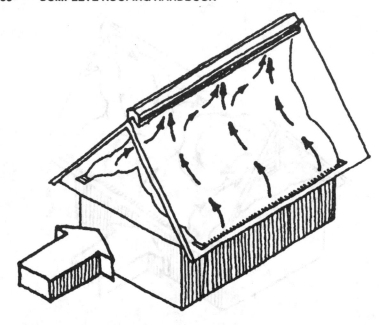

Fig. 4–7. Ridge and soffit vent with wind parallel.

Drip edge vents have 4 square inches or less of NFVA. The number of complaints about ice dams from owners of houses equipped with drip edge vents is rising. Snow can enter the attic through the vents to a distance of 6 or more feet before piling up on the insulation. No current available vent has the baffle that would prevent snow penetration.

Screen soffit vents have screen holes less than $\frac{1}{16}$-inch diameter, which results in reduced NFVA. If you have ever been in a screen tent, have you noticed how hot it is? The mesh is so fine that air flow is reduced by 70 percent or more.

Section 709.1 of the BOCA Code (1987) calls for the use of "corrosion resistant mesh not less than $\frac{1}{4}$-inch nor more than $\frac{1}{2}$-inch in any direction." No one manufactures such a mesh.

The double louvered soffit vent (avoid the single louvered vent because there are two ways to install it and one of them is wrong) should be installed next to the fascia but not in the middle of the soffit.

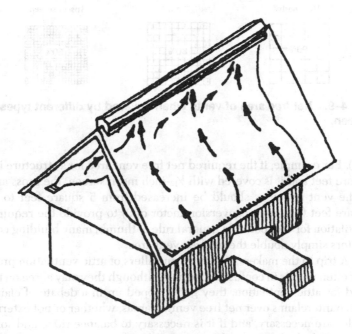

Fig. 4–8. Ridge and soffit vent with wind perpendicular.

As the soffit vent gets closer to the building and to high positive pressure, the possibility of rain and snow intake increases.

Screening—Vents are usually covered with screening or wire cloth to prevent insects or small birds from entering through the opening. Gable vents are covered with both louvers and screening. Covering the vent protects the opening, but it also reduces the net free area and restricts air flow. The amount of reduction will depend on the type of cover used over the vent. The small openings in an insect screen reduce the net free area by approximately 50 percent (Fig. 4–9). The ½-inch mesh screen, on the other hand, reduces the net free area by only 10 percent. To maintain adequate ventilation, the area of the vent openings must be increased to offset the obstructing cover. The amount of the increase is determined by multiplying the required net free vent area by a conversion factor for each type of vent cover (Table

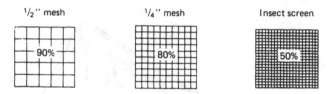

Fig. 4–9. Net free area of vents when covered by different types of screen.

5–1). For example, if the required net free vent area for a structure is 5 square feet and it is covered with $\frac{1}{16}$-inch mesh screen, the gross area of the vent openings should be increased from 5 square feet to 10 square feet (5 times a conversion factor of 2) to provide the required ventilation for the roof. As a general rule of thumb, many building contractors simply double the net free vent area.

A trip to the makers, sellers, or installers of attic ventilation products often results in confusion because, although they may agree on the need for attic ventilation, they get wrapped up in a debate of claims and counterclaims over net free venting areas, whether or not external baffles are necessary, and if it is necessary to balance ridge and soffit vents. With the basic ventilation knowledge you now have, it is possible to cut through the confusion and develop a ventilation scheme that is best for your situation.

The continuous ridge/soffit vent system is not foolproof. Under certain conditions of location: trees, wind and so on, its performance

Table 4–1. Opening Size Adjustments for Different Types of Vent Covers

Type of Covering	Size of Opening
$\frac{1}{4}$″ hardware cloth	1 × net vent area
$\frac{1}{4}$″ hardware cloth and rain louvers	2 × net vent area
$\frac{1}{8}$″ screen	1$\frac{1}{4}$ × net vent area
$\frac{1}{8}$″ screen and rain louvers	2$\frac{1}{4}$ × net vent area
$\frac{1}{16}$″ screen	2 × net vent area
$\frac{1}{16}$″ screen and rain louvers	3 × net vent area

(Courtesy Mineral Insulation Manufacturers Association, Inc.)

could be adversely affected. But overall, it is the best bet among all other so-called attic ventilation schemes.

Ventilation Installation

Soffits or Eave Vents

The overhang of a pitched roof at the eave line is called the *cornice*. The cornice consists of a fascia board nailed to the ends of the rafters, a soffit for a closed cornice, and appropriate moldings. In open-type cornice construction, the bottoms of the rafters in the roof overhang are exposed to view. The underside of the overhang is closed off by a soffit in closed-type cornice construction (Fig. 4–10). The soffit forms a connection between the roof and the sidewall, and is made of wood or plywood.

The inlet vents of a gable roof are installed along the soffit as one continuous slot (Fig. 4–10). A continuous louver vent should be installed near the outer edge of the soffit near the fascia to minimize the

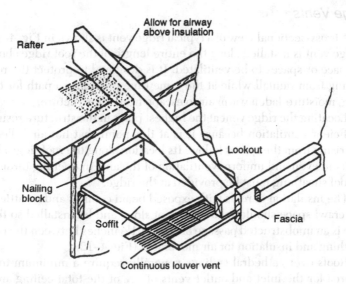

Fig. 4–10. Continuous louver vent.

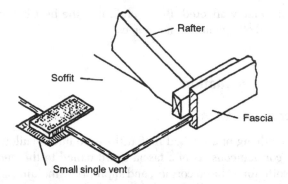

Rafter

Soffit

Fascia

Small single vent

Fig. 4–11. Small vents will not provide enough attic ventilation.

possibility of snow entering. This type of inlet vent may also be used on the extension of flat roofs. If the soffit is made of wood or a wood-type material, these louvered vents can be obtained in most lumberyards or hardware stores and are easy to install. Individual vents usually will not provide enough NFVA. (Fig. 4–11).

Ridge Vents

A cross-sectional view of a typical ridge vent is shown in Fig. 4–12. A ridge vent is installed along the entire length of the roof ridge above the space or spaces to be ventilated. It is designed to protect the roof opening from rainfall while at the same time providing a path for the rising, moisture-laden warm air to escape from the structure.

Locating the ridge vent at the highest point in the structure results in efficient ventilation because it is at this point that hot air collects after rising from the spaces below. Its location along the roof ridge also results in even and uniform ventilation of the air. An opening through the roof sheathing must be provided at the ridge.

The insulation of roofs with exposed beam ceilings and no attic or attic crawl space should follow the roof slope and be installed so that there is an unobstructed passage of at least 1½ inches between the roof sheathing and insulation for air movement (Fig. 4–12).

Roofs over cathedral ceilings generally require a minimum total net area for the inlet and outlet vents of ⅟₃₀₀ of the total ceiling area. When a vapor barrier is used, cross ventilation can be obtained by lo-

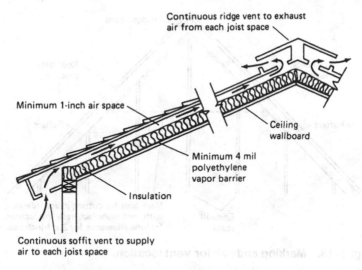

Continuous ridge vent to exhaust
air from each joist space

Minimum 1-inch air space

Ceiling
wallboard

Minimum 4 mil
polyethylene
vapor barrier

Insulation

Continuous soffit vent to supply
air to each joist space

Fig. 4–12. Ventilating roofs over cathedral ceilings with ridge and soffit vents.

cating half the required vent area at each eave. If a vapor barrier is not used, the vent area should be doubled.

Gable Vents

For structures that cannot be vented with a ridge-soffit scheme, a vent opening should be installed in the upper portion of each gable as close to the ridge as possible to provide a cross draft for ventilation. The size of the vent opening will depend on the amount of attic space that must be ventilated. Both sufficient cross draft and correct vent sizing are required if the attic or attic crawl space is to be properly ventilated.

Read the louver manufacturer's installation instructions carefully before beginning work. Mark the area to be covered by the louvers on the outside wall surface of each gable and remove the siding, brick, or other finishing materials, and the insulation so that the wall studs are exposed. Try not to remove or damage the wall materials below the base level of the louver.

After the outer wall covering materials and the insulation have

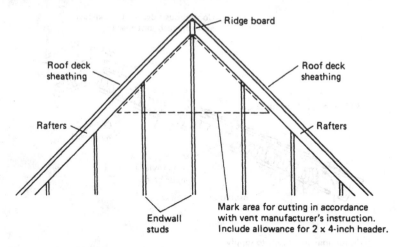

Fig. 4–13. **Marking endwall for vent location.**

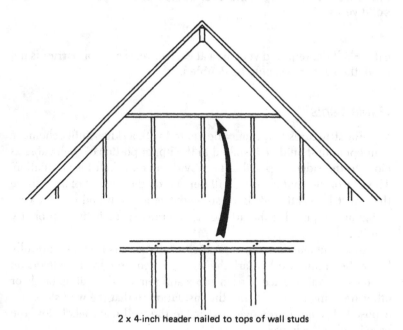

2 x 4-inch header nailed to tops of wall studs

Fig. 4–14. **Cutting opening and installing header.**

been removed, go to the inside of the attic or attic crawl space and cut away the upper sections of the wall studs to provide an opening large enough for the louver. The wall studs must be cut back far enough to provide space for both the louver and header (Fig. 4–13).

The 2 × 4 header provides a base for the louver and a cross sectional nailing support for the wall studs (Fig. 4–14). The header dimensions will depend on the bottom width of the louver. Note also that the header is cut on the oblique at each end to form the desired angle against the face of each end rafter. The angle of the cut will depend on the angle formed by the rafters.

Nail the header to the wall studs with tenpenny (10d) or sixteenpenny (16d) nails. It may be necessary to toenail the oblique butt joint (formed where the ends of the header join the roof rafters). Avoid using too many nails because this may weaken the joint. Fasten the louver in position according to the manufacturer's instructions (Fig. 4–15).

Cover the inside of the louver with a screen to prevent insects and debris from blowing into the attic or attic crawl space. The louver manufacturer will usually supply screening material with the louver, or suggest a suitable type and size to use. Ordinary copper or aluminum window screens are generally preferred.

If you have decided to make your own wood louver, instead of purchasing one ready-made at your local building supply outlet, the following suggestions are offered as a guide:

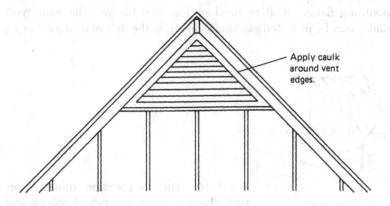

Apply caulk around vent edges.

Fig. 4–15.   Installing louvered gable vent.

1. The size and number of vents is determined by the size of the area to be ventilated.
2. The minimum net open area should be ¼ square inch per square foot of ceiling area.
3. Most louver frames are usually 5 inches wide.
4. The back edge of the louver frame should be rabbeted out for the screening material.
5. Three-quarter-inch slats should be used and spaced about 1¾ inch apart.
6. Sufficient slant or slope to the slats should be provided to prevent rain from entering.
7. The louvers should be placed as near the top of the gable as possible.

Turbine Ventilators

A *turbine ventilator* is a nonpowered roof ventilator that depends on air movement for its operation. This type of roof ventilator is commonly used on agricultural, industrial, and commercial structures where it is usually installed along the ridge line. It is also sometimes used on houses where it is mounted on the roof plane of a rear slope or incline of a pitched roof (Fig. 4–16). When it is mounted on the roof plane of a pitched roof, it is also sometimes called a *roof plane ventilator.*

The principal components of a turbine ventilator are the base, mounting flange, rotating head or top, and blades. The vane-type blades may be mounted externally on top of the unit or designed as an

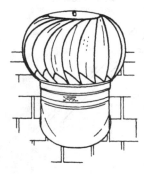

Fig. 4–16. Turbine ventilator mounted on roof slope. *(Courtesy Sears, Roebuck and Co.)*

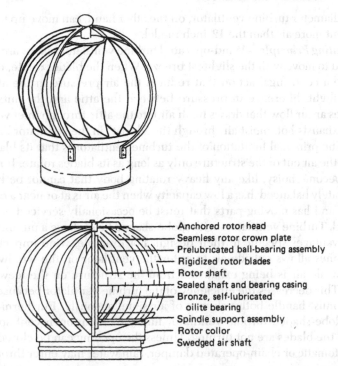

Fig. 4–17. Construction details of typical globe-shaped turbine ventilator. *(Courtesy Penn Ventilator Co., Inc.)*

Anchored rotor head
Seamless rotor crown plate
Prelubricated ball-bearing assembly
Rigidized rotor blades
Rotor shaft
Sealed shaft and bearing casing
Bronze, self-lubricated oilite bearing
Spindle support assembly
Rotor collar
Swedged air shaft

integral part of a globe-shaped head (Fig. 4–17). Globe-shaped turbine ventilators are available with either internal or external metal bracing. External metal bracing provides extra strength for installations in areas with high winds.

Most turbine ventilators are available in 12- and 14-inch diameter sizes. The turbine ventilator size indicates the diameter of the blades, the height of the ventilator above the roof deck, and the size of the cutout in the roof deck. For example, a 12-inch diameter turbine ventilator has blades with a 12-inch diameter, stands 12 inches above the roof deck, and requires a 12-inch diameter cutout in the roof deck for installation.

One 12-inch turbine ventilator with at least 2 square feet of net free inlet opening can ventilate about 600 square feet of attic. A 14-

inch diameter turbine ventilator, on the other hand, can move up to 50 percent more air than the 12-inch model.

Operating Principle—Wind-operated turbine ventilator blades are designed to move with the slightest breeze. When the blades rotate, they create a centrifugal action that reduces the air pressure in the attic. This slight difference in pressure between the attic and the outside creates an air flow that draws fresh air into the attic through inlet vents and exhausts hot, moist air through the "throat" in the ventilator base.

The principal limitation of the turbine ventilator is that its blades draw the air out of the structure only as long as its blades rotate. It may also become noisy, like any heavy rotating body that cannot be kept accurately balanced; has a low capacity when the air is at or near a calm state; and has moving parts that must be occasionally serviced or repaired. Turbine ventilators with globe-shaped heads have a number of disadvantages specific to their design. For example, wind impact on the vanes allows outside air to enter the ventilator on the windward side while air is being exhausted from the structure on the leeward side. This decreases the efficiency of the turbine ventilator because its head must handle both volumes of air simultaneously. Furthermore, the globe-shaped turbine ventilator functions as an open vent space when the blades are not turning. Unless the opening can be closed by an automatic or chain-operated damper, rain water may enter through the opening.

Installing Turbine Ventilators—The number of turbine ventilators installed on a roof will depend on the size of the ventilator and the attic square foot area. The square footage of the attic is determined by multiplying attic length by width in feet. One 12-inch turbine ventilator should be installed for every 600 square feet of attic area and one 14-inch model for every 900 square feet. These are minimums, you may have to increase them to meet your needs. If two turbine ventilators are installed on the roof, locate one ventilator one-fourth of the distance from each end of the structure. Allow a minimum of one square inch of intake vent area for every two square feet of attic floor space. When turbine ventilators are installed in the center of the roof, inlet vents should be around the overhang. Install vents under the eaves if the roof has an overhang and soffit. If the roof has no overhang, install louvers in the gable end walls.

The procedure for installing a turbine ventilator on the roof plane of a pitched roof is illustrated in Fig. 4–18 and may be outlined as follows:

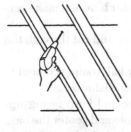

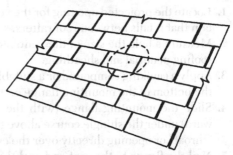

A. Position ventilator base on roof as close to ridge as possible. Measure down from ridge inside attic and drill a hole between two rafters.

B. Draw a circle around the drilled hole equal to the diameter of the turbine ventilator base.

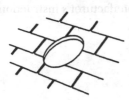

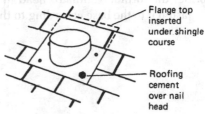

Flange top inserted under shingle course

Roofing cement over nail head

C. Cut out the hole and remove the roofing and roof deck sheathing.

D. Apply roofing cement to the bottom of the base flange and center the base opening over the hole. Nail the base flange to the roof deck and cover the nail heads with roofing cement or caulk.

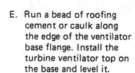

E. Run a bead of roofing cement or caulk along the edge of the ventilator base flange. Install the turbine ventilator top on the base and level it.

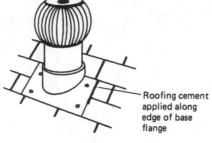

Roofing cement applied along edge of base flange

Fig. 4–18.   Installing turbine ventilator.

1. Locate the projected opening for the ventilator base and make certain that it falls between roof rafters.
2. Measure a hole the size of the ventilator base and cut through the roofing material and sheathing.
3. Apply a caulking compound or a suitable asphalt roofing cement to the bottom of the mounting flange on the ventilator base.
4. Slide the mounting flange (with the connected base facing upward) under the shingle course above the hole and center the base "throat" or opening directly over the cutout hole in the roof.
5. Nail the flange to the roof and seal the edges of the flange with a caulking compound. Cover the nail heads with caulking compound to protect them from corrosion.
6. Set the turbine ventilator head in the base opening, level it, and fasten it to the base according to the manufacturer's instructions.

CHAPTER 5

Skylights and Roof Openings

Dormers, skylights, stairs, chimneys, vent pipes, stacks, and similar types of construction require that openings be cut through the roof deck. Cutting an opening in the roof deck requires basic carpentry skills and some experience because the opening must be carefully planned. All joints must be sealed against rain penetration and any cut rafters or joists must be supported with additional framing to prevent the roof from sagging. Because this type of work affects the external appearance of the roof and offers the possibility of weakening the framework when not done properly, the local building code should be consulted and a building permit obtained if one is required.

Roof openings can be classified as either *major* or *minor*. A major roof opening requires the cutting of one or more rafters, and the installing of headers to which the cut rafters are nailed. Bracing is sometimes added between the rafters near the roof opening to provide additional strength. A minor roof opening can be constructed without cutting any of the rafters. In the case of minor roof openings, the installing of headers or special bracing is often unnecessary.

A roof opening must be carefully flashed to provide protection against leaks. Various methods of applying flashing are described in Chapter 7.

The construction of dormer roof openings is described in Chapter

6. This chapter contains descriptions of the framing and construction details for skylights, stairs, chimneys, vent pipes, and other minor roof openings.

Tools, Equipment, and Materials

Basic carpentry tools are required for cutting and framing roof openings. The tools will include a claw hammer, circular or saber saw, straightedge, combination square, measuring tape, and carpenter's apron. A sturdy ladder is required for access to the top surface of the roof.

Roof openings are framed with 2 × 4s and 2 × 6s. Framing lumber can be purchased at a local lumber yard and the amount purchased will depend on the size of the project. Other materials required for the construction of roof openings include common wire nails for the framing, roofing nails, flashing, roofing felt, and roofing cement.

Skylights

Installing a skylight on the roof is an excellent way to introduce more light into halls, baths, and interior rooms or spaces. Studies have shown the skylights provide three to five times more available light than windows.

The use of skylights has increased dramatically since the introduction of window plastic as a substitute for glass. Acrylic is the most commonly used window plastic for skylights, followed by cellulose acetate butrate (CAB) and Lexan polycarbonate sheet. The use of plastic instead of glass has practically eliminated problems of breakage, leakage, and costly maintenance. Window plastic requires no maintenance. If it must be cleaned, dirt can be easily removed with mild soap and water. Fresh paint, grease, and roofing compounds can be removed with rubbing alcohol or Butyl Cellosolve. Before using these or any other substance to clean window plastic, it is always a good idea to test the product first on an unexposed surface prior to use. *Never use a petroleum-based or abrasive cleaner.*

Clear and colorless window plastic admits the maximum amount of light and heat. When this is not desirable, a translucent white plastic can be used to provide soft, diffused lighting. Window plastic is also

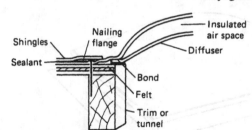

Shingles
Sealant
Nailing flange
Insulated air space
Diffuser
Bond
Felt
Trim or tunnel

Fig. 5–1. Double-dome construction with domes fused together at edges.

available in bronze or gray to further reduce the amount of heat and light passing through the skylight.

Skylight manufacturers have been able to reduce heat loss through window plastic by using double or triple-dome construction. If the domes are fused together at the edges, as in Fig. 5–1, the air space functions as a thermal insulator. These domes are permanently sealed against dirt, air, or insect infiltration. R-values (thermal resistance) and U-values (overall coefficient of heat transmission) for the different types of plastic are listed in Table 5–1.

Condensation may form between the layers of a double or triple dome briefly during extreme weather conditions, but it usually dissipates without any visible sign as weather conditions change. Condensation is most noticeable if all dome layers are clear plastic. Many skylights are designed with condensation gutters and weep holes (or weeping rings) to eliminate moisture (Fig. 5–2).

Types of Skylights

Skylights can be custom-designed or purchased factory-made from the manufacturer. Prefabricated skylights are available in a variety of shapes and sizes for installation on both pitched and flat roofs.

Table 5–1. R-values and U-values for Different Types of Plastic Materials Used in Double-dome Skylight Construction

Plastic Material	U-Factor		R-Factor	
	Winter	Summer	Winter	Summer
Acrylic	.71	.50	1.40	2.00
Cellulose Acetate Butrate (CAB)	.56	.53	1.80	1.88
Lexan Polycarbonate	.288	.281	3.47	3.56

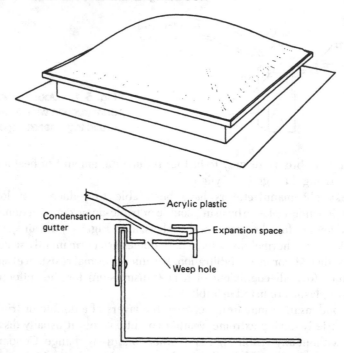

Fig. 5–2. Self-flashing skylight dome with condensation gutter and weep holes.

As shown in Fig. 5–3, skylights are available in many different shapes and styles. The dome, square, and rectangular shaped skylights are the types most commonly used on the pitched roofs of houses. Most skylights are fixed units, but ventilating skylights are also available. Ventilating skylights, which can be operated by hand, pole, or a small electric motor, serve as a source for both additional light and ventilation.

Skylights may be installed over a box frame (or curb) constructed around the roof opening or directly mounted on the plane of a pitched roof. Some curb mounted skylights are manufactured with insulated or noninsulated integral curbs which attach directly to the roof deck (Fig. 5–4). Others require that a curb be constructed on site from wood, concrete, or metal and then secured to the roof deck and flashed (Fig. 5–5). The skylight is then installed over the curb and secured to it with corrosion-resistant fasteners. If the skylight is not designed to be used

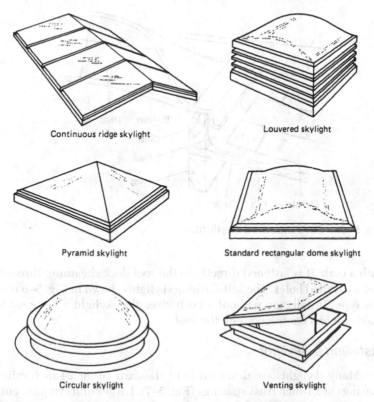

Continuous ridge skylight

Louvered skylight

Pyramid skylight

Standard rectangular dome skylight

Circular skylight

Venting skylight

Fig. 5–3.   Examples of different types of skylights.

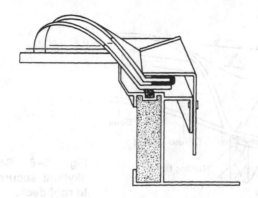

Fig. 5–4.   Skylight with insulated integral curb.

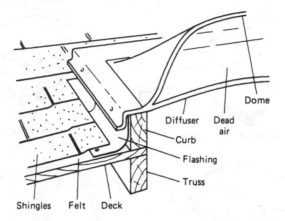

Fig. 5–5. Curb mounted skylight.

with a curb, it is fastened directly to the roof deck sheathing through factory-drilled holes. The self-flashing skylights shown in Fig. 5–6 is of this type. Installation without a curb gives the skylight a low profile making it less conspicuous on the roof.

Installing a Skylight

Many skylights are designed to fit standard 16- or 24-inch rafter spacings or 24-inch truss spacings (Fig. 5–7). Larger units require cutting rafters or trusses and framing the opening. Whether or not rafters or trusses require cutting will depend on the inside dimension (roof

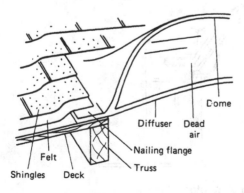

Fig. 5–6. Self-flashing skylight secured directly to roof deck.

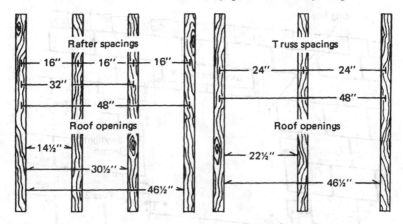

Fig. 5–7. Roof opening dimensions for different rafter and truss spacings.

opening) listed by the manufacturer for the skylight. For example, a 14¼-inch wide skylight can be used on a roof with rafters spaced 16 inches apart without having to cut any of the rafters. The skylight is simply placed over a pair of rafters, and the required opening is cut out of the roof deck area between them. Similarly, a 22¼-inch wide skylight can be installed on a roof with roof rafters spaced 24 inches apart without cutting any of the rafters. When cutting the rafters is necessary in the installation of the skylight, headers must be added and a curb constructed to serve as a base for the unit.

Figs. 5–8 to 5–11 illustrate the procedures used to install a small skylight on a roof with rafters spaced 16 inches on center. Because the skylight is less than 16 inches wide across its inside opening, the rafters do not have to be cut.

Begin by marking the size of the planned opening on the bottom of the roof deck sheathing inside the attic or attic crawl space. The size of the roof opening marked on the sheathing will correspond to the inside opening dimensions of the skylight plus 6 inches added to its length for framing the opening with headers. The 6 inches are distributed so that there are 3 inches at the top of the skylight and 3 inches at the bottom (Fig. 5–8). Skylight sizes commonly begin at 14¼ × 14¼ inches, which is small enough to fit between rafters without cutting. The first dimen-

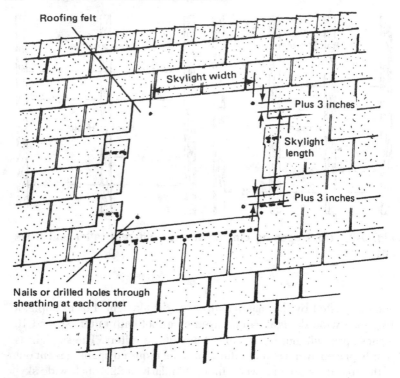

Fig. 5–8. Roofing removed from projected skylight opening.

sion is the width of the skylight inside opening and the second dimension is its length.

Use an electric power drill to drill a hole or drive a tenpenny (10d) nail through the roof at each corner of the square marked on the bottom of the sheathing. Remove the shingles covering the area marked by the four drilled holes or nails plus an additional 5 or 6 inches around them. Do not remove the roofing felt.

Mark the dimensions of the skylight opening on the roofing felt and cut away the roof deck sheathing with a power circular saw. Install doubled 2 × 6-inch headers at both the top and bottom of the roof opening to support the skylight (Fig. 5–9). Cut rectangular pieces of plywood thick enough to lie flush with the sheathing and large enough to

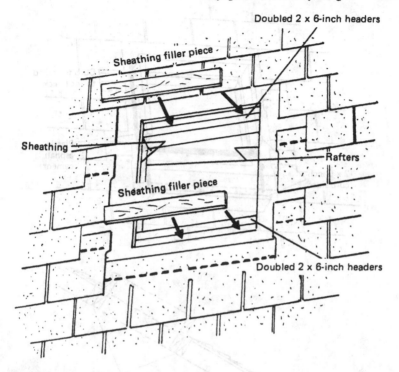

Fig. 5–9. Installing headers.

cover the headers. Nail the plywood pieces in place and construct a curb around the roof opening (Fig. 5–10). Curbs may be constructed of 2 × 4s or 2 × 6s depending on the skylight manufacturer's instructions. In either case, the inside surfaces of the curb must be flush with the inside surfaces of the roof opening framework (i.e., headers and rafters). Coat the surface where the curb will lie with plastic asphalt cement, position the curb, and toenail its sides to the framing and roof deck with nails spaced 4 to 6 inches apart. Fit the metal frame of the skylight over the curb and secure it in place with the required number of fasteners (Fig. 5–11).

A 48-inch wide skylight requires the construction of a major roof opening. As shown in Fig. 5–12, it requires the cutting of two rafters on a roof with rafters spaced 16 inches apart on center. Only one rafter

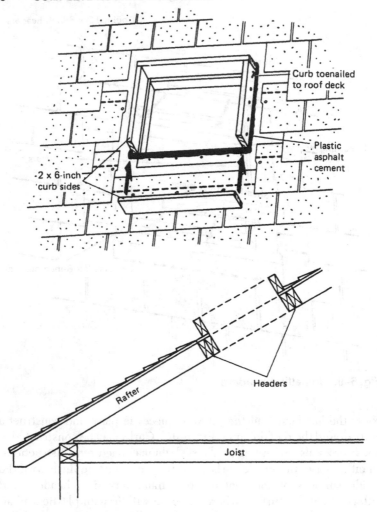

Fig. 5–10. Constructing skylight curb.

needs to be cut when the spacing is 24 inches apart on center. In any event, the roof must be braced around the planned opening *before* any cutting is done, or the roof deck may sag, causing damage to the sheathing and roofing materials (Fig. 5–13).

As with all types of roof openings, an opening constructed for a

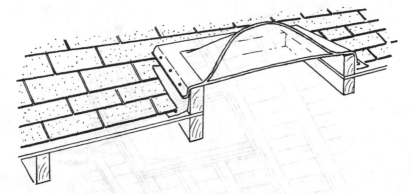

Fig. 5–11.   Cutaway view of skylight installed over curb.

skylight must be carefully flashed to provide protection against leaks. Typical flashing methods are described in Chapter 7.

Roof Access Openings

A roof access opening is sometimes installed on a flat roof, particularly on the roofs of commercial or industrial buildings. When this is the case, a major roof opening must be constructed in the deck. This requires cutting one or more ceiling joists or rafters and installing headers to reinforce the framing around the opening. Construction details for major roof openings are illustrated in Fig. 5–14.

The first step in constructing an access opening for a flat roof is to calculate its exact location. The ceiling joists or rafters should then be marked for cutting and the remaining portions braced with 2 × 4s extending down to the floor below. The bracing will prevent the roof from sagging after the opening has been cut through the framing. If stairs are used and they are to be enclosed by partitions, the studs in the partitions will provide excellent support for the ceiling joists or rafters. If no enclosing partitions are planned, permanent braces should be installed from the framing joists or rafters to the floor below.

Roof scuttles and hatches for either ladders or stairs may be placed over the roof opening and secured with nails or bolts through holes provided in the flanges. Roofing material is inserted under the integral

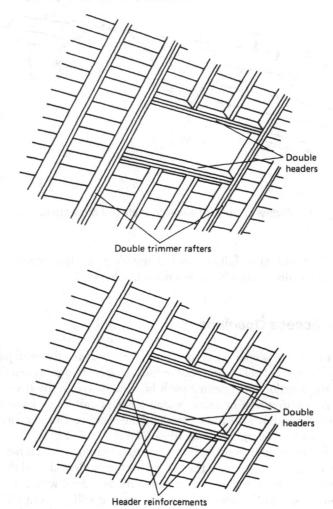

Fig. 5–12. Two methods of framing major roof openings for sky-lights.

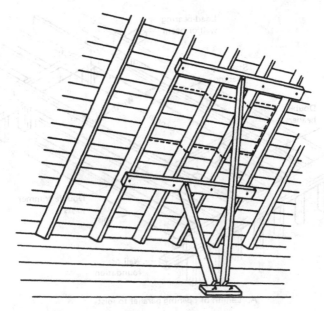

Fig. 5–13.   Bracing rafters before cutting them.

cap flashing and caulked with oakum and brush asphalt. Examples of roof scuttles and hatches are shown in Figs. 5–15 and 5–16. A roof access opening may also be covered by a wood trap door or a small shed-type structure.

Chimney Openings

A chimney that projects through the roof deck usually requires a major roof opening. At least one roof rafter must be cut and headers installed to frame the opening (Fig. 5–17).

At least two inches of air space should be provided between the masonry walls of the chimney and the roof framing and deck materials. This is necessary to protect the wood from the heat radiating from the surface of the chimney wall. Always check the local building code for the minimum allowance because it may vary among localities.

Bridging of the same size as the rafters is usually installed in the

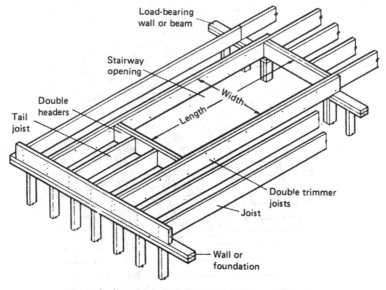

A. Length of opening parallel to joists

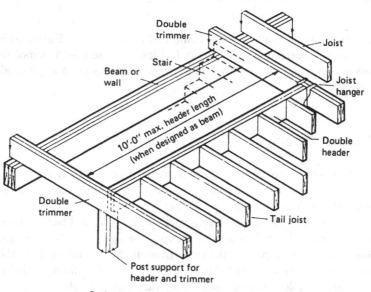

B. Length of opening perpendicular to joists

Fig. 5–14. Two methods of framing stair openings.

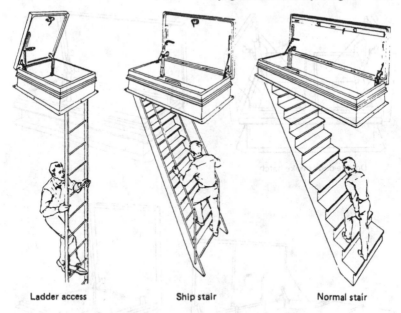

Ladder access   Ship stair   Normal stair

Fig. 5–15.   Various types of roof scuttles.

area near the chimney to provide additional strength for the framing. The bridging runs in the same direction as the headers, that is, at a right angle to the rafters.

Vent Pipe and Stack Openings

Vent and stack openings in the roof deck do not require the cutting of any of the framing members, because the diameter of the opening is invariably less than the 16-inch spacing of the rafters.

If possible, the opening in the roof deck should be placed so that it is directly over the point at which the vent pipe or stack protrudes through the attic floor. The easiest way to mark the opening is to hang a plumb bob from the underside of the roof deck. The point of the plumb bob should hang in the exact center of the vent pipe or stack opening (Fig. 5–18).

If the vent pipe or stack protrudes through the attic floor at a point directly under the roof rafter, a bend or elbow should be installed in

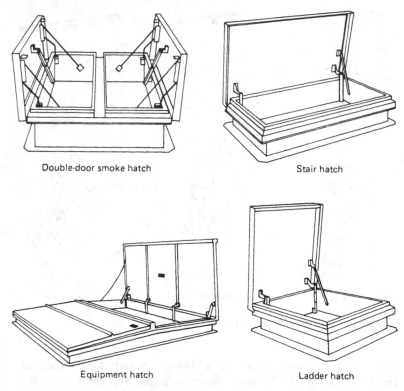

Double-door smoke hatch

Stair hatch

Equipment hatch

Ladder hatch

Fig. 5–16. Examples of different types of roof hatches.

the piping so that the opening in the roof deck will be located between rafters. This is not as satisfactory an arrangement as a straight pipe, because the bend in the piping offers resistance to the flow of the rising gases, but it does avoid the more complicated and time consuming procedure of cutting a rafter and installing headers and trimmers. The pipe bend or elbow should be installed first and then a plumb bob can be hung from the underside of the roof deck to determine the location of the opening.

Any pipes carrying hot gases and other by-products of the combustion process should have an air space between the surface of the pipe and the wood of the roof deck (Fig. 5–19). The size of the air space can be determined by consulting the local building code. The interior of

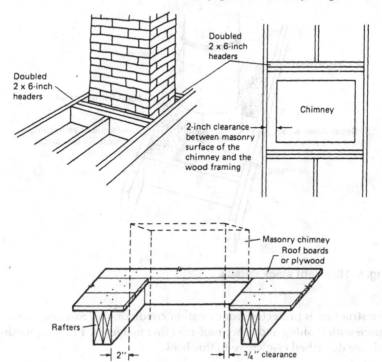

Doubled
2 x 6-inch
headers

Doubled
2 x 6-inch
headers

Chimney

2-inch clearance
between masonry
surface of the
chimney and the
wood framing

Masonry chimney
Roof boards
or plywood

Rafters

2''

¾'' clearance

Fig. 5–17. Chimney opening framing details.

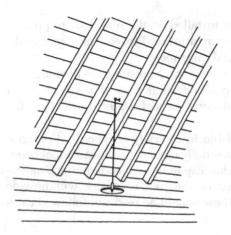

Fig. 5–18. Using plumb bob to mark center of opening on sheathing.

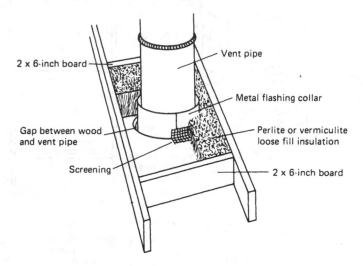

Fig. 5–19. Vent stack details.

the structure is protected from weather conditions by covering the air space with flashing and other roof-covering materials. Flashing methods are described elsewhere in this book.

Roof Ventilators

Ventilators are sometimes installed on the roof deck to provide sufficient ventilation in the attic or attic crawl space. These roof ventilators are used on both flat and sloping roofs.

As is the case with a vent pipe or stack opening, the opening for a roof ventilator is made between the rafters. Because this is a minor roof opening, there is no need to cut any of the roof rafters in order to install the ventilator.

Roof ventilators are available in a variety of types and designs, ranging from those that are essentially nothing more than a roof opening protected by a metal weather cap to those incorporating a motor-driven fan. The manufacturer of the roof ventilator will provide installation instructions, and these should be read and followed carefully.

Interior Roof Drains

Flat roofs on commercial buildings are often equipped with interior roof drains to prevent the ponding of water when it rains, or when ice or snow melt.

Roof drains must be located at the lowest point of the roof to obtain proper drainage. Each roof drain should be provided with a wide flange, expansion joint sleeves, and strainer construction that will allow adjustments for meeting variations in roof level or slope (Fig. 5–20).

Expansion Joints

Structural expansion joints are installed on flat built-up roofs of large expanse to protect the roof membrane from stress forces (Fig. 5–21). The location of a structural expansion joint should be determined by an architect, engineer, or building contractor knowledgeable about the types of stresses found on large roof decks.

Structural expansion joints should be installed at the following locations:

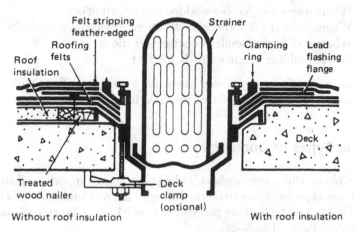

Fig. 5–20. Interior roof drain construction details.

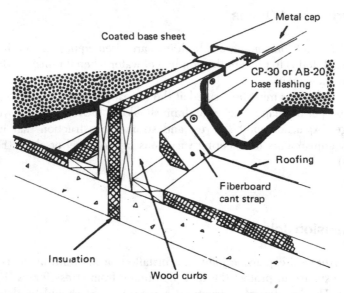

Fig. 5–21. Construction details of typical expansion joint.

1. Where the deck structures change directions as in an *L*-shaped building
2. Where two dissimilar deck materials join
3. Where a new roof has been added to an existing one
4. Where there is a difference of elevation of two adjoining decks
5. Where roof frame members change directions
6. Where building length exceeds 200 feet

The omission of properly placed expansion joints quite often results in roof membrane fracturing caused by stress forces.

Other Types of Openings

Column stubs, sign anchors, railing posts, tank supports, and similar types of projections extending through a flat roof surface should be installed in a metal flashing pan with a 4-inch high collar and a 6-inch wide deck flange (Fig. 5–22).

These types of roof deck openings should be made large enough to

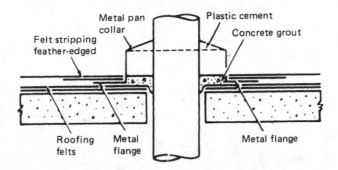

Fig. 5–22. Metal flashing pan with collar and deck flange.

provide a narrow space between the edge of the deck and the object projecting through it. This space provides an allowance for the expansion and contraction of the roof deck materials.

The flanges of the flashing pan should be embedded in plastic asphalt cement over the roofing felts. Two collars of roofing felt should then be cut to fit snugly over the flashing pan to seal the flange. The flange should be overlapped 6 inches and 9 inches respectively on all sides with each of the two roofing felt collars embedded in an application of plastic asphalt cement. The roof-covering materials are installed over the collars.

Fill the bottom of the flashing pan with one inch of concrete grout, and the remainder with plastic asphalt cement. Slope the plastic asphalt cement layer downward to the outside of the pan lips to provide sufficient drainage.

CHAPTER 6

Dormer Construction

A dormer is an enclosed room or space with vertical walls that project through the surface of a sloping roof. The word *dormer* is derived from the Latin verb *dormire* (to sleep) and was originally used to designate a room or space used for sleeping. Today, it identifies virtually any type of enclosed room or space with walls rising vertically from a sloping roof.

Types of Dormers

A dormer has one or more windows, depending on its construction. The most common type of dormer used in residential construction is the small window dormer shown in Fig. 6–1. Two or more of these dormers are generally used on the front slope of a pitched roof. The number used will depend on the length of the roof. When this type of dormer is constructed with a gable roof, it is called a *gable dormer*. Other types of roofs used on the window dormer are the hip roof and shed roof (Figs. 6–2, 6–3). Most window dormers allow both light and ventilating air into the attic. Some dormers, however, are constructed

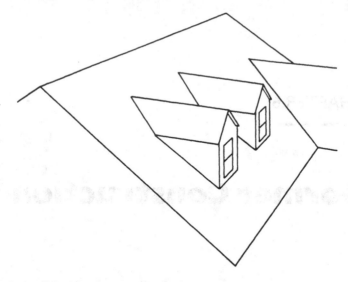

Fig. 6–1. Window dormer with gable roof.

Fig. 6–2. Window dormer with hip roof.

Fig. 6–3. Window dormer with shed roof. *(Courtesy Masonite Corp.)*

for purely decorative purposes. These nonfunctional dormers are attached to the roof without cutting an opening through the roof deck.

Some houses have large dormers that extend almost the entire length of one roof slope. These large dormers may be constructed on either the front slope of the roof, as in Fig. 6–4, or on the rear slope. Often, two or more window dormers will be constructed on the opposite slope of the same structure. These large roof-length dormers are sometimes called *roof dormers* or, because they are covered with shed roofs, *shed dormers*. In addition to providing light and ventilation, a shed or roof dormer greatly increases the attic area, particularly its head room.

Both the window dormer and the shed or roof dormer require major roof openings. A number of roof rafters must be cut and removed and the opening framed with headers and trimmers in order to accommodate the dormer framing. Construction details for both a gable dormer and a shed dormer are included in this chapter.

Fig. 6–4. Shed dormer. *(Courtesy Masonite Corp.)*

Tools, Equipment, and Materials

The tools and equipment required for the construction of a roof dormer are the same as those required for other types of wood frame construction. The needed tools include a claw hammer, circular or saber saw, straightedge, combination square, measuring tape, plumb bob and string, and carpenter's apron. A sturdy ladder or scaffolding is required for access to the top surface of the roof.

All framing lumber should be well seasoned. Lumber 2 inches thick or less should have a moisture content of not more than 19 percent. Second grades of the various softwoods are normally used for the rafters. Existing rafters cut to form the roof opening for the dormer can also be used on the dormer roof.

Use 2 × 4s for wall studs, top plates, soles, roof rafters, and ceiling joists. A 1 × 6 is recommended for the ridge board of a gable dormer roof. The trimmers and headers are cut from 2 × 6s. The attic and dormer subfloors are usually constructed from ¾-inch thick 4 × 8-foot plywood sheathing panels. The dormer roof and wall sheathing are identical to those used on the rest of the structure. The same holds true for the roofing and siding. Obtain fascia and cap molding to match the trim on the rest of the structure.

Framing lumber is ordered by the nominal size and priced in board feet. The nominal size is not the actual lumber measurement or dimension, but instead a designation of the type of frame member. For example, the 2 × 4 commonly used for wall studs actually measures 1½ × 3¼ inches. These size differences should be taken into consideration when ordering lumber for dormer construction. Nominal and actual sizes for framing lumber are listed in Table 6–1.

Windows can be purchased made to order or prefabricated complete with exterior and interior trim. The size of the window will determine the size of the rough opening in the dormer.

Purchase 5 pounds each of 16d, 8d, and 6d common nails for the job. A 20d nail is recommended for nailing the doubled 2 × 4s of the sole plate to the roof deck sheathing.

Attic Preparation

Most of the work involved in the construction of a dormer is performed from inside the attic. An unfinished attic will usually have ex-

Table 6–1. Nominal and Dressed Sizes of Framing Lumber

Nominal Size	Dressed Size
1 × 2	¾ × 1½
1 × 3	¾ × 2½
1 × 4	¾ × 3½
1 × 6	¾ × 5½
1 × 8	¾ × 7¼
1 × 10	¾ × 9¼
1 × 12	¾ × 11¼
2 × 2	1½ × 1½
2 × 4	1½ × 3¼
2 × 6	1½ × 5½
2 × 8	1½ × 7¼
2 × 10	1½ × 9¼
2 × 12	1½ × 11¼

posed floor joists that must be covered before work on the dormer can begin. Covering these joists provides a convenient work surface and protects the ceilings of the rooms below the attic from tools or materials that are invariably dropped during construction. Another important factor to be considered before constructing a dormer or finishing an attic is the type of access involved. Some attics have existing stairways, whereas others are entered through ceiling hatches. If the attic is to be finished as a bedroom, den, or other type of living quarters, the hatch entry should be replaced by a stairway. The stairway opening should be framed and the stairs built before the dormer is constructed.

Attic Subfloor

At this point in the construction of a dormer, only a subfloor is nailed to the joists. The finish boards, or floor covering materials such as carpets, tiles, or linoleum, are applied after the dormer has been framed and enclosed and the attic finished.

Subflooring commonly consists of wood boards or sheet materials such as plywood, oriented strand board, or waferboard. Oriented strand board—often abbreviated OSB—resembles waferboard but differs in the manner in which its wood strands are arranged. The orientation and length of the strands makes a sheet of OSB strong enough to span greater distances under heavier loads than waferboard.

Subfloor sheets are rated by span index. The numbers of a span index indicate the distance the sheet can span. For example, a sheet of $7/16$-inch OSB has a span index of $24/16$, meaning it can span 24 inches when used as roof sheathing and 16 inches when used as subflooring. To span a greater distance between floor joists, a thicker sheet of OSB must be used. Always check the American Plywood Association (APA) stamp on the product for its span capability.

The size of the plywood panels used to construct the attic subfloor may be a problem. Plywood is usually available in 4×8-foot panels and there may be no opening in the attic floor large enough to admit panels this large. If this should be the case, nail four or five 1×6-inch boards across the joists in the projected work area. The plywood panels can then be brought into the attic through the dormer roof opening once it is cut.

A more expensive solution to bringing materials into the attic is to use 1 × 6-inch sheathing boards or planks for the subfloor. This method requires more time because far more cutting and nailing is required than when working with plywood panels. Sheathing boards are usually applied so that they run at right angles to the floor joists (Fig. 6–5).

Plywood panels are applied with their length running parallel to the joists. Butt the panels so that the joints are centered over the joists. Nail them to the joists with common eightpenny (8d) nails placed every 6 to 8 inches (Fig. 6–6).

Before laying the subfloor, bridging should be installed between the joists for reinforcement (Fig. 6–7). As the bridging is being installed, rough in all electrical wiring, plumbing, and ductwork required for the finished attic.

Attic Access

Access openings in attic floors are framed out during the construction of the floor system. If the attic is intended only for storage, the access opening will usually be covered by a hatch which may or may not be hinged to the frame. Construction details of two access openings

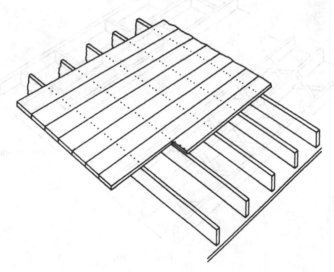

Fig. 6–5. Subfloor of sheathing boards.

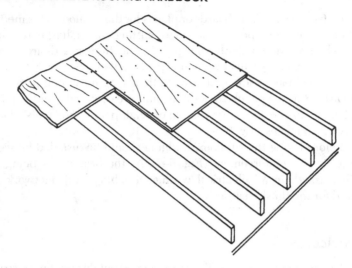

Fig. 6–6. Plywood subfloor.

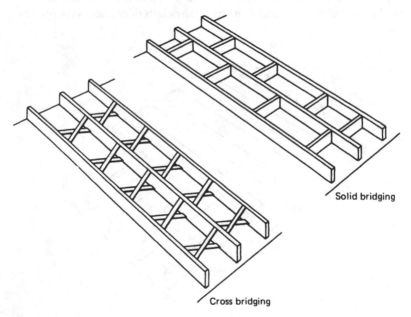

Solid bridging

Cross bridging

Fig. 6–7. Bridging between floor joists.

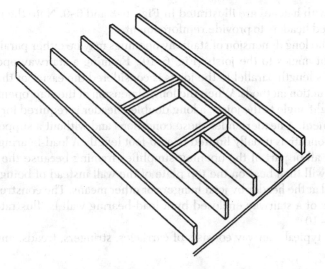

Fig. 6–8. Hatch-type access opening requiring the cutting of one or more floor joists.

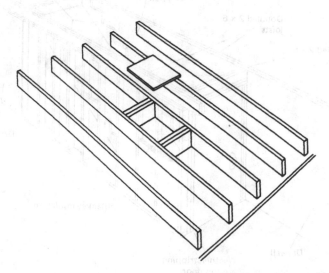

Fig. 6–9. Hatch-type access opening not requiring the cutting of floor joists.

used with hatches are illustrated in Figs. 6–8 and 6–9. Note the use of doubled headers to provide reinforcement.

The long dimension of stairway openings may be either parallel or at right angles to the joists (Fig. 6–10). Framing a stairway opening with its length parallel to the joists is considered the easier of the two construction methods. When the long dimension of the stair opening is at a right angle to the joists, a long doubled header is required for reinforcement. A header under these conditions and without a supporting wall beneath is usually limited to a 10-foot length. A load-bearing wall under all or part of the opening simplifies framing because the floor joists will then bear on the top plate of the wall instead of being supported at the header by joist hangers or other means. The construction details of a stairway enclosed by a load-bearing wall is illustrated in Fig. 6–10.

A typical stairway consists of carriages, stringers, treads, and ris-

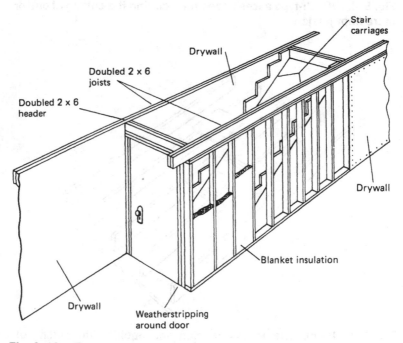

Fig. 6–10. Enclosed stairway construction details.

ers. The *stair carriages* carry the treads and support the load on the stairs. Stair carriages are usually made from 2 × 12-inch planks which are notched to receive the treads. Carriages are placed on each side of the stairs. An intermediate carriage is required at the center of the stairs when the treads are $1\frac{1}{16}$ inches thick and wider than 2 feet 6 inches. Another intermediate carriage is required when the treads are $1\frac{5}{8}$ inches thick and more than 3 feet wide.

The typical layout of a straight-run stairway connecting a lower floor to an attic is illustrated in Fig. 6–11.

If the attic is used primarily for storage or if no space is available for a stairway to a finished attic, hinged or folding stairs are often used and may be purchased ready to install. They are operated through an opening in the ceiling of a hall and swing up into the attic space out of the way when not in use. Where such stairs are to be installed, the attic floor joists should be reinforced for limited floor loading. One common size of folding stairs requires only a 26 × 54-inch rough opening. These openings should be framed out as for normal stair openings.

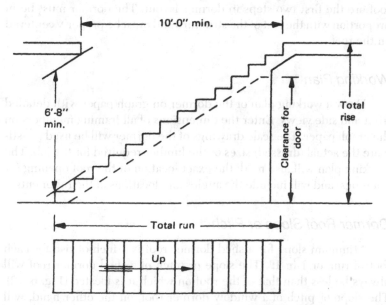

Fig. 6–11. Typical straight-run stairway between attic and lower floor.

Dormer Layout

Always check the local building code before beginning any work to make certain the planned dormer and its specifications meet code requirements. Some building developments or subdivisions also have restrictions on the types of additions that can be made to a house, and these will also have to be checked before going ahead with construction.

A local building permit will have to be obtained before construction begins. A fee is charged for a building permit, which is issued by the city or county building department after the plans or blueprints have been approved. The building permit must be posted in a conspicuous place at the construction site. There will probably be at least one visit from the building inspector while the job is in progress.

Dormer Location

Choosing the type of dormer and determining its location on the roof are the first two steps in dormer layout. The dormer must be in proportion with the rest of the structure and must be properly centered on the roof.

Working Plan

Draw a working plan of the dormer on graph paper with detailed front and side views. Enter the dimensions of all framing members on the graph paper. The scale drawings of the dormer will be used to estimate the actual (dressed) sizes of the lumber required for the job. The working plan will also mark the exact location of the roof opening (or openings) and will include the angles and locations of the rafter cuts.

Dormer Roof Slope or Pitch

Minimum slope for a shed dormer roof is 1 inch of rise for each foot of run, or 1-in-12. The slope or pitch of a shed dormer roof will always be less than that of the roof on which it is located (Fig. 6–12). The slope or pitch of a window dormer roof, on the other hand, will usually be the same as for the house roof (Fig. 6–13).

Read the sections in Chapters 1 and 2 covering slope and pitch

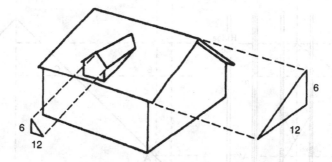

Fig. 6–12.   Window dormer roof slope.

before continuing, because these aspects of roof layout must be clearly understood.

Window Dormer Layout

Draw a cross section of the roof to scale with the rafters at the correct angle and proceed as follows:

1. Locate the top surface of the bottom header (point *A* in Fig. 6–14) at least 2 feet above the attic floor.
2. Run a vertical line from point *A* to point *B*, a scale distance of 5 feet.
3. Draw a horizontal line at a right angle to point *B* and extend it until it meets the roof at point *C*. The horizontal line *BC* repre-

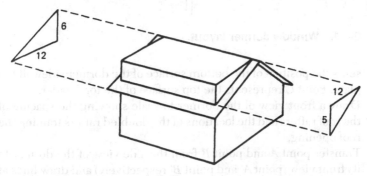

Fig. 6–13.   Shed dormer roof slope.

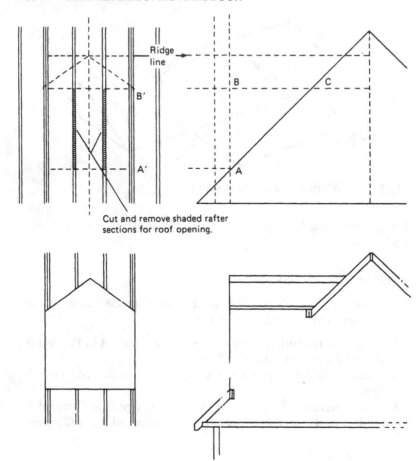

Cut and remove shaded rafter
sections for roof opening.

Fig. 6–14. Window dormer layout.

sents the position of the bottom surface of the dormer sidewall top
plate. Point C represents the top surface of the top header.

4. Draw a front view of the dormer to scale showing the spacing of
   the roof rafters and the locations of the doubled rafters framing the
   roof opening.
5. Transfer point A and point B from the side view of the dormer to
   its front view (point A' and point B' respectively) and draw lines at
   right angles to the rafters at these points.

6. Run a line perpendicular to the horizontal lines at points A' and B' in the exact center of the roof opening to locate the position of the dormer roof ridge.
7. Connect point B' to the perpendicular line at an angle equal to roof slope or pitch.
8. Transfer the height of the dormer roof ridge in the front view drawing of the dormer to its side view drawing, draw a horizontal line to represent the top of the roof ridge, and mark the point where it intersects the roof.

Shed Dormer Layout

Draw a cross section of the roof to scale with the rafters at the correct angle and proceed as follows:

1. Run a vertical line from the bottom of the ridge board (point A in Fig. 6–15) to the attic floor (point B).
2. Draw a line marking the location of the outside surface of the shed dormer front wall (point C if it is flush with the lower exterior wall or point D if it is set back from it) perpendicular to the attic floor.
3. Run a horizontal line from point A at a right angle to line AB until it intersects with vertical lines from points C and D, and label the intersecting points $E1$ and $E2$ respectively.
4. Measure the length of line $AE1$ (flush dormer wall) or line $AE2$ (set back dormer wall) and enter the amount on the scale drawing in feet.
5. Measure down from point $E1$ or point $E2$ a distance in inches *at least* equal to the distance in feet of line $AE1$ or $AE2$ to find the minimum slope (1-in-12) of a shed dormer roof. Label these points $F1$ and $F2$ respectively. If, for example, line $AE1$ was 12 feet, the distance between points $E1$ and $F1$ or between $E2$ and $F2$ would have to be a minimum of 12 inches.
6. Draw a line between point $F1$ or $F2$ and point A to represent the bottom edge of the shed dormer roof rafters.

Cutting and Framing Roof Openings

Never cut the roof opening for the dormer until the floor joists inside the attic have been covered with a subfloor or suitable temporary flooring placed directly below the projected opening in the roof.

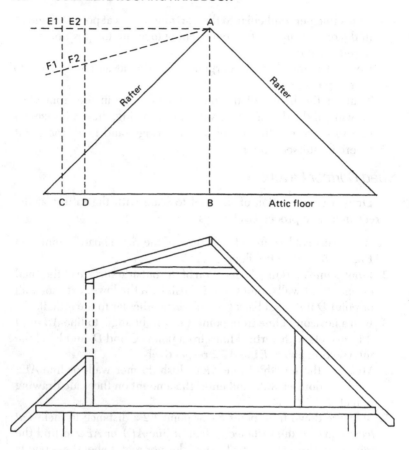

Fig. 6–15. Shed dormer layout.

This precaution will protect the ceiling of the room below from falling debris.

Before cutting through the roof, be certain that some sort of protection has been provided for the attic in case it rains, or during those periods when no work is being done. A canvas, plastic sheet, or tarpaulin large enough to extend at least a foot beyond the roof opening on all sides will provide adequate protection. The top of the cover should be nailed to the roof deck on the other side of the roof ridge. The nails are driven through short lengths of 1 × 4-inch boards to prevent the cover from being torn away from the nails in strong winds. When rain threatens, the cover should be drawn down over the opening and nailed at the bottom and sides in the same manner as along the top.

Dormer layout calculations will provide the basis for estimating the dimensions of the roof opening (see the discussion of dormer layout in this chapter).

The procedure for cutting a dormer roof opening may be summarized as follows:

1. Snap a chalk line along the bottom of the rafters to indicate the lower edge of the roof opening (be sure to provide an allowance for the double header and front wall frame).
2. Mark each rafter with a bevel square or chalk line and plumb bob for the exact angle of the cut (Fig. 6–16).
3. Install 2 × 4 braces under the roof ridge board and under all sections of rafters that will remain after cutting the roof opening.

The rafter and ridge board braces must remain in position until the roof opening is framed and permanently braced. Bracing is required to prevent the roof from sagging. Additional information about the cutting and framing of roof openings is provided in Chapter 5.

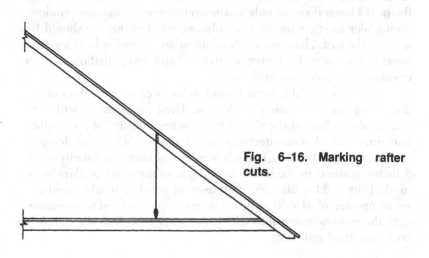

Fig. 6–16. Marking rafter cuts.

Window Dormer Construction

The window for a window dormer should be purchased complete with exterior and interior trim. The window manufacturer will provide instructions for installing the window and the trim. Read these instructions carefully before beginning work. If there are any questions, return to the building supply dealer where the window was purchased for clarification of the installation instructions.

The dormer window should match the style and dimensions of other windows in the structure. An exact match is not necessary, but avoid using window styles and dimensions so different that they make an unpleasant contrast.

Another important consideration is the *number* of window dormers planned for the roof. The main roof of an average size house or building usually has enough space for two or perhaps three window dormers without giving the surface of the roof a crowded appearance. This is usually the case if no minor roof intersects the main roof on the same slope where the window dormers are to be installed. Usually only one or two window dormers should be used on a main roof to which a minor roof is joined.

Window dormers can be installed on a minor roof, but only if the roof is long enough to accommodate them. Once again, a crowded appearance should be avoided. Dormer windows should be spaced the same distance apart as windows would be normally spaced on lower floors. If a lower floor can only accommodate one average size window, then under most circumstances only one window dormer should be used on the roof. These recommendations are offered only as a guide, because there is such a variety in architectural design that one always encounters an exception to the rule.

After the size of the dormer window has been selected, lay out the dimensions of the dormer enclosure. These dimensions will vary widely, depending on the size of the window, the size of the dormer enclosure, and other architectural considerations such as roof design.

In older construction, a rough window opening is ordinarily made 5 inches wider than the width and height of the window glass to be used. Thus, a 24 × 26-inch, double-hung window would require a rough opening of 34 × 36 inches. The extra space is used to accommodate the window frame into which the window sashes (the movable parts) are fitted and hung.

Most windows available today do not require as much space for the window frame. This is particularly true of those with aluminum or other light metal frames. In any event, the size of the rough window opening will be determined by the window dimensions.

Window dimensions also have a direct influence on the width and height of the dormer enclosure. Under normal circumstances, the width of the dormer will not exceed 10 to 12 inches on either side of the window. Thus, a dormer enclosure with a 24-inch wide window should have a maximum width of 44 to 48 inches. Once again, dimensions will vary, depending on the size of the window and other design factors; but the window area should dominate, with the surface of the dormer enclosure functioning simply as a frame for the window.

Fig. 6–17 illustrates the construction details of a typical gable-type window dormer used on a roof with rafters spaced 16 inches on center.

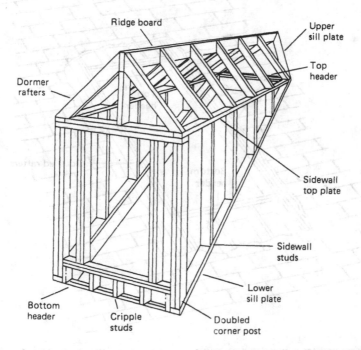

Fig. 6–17. Construction details of typical gable-type window dormer.

These construction details are applicable to any size window dormer enclosure; only the dimensions will change.

As shown in Fig. 6–18, a header usually functions as the base for the window frame as well as a nailing base for the rafter that had to be cut to make room for the dormer opening. Sometimes a plate (or sill header) is nailed across the top of the header, but the bottom of the dormer window usually remains no higher than 5 or 6 inches above the roof surface on roofs with an average slope. However, roofs with steep slopes generally require that lower cripples and a sill header be installed to raise the window opening so that it is centered properly. In larger dormers, it may also be necessary to use cripples above the window opening.

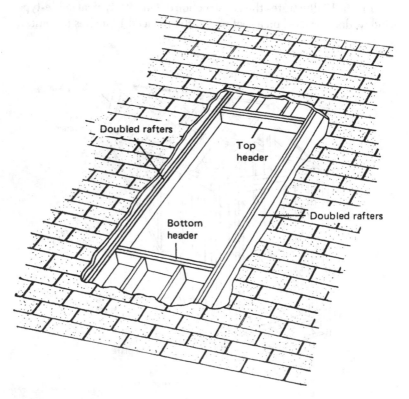

Fig. 6–18. Roofing and sheathing cut back to illustrate framing of roof opening for dormer.

The step-by-step procedure used to construct a gable-type window dormer is illustrated in Figs. 6–19 to 6–24. Framing details of window dormers covered with hip and shed roofs are shown in Figs. 6–25 and 6–26.

The same siding and roofing materials used on the rest of the structure should also be used to cover the window dormer. Installation instructions are included in the various chapters covering specific types of siding and roofing materials. For a description of how to finish the interior, see the discussion of finishing attics in this chapter.

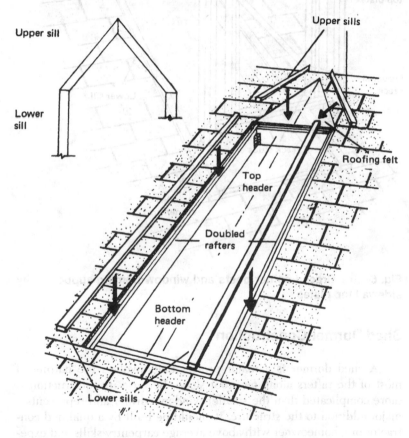

Fig. 6–19. Nailing dormer sill plates to rafters and roof deck.

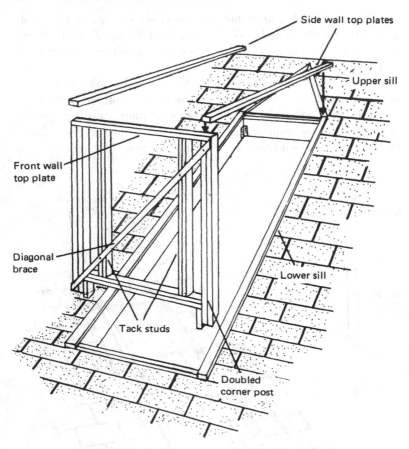

Fig. 6–20. Braced corner posts and window framing supported by sidewall top plates.

Shed Dormer Construction

A shed dormer is a major project that involves the cutting of most of the rafters along one roof slope. Because its construction is more complicated than the smaller window dormer and represents a major addition to the structure, it should be built by a qualified contractor or a homeowner with above average carpentry skills and experience.

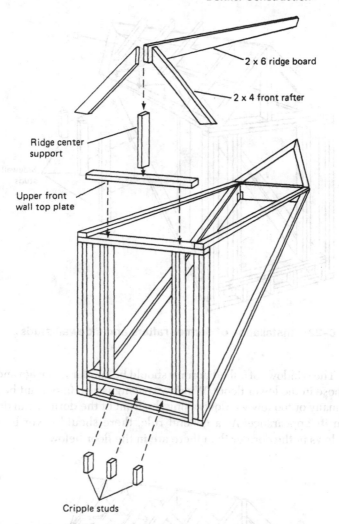

Fig. 6–21. Addition of cripple studs, ridge board, and ridge board support framing.

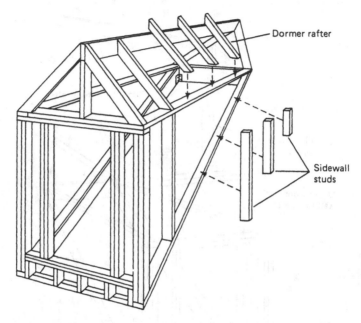

Fig. 6–22. Installation of dormer rafters and sidewall studs.

The windows of a shed dormer should be similar in design and size to those in the lower floors. Window spacing is also important because too many or too few windows along the front of the dormer can detract from its appearance. As a general rule, there should never be more windows in the dormer than there are in the floor below.

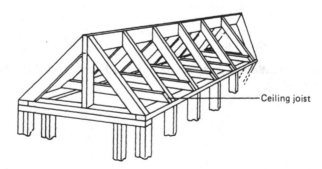

Fig. 6–23. Addition of ceiling joists.

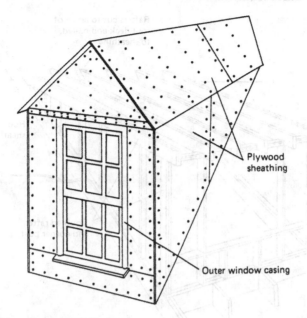

Plywood
sheathing

Outer window casing

Fig. 6–24. Finishing dormer.

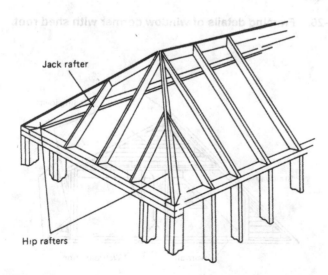

Jack rafter

Hip rafters

Fig. 6–25. Framing details of window dormer with hip roof.

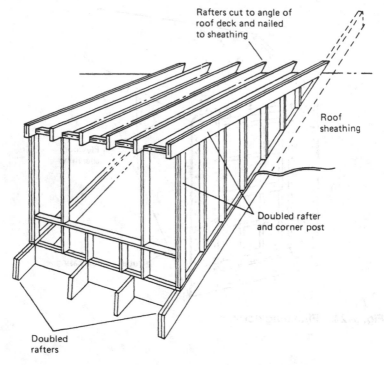

Fig. 6–26.   Framing details of window dormer with shed roof.

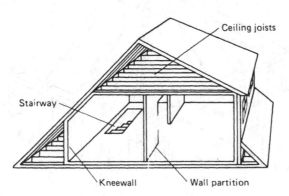

Fig. 6–27.   Cross section of typical shed dormer.

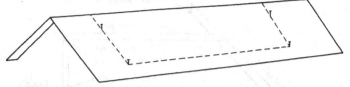

1. Mark roof opening dimensions on bottom surface of roof, drive nails through roof deck, and snap chalk lines between nails. Extend to the roof ridge both chalk lines running parallel with roof rafters.

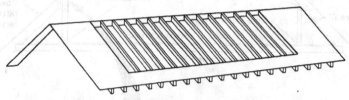

2. Remove all roofing materials within the lines, temporarily brace the roof ridge board and the lower rafter sections, and cut the rafters.

Fig. 6–28. Marking and cutting roof opening.

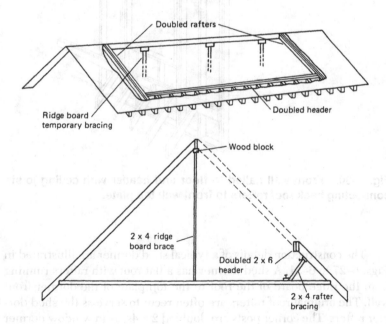

Doubled rafters

Ridge board
temporary bracing

Doubled header

Wood block

2 x 4 ridge
board brace

Doubled 2 x 6
header

2 x 4 rafter
bracing

Fig. 6–29. Framing roof opening.

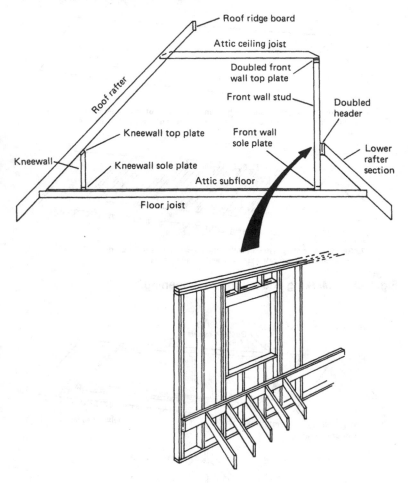

Fig. 6–30. Front wall nailed to floor and header with ceiling joists connecting back roof rafters to front wall top plate.

The construction details of a typical shed dormer are illustrated in Figs. 6–27 to 6–31. A shed dormer has a flat roof with rafters running from the ridge board of the roof to the top plate of the dormer front wall. The original roof rafters are often recut to serve as the shed dormer rafters. The corner posts are doubled 2 × 4s, as in window dormer

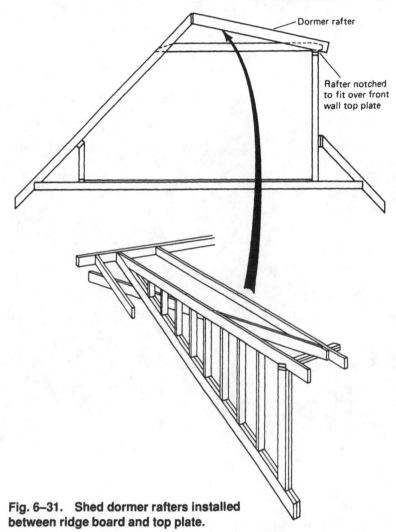

Dormer rafter

Rafter notched
to fit over front
wall top plate

**Fig. 6–31. Shed dormer rafters installed
between ridge board and top plate.**

construction, and are usually located over the third rafter from each
end of the roof. The third rafter is always doubled to reinforce the roof
opening.

The methods used to finish the exterior of a shed dormer are the
same as those described for a window dormer.

CHAPTER 7

Roof Flashing Details

An important preliminary step to the application of the roofing materials is the installation of flashing. The roof must be protected with flashing at every point in its construction where leaks might occur, especially where there may be a particularly heavy concentration of water runoff, such as in roof valleys or along the roof eaves.

Leaks frequently occur at the joint where a minor roof intersects a main roof or where the roof deck meets a vertical wall. On wood roof decks, the sheathing is subject to a certain amount of expansion and contraction under varying weather conditions, which may cause the sheathing to pull away from another surface and provide an entry point for water. The point at which a chimney projects through a roof is also susceptible to leakage. The flashing installed around a chimney must maintain a watertight seal while at the same time allowing the separate movement of the roof deck and chimney, which are built on separate foundations to avoid any stress and distortion caused by uneven settling between the two. Similar opportunities for leakage occur around projections through the roof deck, such as those made by a vent pipe, soil stack, flue, or roof ventilator.

The various types of flashing materials, applications, and application methods for pitched-roof construction are described in this chapter. The flashing of flat or low-pitched roofs is described in Chapter 11.

Flashing Applications

As shown in Fig. 7–1, the flashing for a typical pitched roof is applied along the roof valleys, rakes, and eaves; around the bases of chimneys, soil stacks, vent pipes, flues, and roof ventilators; over chimney crickets; along the sides of dormers; around skylight curbs and other square or rectangular openings; at the juncture between a roof and a vertical wall; and along roof ridges and the joint formed by the joining of two sloping roof planes.

Flashing is applied in several forms, depending on its location on the roof. Long, wide sheets of flashing material are used in the roof valleys. Narrower widths of material are applied along the roof eaves, rakes, and ridges. Base, step, and counter flashing are used elsewhere on the roof.

Base flashing is applied along the front and sides of a chimney, and along the joint formed between a roof slope and a vertical wall. *Step flashing* may be used instead of base flashing along the sides of chimneys, dormers, and junctures formed by a vertical wall and a roof slope. It is sometimes called *shingle flashing* because it is applied in the form of individual, overlapping pieces. *Counter flashing* or *cap flashing* is used to cover base or step flashing where considerable movement can

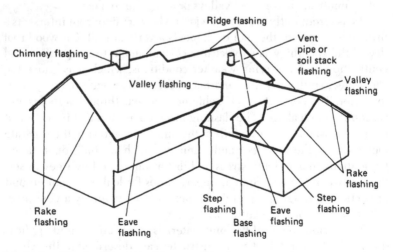

Fig. 7–1. Typical roof flashing applications.

be expected to occur between adjoining surfaces by overlapping (or "capping") it from above.

A flashing flange is usually installed around the base of a soil stack, vent pipe, flue, or other types of small, circular protrusions through the roof surface. A form of base flashing may be used to flash skylight curbs or other types of square or rectangular roof openings. Cricket flashing is a single piece of flashing cut to cover the top of a chimney cricket.

Flashing Materials

Galvanized steel, copper, aluminum, mineral surface roll roofing, and roofing felt are all used as flashing materials on both residential and nonresidential structures. Metal alloys, such as Terne metal or Terne-coated stainless steel, are also used for flashing, primarily on the roofs of public buildings or commercial structures.

In residential construction, metal flashing is generally recommended for roofs covered with wood shingles or shakes, slate, or tile, but it can also be used with asphalt shingles and other types of roofing. Mineral surface roll roofing is commonly used on asphalt shingle roofs. Roll roofing is sometimes used as a flashing reinforcement along roof eaves in combination with a metal drip edge.

Metal flashing is available in rolls or preformed units (Fig. 7–2). The performed units include drip edges and rake edges for mineral fiber, asphalt, slate, and wood shingle and shake roofs (Fig. 7–3), and metal gravel stops for built-up roofs. They are generally available in 10-foot lengths, which can be cut to size with ordinary tinsnips. Preformed metal valleys with a 1-inch crimp standing seam are produced in 14-inch and 20-inch widths with other sizes available on special order (Fig. 7–4). Valley flashing or roll flashing is also available in 12- to 18-inch rolls with most rolls containing 50 linear feet of flashing metal.

Galvanized steel is the most widely used flashing metal. It is cheaper than copper flashing, will not react to contact with masonry or cement, and is easily cut with ordinary tinsnips. The galvanized steel used for flashing is a 26-gauge metal. It must be painted with a protective finish before it is applied or it will rust and deteriorate. Even with a protective finish, a scratch or dent may cause the metal to deteriorate.

Aluminum is an ideal flashing metal for structures located near

Fig. 7–2. Aluminum and galvanized roll valley flashing. *(Courtesy Howmet Aluminum Corp.)*

bodies of water or in areas experiencing high humidity or heavy or frequent rains, because moisture will not rust aluminum or cause it to deteriorate. Aluminum flashing is available unfinished or finished with a thermally-set paint applied in full coat to both sides of the metal. The paint colors are either white or brown, which allows the flashing to be used with a wide range of roofing material colors. An unfinished or natural metal finish will weather to a soft gray appearance and will not rust or discolor. Aluminum flashing should have a .019-inch minimum thickness. If it is used in contact with masonry, cement, or dissimilar metals, a coat of bituminous paint should be applied between the surfaces to guard against electrolytic action.

Copper flashing must be a 16-ounce metal with a .020-inch minimum thickness. Copper flashing is expensive and seldom used except in quality construction. On some roofs, copper may be used to flash the

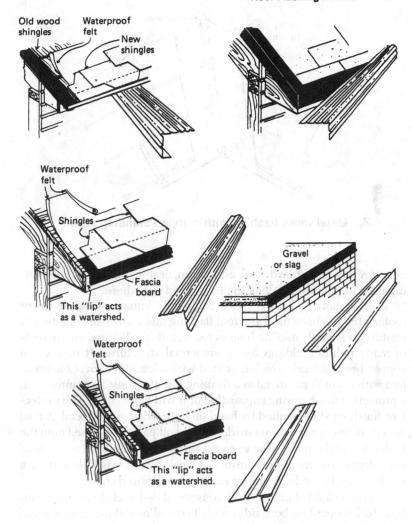

Fig. 7–3. Preformed eave and rake flashing. *(Courtesy Howmet Aluminum Corp.)*

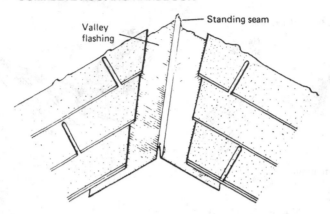

Fig. 7–4. Metal valley flashing with crimped standing seam.

chimney with galvanized steel, aluminum, or mineral surface roll roofing used in the valleys, along the eaves, and elsewhere.

Terne metal, which is more commonly known as valley tin or roofer's tin, has been used as a roof flashing metal since the eighteenth century. It is rarely used on houses today, but is still found on the roofs of many public buildings and commercial structures. It consists of copper-bearing steel, which is coated with a lead-tin alloy. One problem with using Terne metal as a flashing is that it must be painted with a protective finish during application or it will deteriorate. The protective finish must be applied by hand on both sides of the metal. A coat of very slow-drying red iron oxide-linseed oil paint is brushed onto the underside of the metal. The exposed surface receives a primer coat of slow-drying red iron oxide-linseed oil paint, which is followed by a good quality linseed oil exterior paint of any desired color.

Terne-coated stainless steel consists of 3–4 nickel chrome stainless steel covered on both sides with Terne alloy (80 percent lead and 20 percent tin). It requires no protective finish and never needs maintenance if properly installed. Its unfinished surface will eventually weather to a uniform dark gray. Terne-coated stainless steel is used as flashing primarily on public buildings or commercial structures, rarely on residential roofs due to the high initial cost.

Mineral surface roll roofing has the same composition as asphalt shingles. It will not last as long as metal flashing, but it is far less expensive and easier to install. Mineral surface roll roofing is commonly available in 36-inch wide rolls containing 36 linear feet of material. It is surfaced with mineral granules to produce a wide range of different colors. Some roofing manufacturers also produce an 18-inch wide roll for valley applications.

In order to prevent chemical reactions between dissimilar metals, copper nails should be used with copper flashing, aluminum nails with aluminum flashing, and hot galvanized nails with galvanized steel flashing. Hot galvanized steel roofing nails are also used to apply mineral surface roll roofing and roofing felt.

Eave and Rake Flashing

A metal drip edge should be nailed along each roof eave and rake. The drip edge is normally fastened along the eaves *before* the underlayment is applied and along the rakes *after* it is applied (Figs. 7–5 and 7–6). Variations of eave and rake flashing applications are shown in Fig. 7–3.

Metal drip edges are available in 10-foot lengths of corrosion resistant, nondiscoloring metal. Preformed metal drip edges of galvanized steel, aluminum, or copper are the most commonly used types of eave or rake flashing used in roofing. Other materials may be used if ap-

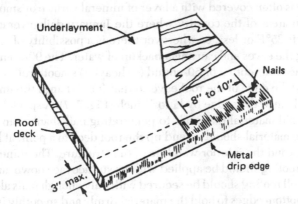

Fig. 7–5. Metal drip edge along eave. (Courtesy Asphalt Roofing Manufacturers Association)

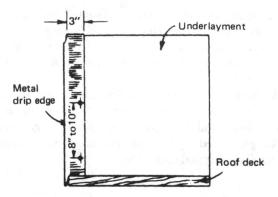

Fig. 7–6. Metal drip edge along rake. *(Courtesy Asphalt Roofing Manufacturers Association)*

proved by the roofing manufacturer. As shown in Figs. 7–5 and 7–6, the metal drip edge is secured to the roof deck with nails spaced 8 to 10 inches apart along the inner edge; it should not extend back from the edge of the roof deck more than 3 inches.

Metal drip edges along the eaves and rakes allow the water to run off the roof without coming into contact with the underlayment or roof deck sheathing. They also prevent or minimize the possibility of rain water being blown back under the roofing. The drip edge along the roof eave is often covered with a layer of mineral surface or smooth roll roofing in areas of the country where the January daily average temperature is 25°F or less, or wherever there is a possibility of ice forming along the eaves and causing a backup of water. The 90-pound mineral surface roll roofing or 50-pound (or heavier) smooth roll roofing is laid along the edge of the roof eave so that it overhangs the underlayment and metal drip edge by ¼ to ⅜ inch (Fig. 7–7). A special ice and water shield material that seals to penetrating nail shanks can also be used. The material should extend up the roof deck to a point at least 12 inches beyond the interior wall line of the structure. The mineral surface roll roofing should be applied with its surface face down, and both types of roll roofing should be secured with only enough nails along the top and bottom edges to hold the material firmly and smoothly in place. If the overhang at the eaves requires material wider than 36 inches to meet the minimum 12-inch extension beyond the interior wall line,

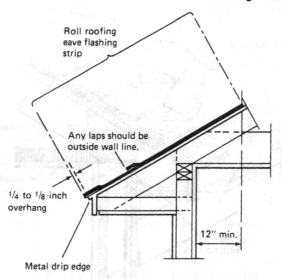

Roll roofing
eave flashing
strip

Any laps should be
outside wall line.

1/4 to 1/8-inch
overhang

12" min.

Metal drip edge

Fig. 7–7. Roll roofing eaves flashing strip. *(Courtesy Asphalt Roofing Manufacturers Association)*

two courses of material should be laid with the top course overlapping the underlying one at a point midway between the eave line and the exterior wall line. The lap must be sealed with asphalt cement when roll roofing is used.

On decks with a slope of less than 4-in-12, but not less than 2-in-12, or in areas of the country where severe ice formation is encountered (Fig. 7–8), the second course of material should extend up the roof deck to a point not less than 24 inches inside the inner wall line of the structure (Fig. 7–9). The underlaying material should be covered with a layer of asphalt cement before the top layer is applied when roll roofing is used.

Valley Flashing

A roof valley is formed where two sloping roof planes meet to form an internal angle. During heavy rains, large volumes of water will flow along the joint forming the valley. As a result, it is absolutely necessary

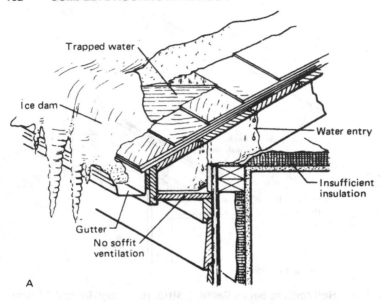

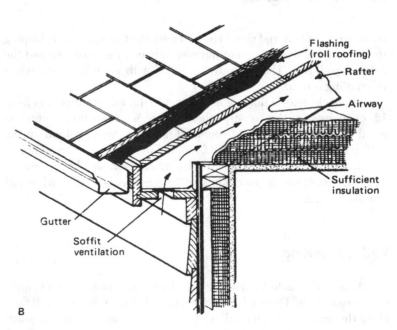

Fig. 7–8. Prevention of ice dam formation along eaves.

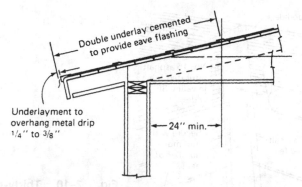

Fig. 7–9. Eave flashing on low roof slope. *(Courtesy Asphalt Roofing Manufacturers Association)*

to insure that this roof joint is capable of handling the excess water without leaking a portion of it into the structure. This is accomplished by lining the valley with a suitable flashing material.

Metal flashing, mineral surface roll roofing, or smooth surface roll roofing can be used as a flashing material in roof valleys. Because mineral surface roll roofing is less expensive than metal flashing, the valleys of most asphalt shingle roofs are flashed with this material. Metal flashing is still used in more expensive, higher quality construction, or on roofs covered with wood shingles or shakes, slate, or tile. Smooth surface roll roofing is used as a valley flashing on asphalt shingle roofs constructed with woven or closed valleys.

Valley flashing must be applied after the roofing felt underlayment is nailed to the roof deck. As shown in Fig. 7–10, a 36-inch wide sheet of No. 15 asphalt-saturated roofing felt is centered in the valley and attached with enough nails to secure it in place before the underlayment courses are laid across the roof deck.

Metal Flashing

Metal valley flashing can be cut to fit the shape of the roof valley or purchased preformed from a supplier. Preformed metal flashing is available with a 1-inch crimped standing seam down its center. Flashing with a standing seam is required for applications where adjoining roofs have different slopes. The standing seam prevents the heavier

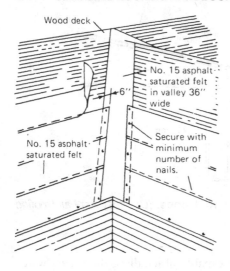

Fig. 7–10. Thirty-six-inch sheet of roofing felt centered in valley. *(Courtesy Asphalt Roofing Manufacturers Association)*

volume of water flowing down the steeper slope from overrunning the valley and being forced under the shingles or roofing material on the adjoining slope. The valley should be wider at the bottom to cope with the larger volume of water drainage at lower levels on the roof slope (Fig. 7–11).

Metal valley flashing is produced with either flat or crimped edges. Like the standing seam down the center of the valley, the crimped edges are also designed to prevent a heavy flow of water from overflowing the valley flashing and running up under the roofing material.

Metal flashing for valleys should not be less than 12 inches wide for roof slopes of 7-in-12 or more, 18 inches wide for slopes of 4-in-12 to less than 7-in-12, or 24 inches wide for slopes less than 4-in-12. The flashing should be secured with nails spaced approximately 12 inches apart in a row 1 inch from each outer edge of flat-edge flashing or driven through cleats attached every 12 inches to crimped-edge flashing. Nails should never be driven near the center of the valley because water may leak through the flashing at the nail holes. Use only nails of a metal compatible with the flashing to prevent corrosion, discoloration, or other reactions. Galvanized steel nails should be used with galvanized steel flashing, aluminum nails with aluminum flashing, and copper nails with copper flashing.

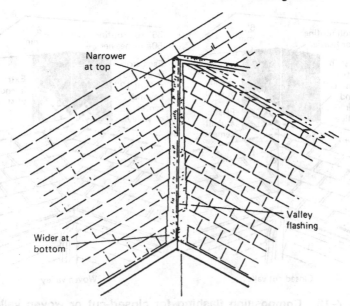

Fig. 7–11. Valley made wider at bottom for more efficient drainage.

Galvanized steel flashing must be painted before it is applied to protect it from corrosion and deterioration. Clean the metal with a suitable solvent, apply a zinc-based metal primer to both sides of the flashing, and spray or brush the top (exposed) surface with two coats of paint in a color that will match or contrast with the roofing material.

Composition Flashing

The application of a composition flashing material, such as mineral surface or smooth surface roll roofing, depends on whether the roof has open, woven, or closed valleys. If the roof has either woven or closed valleys, a type of construction limited to strip-type asphalt shingle roofs, 50-pound or heavier smooth surface roll roofing is used as the flashing material. A 36-inch wide sheet of the material is nailed over the underlayment so that it extends at least 12 inches beyond the center of the valley on either side (Fig. 7–12). The roll roofing sheet should be nailed only along its outer edges (*never* in the center of the valley)

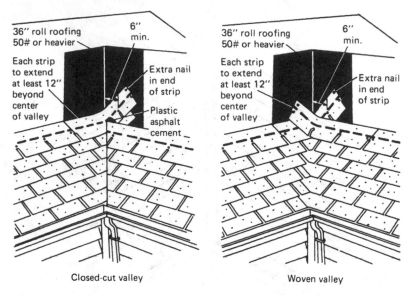

Closed-cut valley   Woven valley

Fig. 7–12. Composition flashing for closed-cut or woven valley.
(Courtesy Asphalt Roofing Manufacturers Association)

with enough nails to hold it firmly in place. The strip shingles are then applied so that they cover the valley flashing either by being woven across the center of the valley or by cutting them to form a seam down the valley center.

An open valley is flashed with two layers of mineral surface roll roofing over the underlayment (Fig. 7–13). The first layer consists of an 18-inch wide sheet which is centered in the valley with its surfaced side face down and its lower edge cut to fit flush with the eave lines of the two intersecting roofs. The material should be secured to the valley with a row of nails placed 1 inch from each outer edge of the sheet. Use only enough nails to hold the sheet firmly and smoothly in place. Push the material firmly into the center of the valley before nailing the second edge. Nail a second 36-inch wide sheet of mineral surface roll roofing over the first sheet with its surface side up. Make certain that the sheet is centered in the valley and nail it in the same manner as the underlaying sheet.

Whenever it is necessary to use more than one sheet of roll roofing

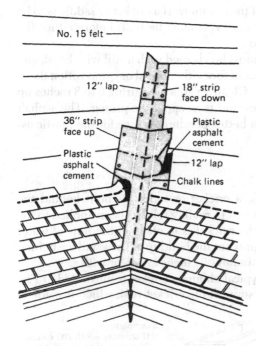

No. 15 felt ——

12" lap

18" strip
face down

36" strip
face up

Plastic
asphalt
cement

Plastic
asphalt
cement

12" lap

Chalk lines

Fig. 7–13. Composition flashing for open valley. *(Courtesy Asphalt Roofing Manufacturers Association)*

in the valley, make sure the upper sheet overlaps the underlying one by at least 12 inches and seal the lap with a coating of plastic asphalt cement between the two sheets.

Chimney Flashing

A chimney is built on a separate foundation from the one that supports the structure. Because there is always movement between the chimney and the structure, resulting to a large extent from the different rates at which the two separate foundations settle, a chimney requires special flashing treatment. The chimney flashing must be able to absorb movement without breaking its seal and allowing water entry. This is accomplished in most cases by allowing the flashing attached to the chimney to overlap or cap the flashing attached to the roof.

Before flashing can be applied, all shingle courses must be completed up to the front face of the chimney. If a cricket or saddle is to be added to the back of the chimney, it must be built before either the roofing or flashing is applied.

The chimney masonry must be cleaned with a stiff wire brush and coated with an asphalt primer before either metal or composition flashing is applied to the surface. Clean an area measuring 6 to 8 inches up from the base of the chimney and then apply the primer. The asphalt primer will produce a much better bonding surface for the plastic asphalt cement.

Cricket Construction

On some pitched roofs, a cricket or saddle is built behind the chimney after the underlayment of roofing felt has been applied to the roof deck. The purpose of the cricket is to channel the flow of water around the chimney and to prevent the buildup of snow and ice behind it. Crickets are used with chimneys 24 inches wide or wider.

The construction of a typical chimney cricket is illustrated in Fig. 7–14. The frame is used to support the cricket roof, which is made of

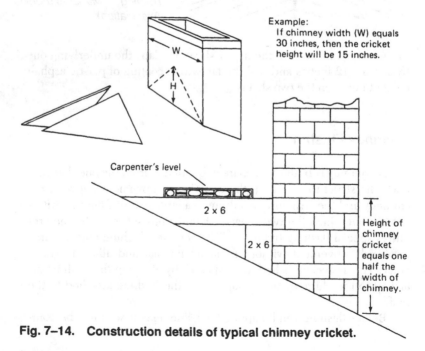

Example:
If chimney width (W) equals 30 inches, then the cricket height will be 15 inches.

Carpenter's level

2 x 6

2 x 6

Height of chimney cricket equals one half the width of chimney.

Fig. 7–14. Construction details of typical chimney cricket.

½-inch thick plywood. The cricket roof is cut to fit the angle of the sloping main roof deck.

As shown in Fig. 7–14, the outer edges of the cricket are designed so that they meet but do not extend beyond the two corners of the chimney. When they do extend beyond, it is usually because a cant strip is installed down both sides of the chimney (Fig. 7–15). The cant strip protects the joint formed between the roof and chimney by channeling the water runoff. If a cant strip is used, the cricket will extend no further than the width of the cant strip.

Metal Flashing

Flashing is always applied to the front of the chimney first. A single piece of base flashing is laid across the front of the chimney so that it extends over the last shingle course. A typical base flashing pattern for the front of the chimney is shown in Fig. 7–16. It is bent along the dotted line in the illustration to conform to the angle formed by the chimney and roof. Lay the lower section over the shingles in a bed of plastic asphalt cement and secure the upper vertical section to the chimney surface with plastic asphalt cement and masonry nails. Drive the nails into the mortar joints. Bend the triangular ends of the upper section around the corners of the chimney and secure them with plastic asphalt cement.

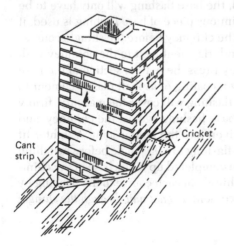

Cant strip

Cricket

Fig. 7–15. Chimney cricket with cant strip. (Courtesy Asphalt Roofing Manufacturers Association)

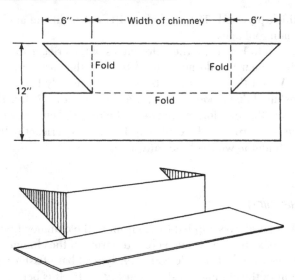

Fig. 7–16. Typical base flashing pattern.

Either base flashing or step flashing may be used along the sides of the chimney. Side base flashing is a single, continuous section of metal. The base section pattern shown in Fig. 7–17 must be bent to conform to the vertical surface of the chimney, the roof slope, and the cant strip attached to the base of the chimney (see dotted lines in Fig. 7–17). If a cant strip is not used, the base flashing will only have to be bent once. When a single, continuous piece of base flashing is used, it is always applied to the side of the chimney before the shingle courses are laid. Cover the roof deck underlayment next to the chimney with plastic asphalt cement and firmly press the bottom section of the base flashing into the cement. Apply a coating of asphalt plastic cement to the chimney surface where the flashing will cover the masonry, firmly press the upper section of the base flashing against the chimney, and secure it with masonry nails driven into the mortar joint. A tighter fit can be obtained by bending the flashing to a 95° angle before installing it. When the shingle courses are completed up to the base of the chimney, secure the end of the last shingle in each course to the flashing by embedding it in plastic asphalt cement. *Do not nail through the shingle and the flashing.*

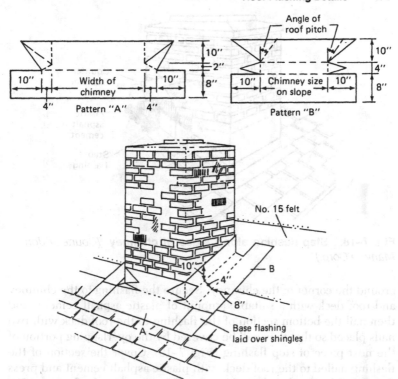

Fig. 7–17. Continuous base flashing strip along sides of chimney.
(Courtesy Asphalt Roofing Manufacturers Association)

If step flashing is used along the sides of the chimney, the shingle courses and the step flashing are alternately applied as the work progresses up the roof slope (Fig. 7–18). Step flashing consists of individual, rectangular pieces of flashing metal cut double the shingle exposure dimension and then bent in half to conform to the angle formed by the chimney and roof. For example, a pitched roof with a 5-inch shingle exposure would require cutting 7 × 10-inch flashing metal pieces and then bending them to provide 5 inches on each side. Each piece would be 7 inches wide and would extend 5 inches up the chimney wall and 5 inches onto the roof deck when installed.

The first step flashing piece should be laid so that one edge wraps

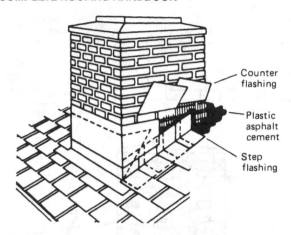

Fig. 7–18. Step flashing along sides of chimney. *(Courtesy Johns-Manville Corp.)*

around the corner of the chimney. Secure the flashing to the chimney and roof deck with a suitable amount of plastic asphalt cement, and then nail the bottom section of the flashing to the roof deck with two nails placed so that they will be covered by the overlapping portion of the next piece of step flashing (Fig. 7–19). Cover the section of the flashing nailed to the roof deck with plastic asphalt cement and press the end of the last shingle in the course firmly against the flashing (Fig. 7–20). *Do not nail through the shingle and flashing.*

Apply the second piece of step flashing so that it overlaps the first one by 2 inches and covers the nails that attach it to the roof deck. Use the same method to secure it to the chimney and roof as described for applying the first piece of flashing. Continue to apply alternate courses of first the flashing and then the last shingle in the shingle course until the work is completed along the side of the chimney. Make sure each piece of flashing overlaps the underlying one by the required 2 inches for a 5-inch shingle exposure. Do not nail the step flashing to the chimney or it will be too rigid to allow for any movement between the roof and the chimney.

If a cricket has *not* been built against the back of the chimney, the base of the chimney will have to be covered with a section of flashing. This is done after the sides of the chimney have been flashed. A suggested pattern for the flashing at the back of the chimney is illustrated in Fig. 7–21. The lower section is secured to the underlayment with

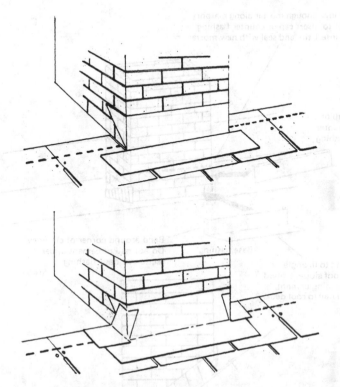

Fig. 7–19. Step flashing pieces installed at front chimney corners.

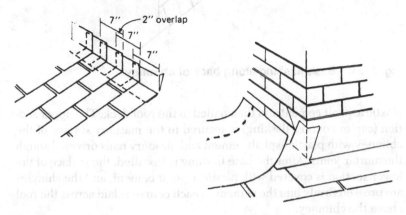

Fig. 7–20. Installing step flashing along sides of chimney.

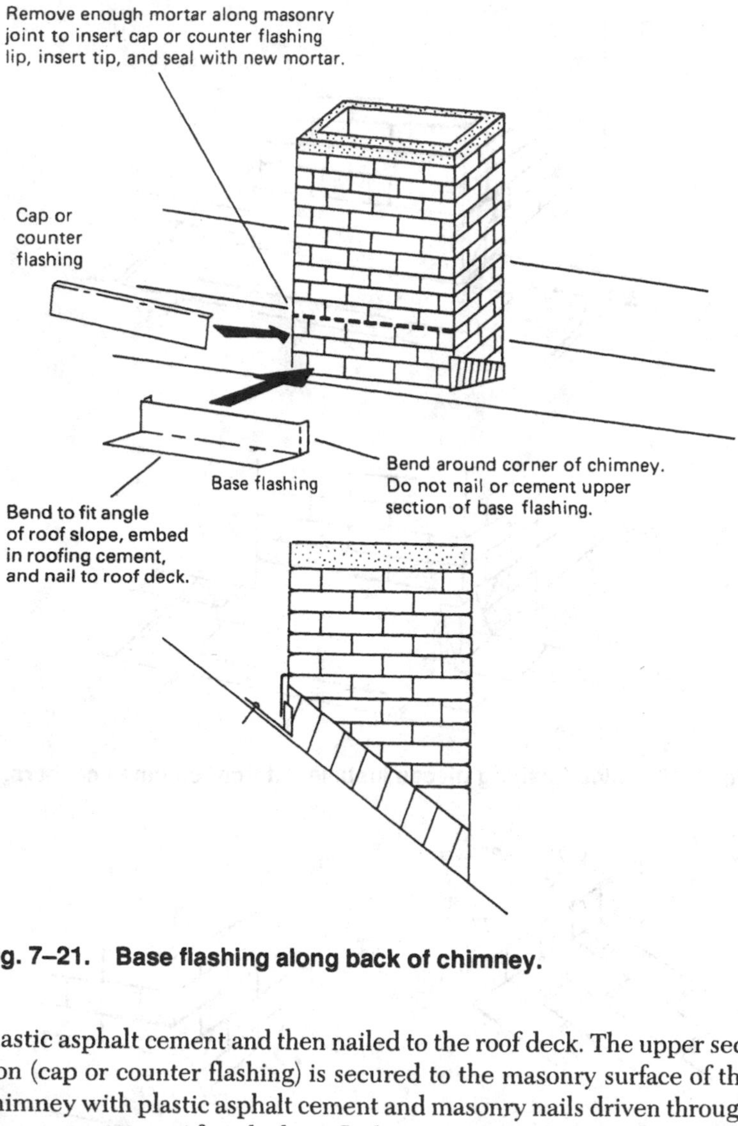

Remove enough mortar along masonry
joint to insert cap or counter flashing
lip, insert tip, and seal with new mortar.

Cap or
counter
flashing

Base flashing

Bend to fit angle
of roof slope, embed
in roofing cement,
and nail to roof deck.

Bend around corner of chimney.
Do not nail or cement upper
section of base flashing.

Fig. 7–21. Base flashing along back of chimney.

plastic asphalt cement and then nailed to the roof deck. The upper section (cap or counter flashing) is secured to the masonry surface of the chimney with plastic asphalt cement and masonry nails driven through the mortar joints. After the base flashing is installed, the surface of the lower section is covered with plastic asphalt cement, and the shingles are pressed firmly into the cement as each course is laid across the roof above the chimney.

The base flashing at the front of the chimney is covered by a sheet of counter flashing or cap flashing. Counter flashing may also be used to cover the base flashing or step flashing along the sides of the chimney.

Counter flashing is secured to the chimney by bending a 1½-inch lap or reglet along its top edge and inserting it into a mortar joint from which the mortar has been removed. The joint is refilled with either plastic asphalt cement or mortar mixed to the desired consistency (Fig. 7-22).

The counter flashing along the front of the chimney is one unbro-

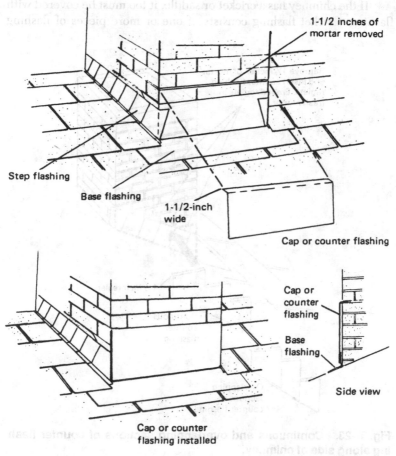

Fig. 7-22. Installing cap or counter flashing along front of chimney.

ken piece. Along the sides of the chimney the counter flashing may be either a continuous piece or several overlapping pieces cut to conform to brick dimensions (Fig. 7–23).

If a single, continuous piece is used along the sides, the bottom end should be bent 2 inches to overlap the counter flashing on the front of the chimney. Embed the 2-inch lap in plastic asphalt cement to bond it to the front counter flashing. Individual counter flashing pieces along the sides of the chimney should overlap at least 3 inches and plastic asphalt cement should be applied between the overlapping sections to seal the joint.

If the chimney has a cricket or saddle, it too must be covered with flashing. Cricket flashing consists of one or more pieces of flashing

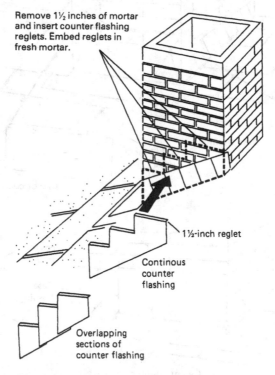

Fig. 7–23. Continuous and overlapping sections of counter flashing along side of chimney.

metal cut to the required dimension and nailed to the cricket roof. If more than one piece of flashing is required to cover the cricket, they should be soldered together after they have been nailed in place.

Applying flashing to a cricket can be made easier by first cutting and fitting a pattern similar to the one shown in Fig. 7–24. A large rectangular sheet of roll roofing or stiff art paper can be used for the pattern material. Fit the pattern material to the cricket and trim it to the required dimensions. Transfer the pattern to a sheet of metal flashing, and then cut and bend the metal flashing to fit the cricket.

The cricket flashing should be large enough to cover the cricket

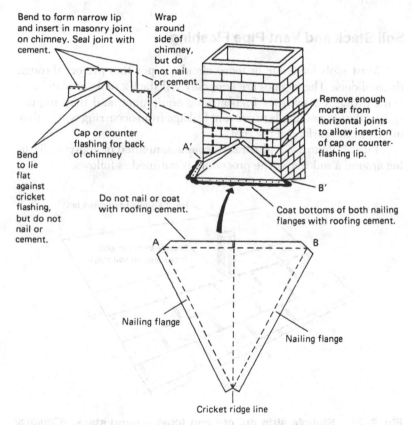

Bend to form narrow lip and insert in masonry joint on chimney. Seal joint with cement.

Wrap around side of chimney, but do not nail or cement.

Cap or counter flashing for back of chimney

Remove enough mortar from horizontal joints to allow insertion of cap or counter-flashing lip.

Bend to lie flat against cricket flashing, but do not nail or cement.

Do not nail or coat with roofing cement.

Coat bottoms of both nailing flanges with roofing cement.

Nailing flange

Nailing flange

Cricket ridge line

Fig. 7–24. Cricket flashing.

roof and extend a short distance onto the roof deck and up the back surface of the chimney. Cover the cricket with asphalt primer and plastic asphalt cement, and then install the flashing. The edge flap lying against the surface of the chimney is secured in place by embedding it in plastic asphalt cement. The edges lying flat against the roof surface should be embedded in plastic asphalt cement and nailed to the roof deck. Run a ribbon of plastic asphalt cement along the edges of the cricket flashing to provide additional protection against leakage. Counter flashing is sometimes used to cover the back of the chimney after the cricket flashing has been applied.

Soil Stack and Vent Pipe Flashing

Most roofs have one or more circular pipes that project through the roof deck. These pipes are generally soil stacks, roof ventilators, or vents (exhaust stacks) for fuel-burning equipment, and they require special flashing methods to prevent leakage from occurring where they interrupt the surface of the roof.

Figs. 7–25 to 7–28 illustrate a common method of applying flashing around a soil stack. The procedure is outlined as follows:

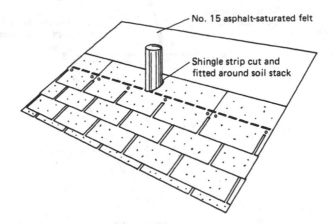

Fig. 7–25. Shingle strip cut out and fitted around stack. *(Courtesy Asphalt Roofing Manufacturers Association)*

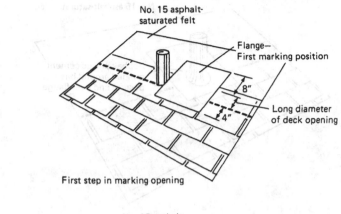

First step in marking opening

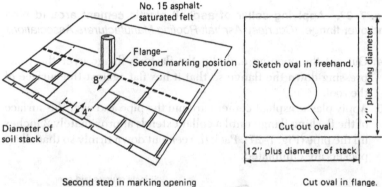

Second step in marking opening · Cut oval in flange.

Fig. 7–26. Marking and cutting flange opening. *(Courtesy Asphalt Roofing Manufacturers Association)*

1. Complete the shingle course through which the pipe will project. As shown in Fig. 7–25, one of the strip shingles will have to be cut and fitted to the diameter of the pipe.
2. Cut a rectangular flange from mineral surface roll roofing, or 50-pound or heavier smooth surface roll roofing, to fit over the pipe. The roll roofing flashing flange should be large enough to extend 4 inches below the pipe, 8 inches above the pipe, and 6 inches on either side of the pipe.
3. Locate the exact opening on the flange for the circular pipe following the steps illustrated in Fig. 7–26.

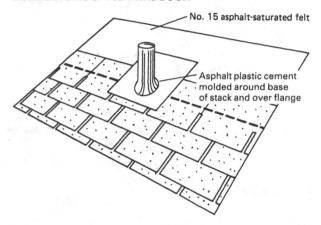

Fig. 7–27. Applying collar of asphalt plastic cement around pipe and over flange. *(Courtesy Asphalt Roofing Manufacturers Association)*

4. Cut out the opening and slip the flashing flange over the pipe, pressing down the flange so that it lies flat against the surface of the roof.

5. Apply plastic asphalt cement around the pipe and over the surface of the flashing flange until a collar extends approximately 2 inches up the pipe (Fig. 7–27). Pack the cement down firmly so that all air pockets are eliminated.

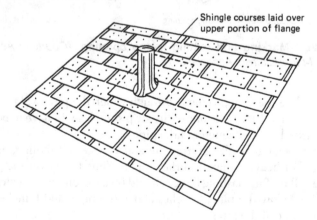

Fig. 7–28. Shingling above the soil stack. *(Courtesy Asphalt Roofing Manufacturers Association)*

6. Complete the next shingle course across the roof, but do *not* nail through the flashing flange. One of the strip shingles may have to be cut to fit around the upper portion of the circular pipe. Any shingle or portion of a shingle overlying the flashing flange must be embedded in plastic asphalt cement (Fig. 7–28).

If it is necessary to cut the rectangular flashing flange from two sections of roll roofing, the upper section should overlap the lower one by approximately 2 inches. Cement the two sections together with plastic asphalt cement.

Flashing around a circular pipe located near the roof ridge is applied in the same manner as described in steps 1–6 above with the following exceptions:

1. Bend the flashing flange over the ridge and allow it to lap the roof shingles at all points.
2. Cover the flashing with hip shingles embedded in plastic asphalt cement.

Some soil stacks, vent pipes, and similar kinds of circular pipes are sold with adjustable metal flange collars designed to fit any roof slope. These are as effective as any other kind of flashing when the manufacturer's installation instructions are carefully followed.

Vertical Wall Flashing

On some structures, the rake of a sloped roof may butt against a vertical wall (Fig. 7–29). The joint formed where the rake and vertical wall meet must be protected with flashing or leaks may develop. The recommended flashing method is to use individual metal flashing pieces applied with a 2-inch side lap. The application procedure is the same as the one used to apply step flashing to the side of a chimney.

In new construction, the underlayment should extend up the vertical wall about 4 inches to cover the joint formed by the wall and roof (Fig. 7–30). If the underlayment has already been applied to the roof deck and the edge of each strip abuts at the vertical wall, lay an 18-inch wide strip of No. 15 asphalt-saturated roofing felt parallel to the rake. Allow 4 inches of the material to extend up the vertical wall, firmly press the roofing felt into the joint, and nail the 14-inch wide

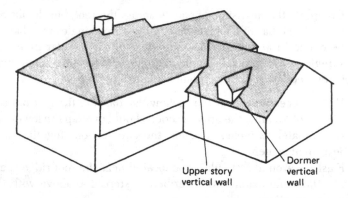

Fig. 7–29. Common vertical wall flashing applications.

section to the roof deck with as few evenly spaced roofing nails as possible. Nail the 4-inch wide portion to the vertical wall sheathing in the same way.

Each piece of metal flashing must be provided with a 2-inch side lap. This dimension and the amount of shingle exposure will determine the width of the metal flashing piece. For example, a 5-inch shingle exposure will require a piece of metal flashing 7 inches wide (5-inch exposure plus 2-inch side lap).

Each metal flashing piece should be long enough to extend 4 inches up the vertical wall, and 2 inches onto the roof deck. Taken together, the width and length dimensions require that each metal flashing piece measure 6 × 7 inches.

Begin at the eave and nail the first metal flashing piece to the vertical wall sheathing with a single roofing nail placed in the upper top corner. Use a second roofing nail to nail the 2-inch wide portion of the metal flashing piece to the roof deck. Both nails should be placed where they will be covered by the 2-inch side lap of the next flashing piece (Fig. 7–31). Coat the head of each nail with plastic asphalt cement to prevent any possibility of a leak developing.

Complete the first course of shingles along the roof eave, and secure the end of the last shingle to the metal flashing piece with plastic asphalt cement. *Do not nail through the shingle and metal flashing.*

Apply the second metal flashing piece to the vertical wall sheathing and the roof deck using the same method described for the first

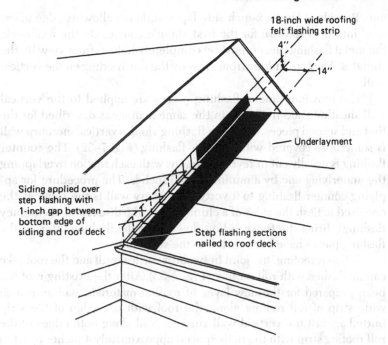

18-inch wide roofing
felt flashing strip

4"

14"

Underlayment

Siding applied over
step flashing with
1-inch gap between
bottom edge of
siding and roof deck

Step flashing sections
nailed to roof deck

Fig. 7–30. Eighteen-inch wide flashing strip applied over underlayment.

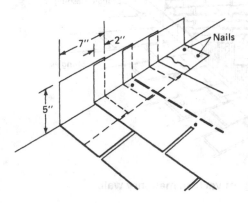

7"   2"

Nails

5"

Fig. 7–31. Flashing shingle details.

one. Provide at least a 2-inch side lap, but do not allow its edge to extend into the exposure for the first shingle course. On the roof deck, the metal flashing pieces must be completely hidden from view by the shingles. They are hidden from view by the finish siding on the vertical wall.

The remaining metal flashing pieces are applied to the vertical wall sheathing and roof deck in the same manner as described for the first and second pieces. The base flashing along a vertical *masonry* wall is sometimes covered with counter flashing (Fig. 7–32). The counter flashing is usually cut in several sections with each section overlapping the underlying one by a minimum of 3 inches. The procedure for applying counter flashing to a vertical masonry wall is the same as the one used to flash the sides of a chimney (see the discussion of chimney flashing). Bring the finish siding down over the flashing after all the flashing pieces have been nailed to the sheathing.

When reroofing, the joint between a vertical wall and the roof rake can be flashed with roll roofing (Fig. 7–33). After the existing roof has been prepared for the new layer of roofing materials, nail an 8-inch wide strip of roll roofing along the roof with one edge of the strip butted against the vertical wall surface. Nail along both edges of the roll roofing strip with the nails spaced approximately 4 inches apart in rows 1 inch from each edge. As each course of shingles is laid, cover the roll roofing flashing strip with plastic asphalt cement and firmly press the end of the last shingle of each course into the cement. *Do not nail the shingle to the flashing strip.*

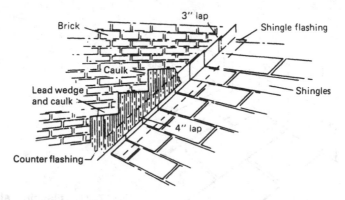

Fig. 7–32. Counter flashing on vertical masonry wall.

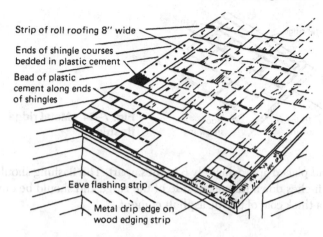

Strip of roll roofing 8" wide

Ends of shingle courses bedded in plastic cement

Bead of plastic cement along ends of shingles

Eave flashing strip

Metal drip edge on wood edging strip

Fig. 7–33. Vertical wall flashing details when reroofing.

Other Types of Roof Flashing

Flashing is also used around dormers, skylights, scuttles, and other square- or rectangular-shaped roof openings; along the joint of the external angle or juncture formed by a change of pitch on a gambrel roof; and sometimes along the roof ridge. Flashing details for skylights, scuttles, and similar types of roof openings are provided in Chapter 5. Dormer flashing is discussed in Chapter 6.

Ridge flashing is not used on most roofs because they are adequately protected by a row of ridge shingles that protect the joint by deflecting the water. However, ridge flashing is recommended for a wood shingle or shake roof with Boston-type ridge construction to prevent water entry. The ridge is flashed with a strip of roll roofing extending 3 inches down from both sides of the ridge and nailed along its outer edges. The nails should be driven in rows 1 inch from each edge and spaced approximately 4 inched apart. The wood shingles or shakes are installed over the ridge flashing.

A metal flashing strip can be used along the ridge of either a wood shingle or shake roof, or an asphalt shingle roof instead of the usual row of ridge shingles (Fig. 7–34). The 6-inch wide metal flashing strip is installed over the ridge *after* the roofing has been applied. Use nails of a compatible metal driven in rows 1 inch from each outer edge of the

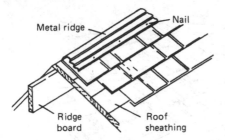

Fig. 7–34 Metal ridge flashing.

strip and placed approximately 6 inches apart. The flashing should extend 3 inches down from each side of the ridge and should be embedded in a thick coat of plastic asphalt cement.

Roof Juncture Flashing

A gambrel roof will have one or two changes of pitch on each slope, depending on its design. These changes of pitch form roof junctures that must be covered with a flashing material. The flashing details of a typical gambrel roof are shown in Fig. 7–35. Note that a cant strip is required under the shingle course immediately above the roof

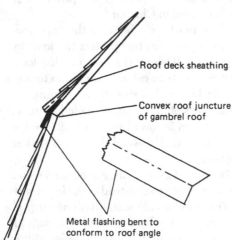

Fig. 7–35. Flashing slope changes on gambrel roofs.

juncture. The shingle course immediately below the juncture is doubled, and the flashing is inserted between them.

Flashing Repairs

Carefully inspect metal flashing for cracks in the plastic asphalt seals, loose or missing nails, small holes, or small rust spots. Cracked seals can be repaired by scraping away the old material and recoating the surface with plastic asphalt cement or a suitable caulking compound. Cracks around the embedded edges of chimney flashing should be freed of loose mortar, cleaned with a stiff wire brush, and then sealed. Never apply the cement or caulking compound when temperatures are below 40°F. If the seal *beneath* the flashing has deteriorated and is causing the leak, carefully pull the flashing back from the surface, remove the old cement, and apply a thick coat of new plastic asphalt cement with a putty knife.

Small holes in metal flashing can be repaired with a spot of plastic asphalt cement or a patch covered with cement. Replace the flashing if the holes are larger than ½ inch across.

Small rust spots should be cleaned down to the bare metal with a suitable solvent and wire brush and then covered with a metal patch and plastic asphalt cement.

Composition flashing can be repaired by first cleaning the surface and then covering the damaged area with plastic asphalt cement or other suitable sealant. Do not attempt to repair composition flashing if the material shows signs of extensive deterioration.

Reroofing

Old metal flashing can be left in place and reused when reroofing if it will remain in a serviceable condition for the life of the new roof. If it is badly deteriorated, it should be removed and replaced with new flashing. If the old flashing is asphalt-saturated roofing felt and the roof covering material (asphalt strip shingles, individual shingles, etc.) is to be completely removed, the general rule is to remove and replace the flashing as well.

CHAPTER 8

Applying Strip Shingles

Asphalt shingles are the most popular type of roofing material in both new construction and reroofing applications. It is estimated that asphalt shingles have been used for roofing on almost three-fourths of the houses in the country (Fig. 8–1). They have also been used on commercial and, to a lesser extent, industrial structures. The popularity of asphalt shingles may be attributed to the fact that they are relatively inexpensive, widely available, and produced in a variety of colors, shapes, and styles. They are also easy to install, maintain, and repair, and they have a service life of at least 15 years.

Asphalt shingles are available in strip form or as individual shingles. The application of individual and hex strip shingles is discussed in Chapter 9. The application of asphalt strip shingles, the most widely used form of roofing material, is described in this chapter.

Asphalt Strip Shingles

Asphalt shingles are sometimes called composition shingles because they are composed of more than one material. Each shingle has a base covered on both sides with an asphalt coating to provide a wa-

Fig. 8–1. Asphalt shingle roof.

terproofing seal (Fig. 8–1). Ceramic or mineral granules of various colors are embedded in the top coating of asphalt to provide an attractive, protective surface. The shingle base may be either an organic mat of wood and paper fibers or a fiberglass mat. Shingles formed around an organic mat are frequently called *asphalt shingles*. If a fiberglass mat is used, they are called *fiberglass shingles* (Fig. 8–2). Fiberglass shingles are slightly more expensive than organic base shingles, but they provide greater fire and wind resistance, and have a longer service life. The application instructions provided in this chapter can be used for both asphalt and fiberglass shingles.

Asphalt strip shingles are generally available in standard 12 × 36-

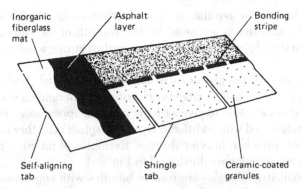

Fig. 8–2. Cutaway view of fiberglass shingle. *(Courtesy Johns-Man-ville Sales Corp.)*

inch lengths, but some roofing manufacturers also produce strip shingles in slightly narrower or wider widths and longer lengths to provide a rustic or textured roof appearance, or deep shadow lines.

The most popular type of asphalt strip shingle is the self-sealing three-tab unit (Fig. 8–3). These are sometimes called *three-tab shingles* or *square tab shingles.* Each self-sealing shingle has a strip of special

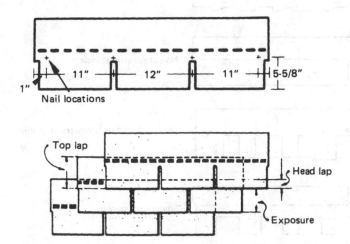

Fig. 8–3. Typical three-tab self-sealing asphalt strip shingles. *(Courtesy Asphalt Roofing Manufacturers Association)*

thermoplastic adhesive that is spaced at intervals on each shingle. When the adhesive is activated by the warmth of the sun, it bonds shingle to shingle to hold them in place against strong winds, rain, and snow.

Asphalt strip shingles are also available with no tabs and cutouts, with staggered or random edge butts, with butts of random width, and other design variations to give a distinctive roof appearance. Some are also manufactured with additional layers of asphalt and other materials to produce a thicker, heavier shingle. Examples of different types of asphalt strip shingles are illustrated in Fig. 8–4.

Asphalt strip shingles are sold in bundles with approximately 27 shingle strips to the bundle. In most applications, three bundles will cover a roofing square (100 square feet of roof surface). Bundles should

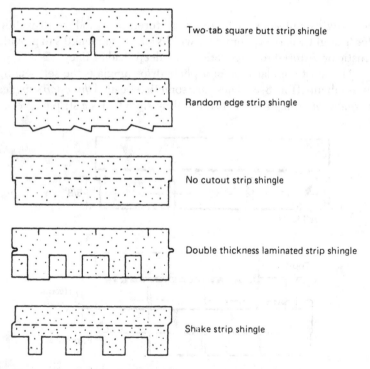

Two-tab square butt strip shingle

Random edge strip shingle

No cutout strip shingle

Double thickness laminated strip shingle

Shake strip shingle

Fig. 8–4. Examples of different types of strip shingles.

be stored flat so that the strips will not curl when they are opened. When possible, store the bundles or shingles indoors until they are to be applied to the roof. If the bundles or shingles cannot be stored indoors, stack them on wood planks or 2 × 4s laid close together and clear of the ground. When not working, cover the bundles or shingles with a large sheet of polystyrene or other equally suitable waterproof material for protection from rain or snow. Avoid working when it is extremely cold because the shingles are brittle and will not bend easily. If working in cold weather cannot be avoided, warm the shingles before bending them.

Tools and Equipment

One or more sturdy ladders or scaffolding will be needed when shingling a roof. The tools required for this type of work include a claw hammer, a carpenter's apron to hold the roofing nails, a chalk line, a utility knife, a tape measure or folding rule, and a putty knife or pointed trowel for applying roofing cement. Additional information about the tools and equipment used in asphalt shingling is presented in Chapter 2. Be sure to read and follow the work safety rules included in that chapter before beginning work.

Roof Deck Preparation

A wood roof deck must be completely dry and tightly built of good quality, well-seasoned lumber or plywood. It must also have adequate support and the rafters should meet the requirements of their span in size and spacing. If these conditions have been satisfactorily met, the deck can then be covered with the required underlayment and flashing (new construction), the existing roof can be repaired to provide a smooth, flat surface for the new roofing materials (reroofing), or the existing roofing materials can be stripped and new materials applied directly to the roof deck (reroofing). Detailed instructions on all aspects of roof deck preparation are provided in Chapter 3.

Underlayment

In new construction or when old roofing materials have been removed, wood decks or roofs with slopes of 4-in-12 to 7-in-12 may be covered with overlapping strips of No. 15 asphalt-saturated roofing felt. The roofing felt should be lapped 2 inches horizontally and 4 inches at the ends. It should also be carried 6 inches from both sides around hips, ridges, and corners. Low-pitched roofs and roofs with slopes greater than 7-in-12 require special treatment. Detailed instructions for the application of the underlayment are included in Chapter 3.

Flashing

Chimneys are usually flashed with galvanized steel. Metal drip edges are used along the rakes and eaves. On most asphalt shingle roofs, the valleys are flashed with mineral-surface roll roofing. See Chapter 7 for detailed instructions on installing flashing.

Shingling Patterns and Exposures

Asphalt shingles are laid with the upper portion of each shingle covered by an overlapping one in the next course. Because the uncovered portion of each shingle from its bottom edge to the bottom edge of the overlapping shingle is exposed to the weather, it is sometimes called the weather exposure of the shingle or the shingle exposure. On most roofs, asphalt shingles are laid with a 5-inch exposure.

When applying asphalt shingles, the first course is started with one full size shingle at either rake. The shingle sizes used to begin the second and succeeding courses will depend on the shingling pattern desired for the roof. The 4-inch, 5-inch, and 6-inch patterns are used when shingling with standard three-tab asphalt strip shingles.

The 6-inch shingling pattern is the one used on most roofs and its application is described in the next section (see the discussion of applying strip shingles). The 6-inch shingling pattern requires that the second course be started with a shingle from which 6 inches (one half tab) have been removed. Succeeding courses are started with shingles from which 12 inches, 18 inches, 24 inches, and 30 inches have been removed. The seventh course is started with a full size shingle and the pattern is repeated until the ridge is reached (Fig. 8–5). The starting shingles of this pattern are trimmed in 6-inch increments and are applied so that their cutouts break joints on halves. The cutouts of each shingle are centered on the tabs of the underlying shingles. Vertical chalk lines must be snapped on the underlayment as work progresses to keep the cutouts aligned up the roof. The shingle edges or cutouts in every other course must be aligned within ¼ inch. Because of minor differences in shingle dimensions, this alignment is difficult to maintain without the use of vertical chalk lines to serve as guides.

The 4-inch shingling pattern is recommended for low-pitched roofs with slopes of 2-in-12 to less than 4-in-12. When using the 4-inch shingling pattern, the second course is started with a shingle from which 4 inches have been removed. Succeeding courses are started with shingles from which 8 inches, 12 inches, 16 inches, 20 inches, 24 inches, 28 inches, and 32 inches have been removed. The tenth course is started with a full-size shingle and the pattern is repeated until the roof ridge is reached. With this shingling pattern, the cutouts break

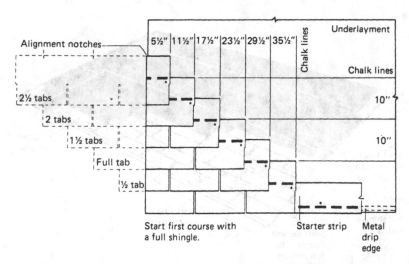

Fig. 8–5. Six-inch shingling pattern. *(Courtesy Johns-Manville Sales Corp.)*

joints on thirds and create a diagonal alignment. A variation of the 4-inch shingling pattern is to repeat the 4-inch, 12-inch, and 16-inch offsets in the fifth, sixth, and seventh courses (Fig. 8–6).

Like the 4-inch shingling pattern, the 5-inch pattern also produces a diagonal shingle alignment. The second course is started with a shingle from which 5 inches have been removed. The first shingles in succeeding courses are trimmed in 5-inch increments (10 inches, 15 inches, 20 inches, etc.) until only a 6-inch section remains (30 inches removed from a 36-inch shingle). The pattern is then repeated to the roof ridge.

A 9-inch shingling pattern is recommended for two-tab and no cutout asphalt strip shingles (Fig. 8–7). The second course starts with a shingle from which 9 inches have been removed. The first shingles in succeeding courses are trimmed in 9-inch increments. A full-size shingle is used to start the fifth course and the pattern is repeated until the roof ridge is reached.

Some strip shingles, particularly those designed to produce a rustic, thatch, or other distinctive appearance, must be applied in accordance with the roofing manufacturer's suggested shingling pattern. Al-

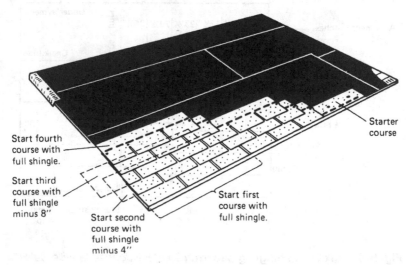

Start fourth course with full shingle.

Start third course with full shingle minus 8"

Start second course with full shingle minus 4"

Start first course with full shingle.

Starter course

Fig. 8–6. Four-inch shingling pattern repeated in the fifth, sixth, and seventh courses. *(Courtesy Asphalt Roofing Manufacturers Association)*

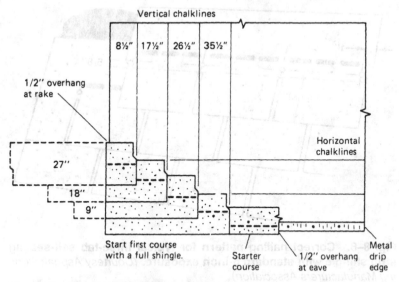

Fig. 8–7. Nine-inch shingling pattern.

ways read and carefully follow the roofing manufacturer's application instructions for these shingles.

Shingles may also be applied in random patterns by varying shingle offsets (see the discussion of random shingle spacing in this chapter).

Nailing Shingles

Properly nailing down the shingles is fundamental to any good roofing procedure. This requires the selection of the right nails, using enough nails, and placing them in the most suitable pattern. The shingle manufacturer will recommend the best nailing pattern to use and will include these recommendations on each bundle of shingles. Typical nailing patterns for strip shingles are shown in Figs. 8–8 and 8–9.

The nails used to apply asphalt strip shingles are 11- or 12-gauge hot galvanized roofing nails with heads at least ⅜ inch in diameter. The nails should be long enought to penetrate at least ¾ inch into solid

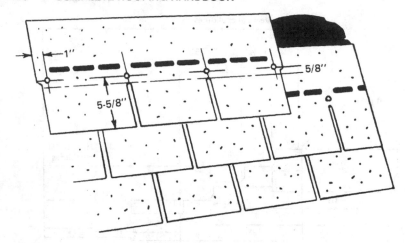

Fig. 8–8. Correct nailing pattern for typical three-tab self-sealing strip shingle with standard 5-inch exposure. *(Courtesy Asphalt Roofing Manufacturers Association)*

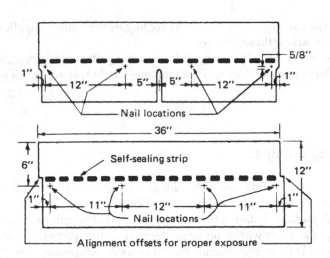

Fig. 8–9. Correct nailing patterns for two-tab and no cutout self-sealing strip shingles. *(Courtesy Asphalt Roofing Manufacturers Association)*

Table 8–1. Recommended Nail Lengths for Different Roofing Applications

Application	Nail Length
Over new deck	1 1/4 inches
Over roll roofing	1 1/4 inches
Over asphalt shingles	1 1/2 inches
Over wood shingles	1 3/4 inches

wood board sheathing and through plywood sheathing. Recommended nail lengths for different roofing applications are listed in Table 8–1.

On most pitched roofs, four roofing nails are used to secure each strip shingle. Use six nails on roofs with steep slopes or roofs located in areas where strong winds are common and place a spot of roofing cement under each shingle tab for additional holding power.

Carefully align each shingle before nailing it and check to make certain that no end joint or cutout is less than two inches from any nail in an underlying course. Check the first course to make sure it has been laid perfectly straight before applying any of the succeeding courses. Check the alignment of shingles and shingle courses at regular intervals against the horizontal chalk lines. Make certain that all cutouts are aligned from the roof eaves to the ridge.

Nail each shingle at the end closest to the shingle just laid and then proceed to nail across the shingle to its opposite end. Do not nail both ends of the shingle first and then the middle. Doing so may cause the shingle to buckle in the middle. If a self-sealing strip shingle is being used, do not nail above or through the adhesive strip. The nails should be driven through the shingle at a point 5/8 inch above the cutouts on tab-type shingles or 6 inches up from the bottom edge on no cutout shingles.

Drive the nails as straight as possible. If a nail is driven crookedly or at a slant, the edge of the head may cut into the shingle. Drive the nails until their heads contact the surface of the shingle. Do not drive the nail heads into the shingle material.

Shingles with the standard 5-inch exposure should be nailed 5⅝ inches above the bottom edge so that the nail heads are covered by the shingles in the next course (Fig. 8–8). The nails at either end of the shingle should be located approximately 1 inch up from the long side edge. If an exposure other than 5 inches is used, the nail heads must be placed so that they are covered by the shingles in the next course.

Applying Strip Shingles

Gable roofs shorter than 30 feet long are generally shingled by starting at either rake and shingling toward the opposite one. Fig. 8–5 illustrates the shingling procedure used when starting at a left-hand rake with three-tab asphalt strip shingles.

In new construction, after the underlayment, drip edge, and any other required flashing have been nailed or otherwise secured to the roof deck, snap vertical and horizontal chalk lines to provide an alignment guide for shingles and shingle courses during application. The vertical chalk lines serve as guides for the shingle cutout. The vertical chalk lines should be spaced at 5½-inch, 11½-inch, 17½-inch, 23½-inch, 29½-inch, and 35½-inch intervals, parallel to the roof edge when using three-tab, square-butt asphalt strip shingles applied with a 6-inch offset (Fig. 8–5).

Applying Starter Course

Lay a starter course along the edge of the roof with a ½-inch overhang at both the rakes and the eaves. The starter course normally consists of a row of shingles with the tabs cut off (Fig. 8–10). If tabless shingles are used, the upper 3 inches of each shingle should be cut off in a straight line parallel with its bottom edge. Nail the shingle down on the roof deck, allowing for a ½-inch overhang along the rakes and eaves of the roof. The nails should be placed approximately 4 inches above the eave line (Fig. 8–11).

Two alternative methods of applying the starter course are to use full shingles with the tabs reversed or a continuous starter strip of 9-inch wide or wider mineral surfaced roll roofing (Fig. 8–12).

A stiffer backing for the shingles along the roof eaves can be provided by using a course of wood shingles under the starter course (Fig. 8–13). Each shingle is laid edge to edge and secured with nails driven 1 inch from each of its side edges and approximately 6 inches up from

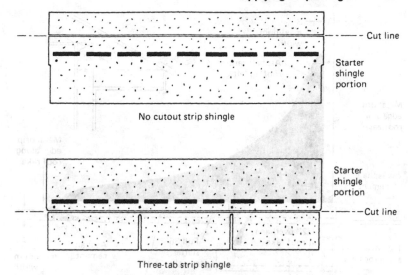

No cutout strip shingle

Three-tab strip shingle

Fig. 8–10. Starter course shingles.

its bottom edge. The course of wood shingles is then covered by the starter course and first course of asphalt strip shingles.

Applying First Course

The first course of shingles is started with a full strip shingle laid directly over the starter strip. The butt edges of the first course shingles should be laid flush with the bottom edge of the starter strip at the left-hand rake of the roof. Align the end of the 36-inch shingle with the 35½-inch chalk line on the roof deck (Fig. 8–14). The ½-inch difference is accounted for by the overhang at the rake.

Applying Second and Succeeding Courses

Cut 6 inches, or one half of a tab, from a shingle strip and begin the second course with it. Align one end of the shingle strip with the 29½-inch chalk line and allow the other end to overlap the rake by ½ inch. The amount of shingle exposure allowed on the first course will de-

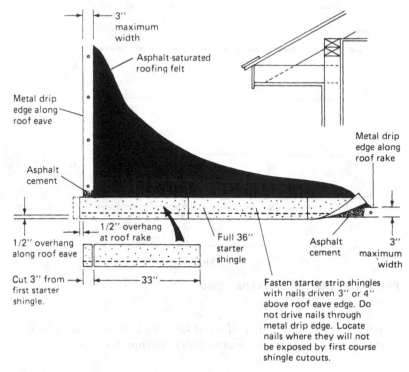

Fig. 8–11. Applying starter course shingles.

pend on the shingle manufacturer's recommendations. For this type of shingle it will generally be about 5 inches (Fig. 8–15).

Start the third course by cutting off 12 inches, or a full tab, from the shingle strip. Nail the shortened shingle strip in position with the far edge of the shingle aligned with the 23½-inch chalk line (Fig. 8–16). Allow the same ½-inch overhang at the rake and the same exposure for the second course of shingles as was provided for the first course.

Cut off 18 inches, or a tab and a half, from a shingle strip and use it to begin the fourth course. One end of the shingle should be aligned with the 17½-inch chalk line, and the other end should extend ½ inch beyond the edge of the rake (Fig. 8–17). Allow the same shingle exposure in the third course as you did for the first and second courses.

Cut 24 inches (two tabs) from a shingle strip and begin the fifth

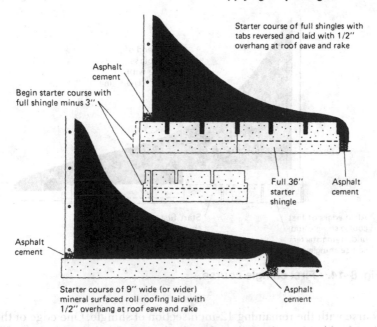

Starter course of full shingles with
tabs reversed and laid with 1/2"
overhang at roof eave and rake

Asphalt
cement

Begin starter course with
full shingle minus 3".

Full 36"
starter
shingle

Asphalt
cement

Asphalt
cement

Starter course of 9" wide (or wider)
mineral surfaced roll roofing laid with
1/2" overhang at roof eave and rake

Asphalt
cement

Fig. 8–12. Alternative types of starter courses.

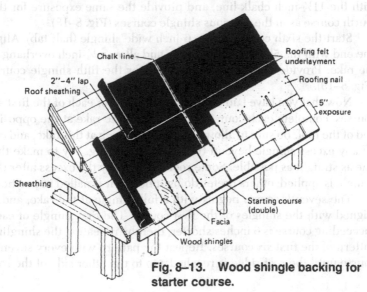

Chalk line

Roofing felt
underlayment

2"–4" lap
Roof sheathing

Roofing nail

5"
exposure

Sheathing

Starting course
(double)

Facia

Wood shingles

**Fig. 8–13. Wood shingle backing for
starter course.**

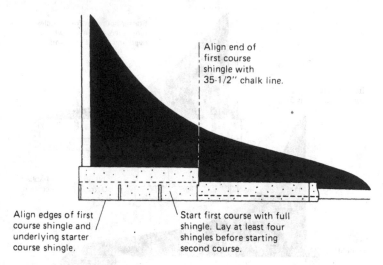

Align end of
first course
shingle with
35-1/2" chalk line.

Align edges of first
course shingle and
underlying starter
course shingle.

Start first course with full
shingle. Lay at least four
shingles before starting
second course.

Fig. 8–14. First shingle course.

course with the remaining 12-inch section of shingle. One edge of the shingle should overhang the rake edge by ½ inch. Align the other edge with the 11½-inch chalk line, and provide the same exposure for the fourth course as in the previous shingle courses (Fig. 8–18).

Start the sixth course with a 6-inch wide shingle (half tab). Align one end with the 5½-inch chalk mark, and allow a ½-inch overhang at the rake. Provide the required exposure for the fifth shingle course (Fig. 8–19).

Now that you have laid the starter shingles for each of the first six courses, complete each course all the way to the rake at the opposite end of the roof. Be sure to allow a ½-inch overhang at this rake, and cut off any excess shingle beyond the ½-inch overhang. Try to make this line as straight as possible. Some roofers cut away the excess after the shingle is applied; others prefer doing it before it is nailed in position.

The seventh course begins with a full shingle at the rake, and is aligned with the shingles of the first course. The first shingle of each succeeding course is 6 inches shorter, thereby repeating the shingling pattern of the first six courses. Repeat the pattern with every seventh course until the roof ridge is reached. Go to the other side of the roof

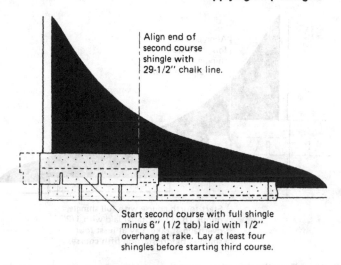

Align end of
second course
shingle with
29-1/2" chalk line.

Start second course with full shingle
minus 6" (1/2 tab) laid with 1/2"
overhang at rake. Lay at least four
shingles before starting third course.

Fig. 8–15. Second shingle course.

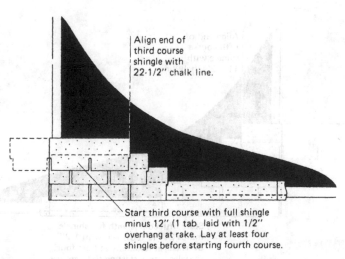

Align end of
third course
shingle with
22-1/2" chalk line.

Start third course with full shingle
minus 12" (1 tab) laid with 1/2"
overhang at rake. Lay at least four
shingles before starting fourth course.

Fig. 8–16. Third shingle course.

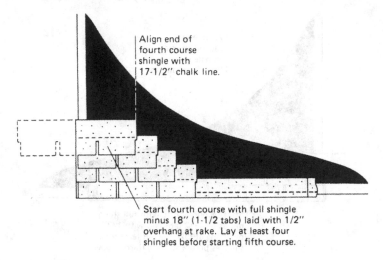

Align end of
fourth course
shingle with
17-1/2'' chalk line.

Start fourth course with full shingle
minus 18'' (1-1/2 tabs) laid with 1/2''
overhang at rake. Lay at least four
shingles before starting fifth course.

Fig. 8–17. Fourth shingle course.

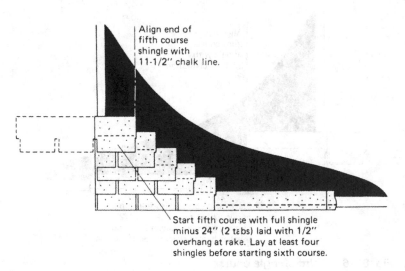

Align end of
fifth course
shingle with
11-1/2'' chalk line.

Start fifth course with full shingle
minus 24'' (2 tabs) laid with 1/2''
overhang at rake. Lay at least four
shingles before starting sixth course.

Fig. 8–18. Fifth shingle course.

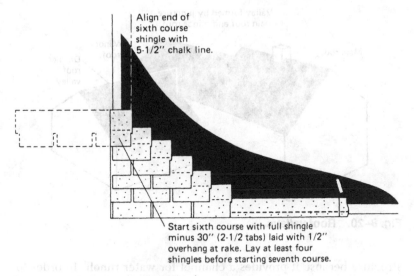

Align end of sixth course shingle with 5-1/2" chalk line.

Start sixth course with full shingle minus 30" (2-1/2 tabs) laid with 1/2" overhang at rake. Lay at least four shingles before starting seventh course.

Fig. 8–19. Sixth shingle course.

and apply shingles in the same manner. The roof ridge is shingled last (see the discussion of shingling hips and ridges).

The dimensions given in the preceding paragraphs are for shingling with 36-inch self-sealing three-tab, square-butt asphalt strip shingles applied with a 6-inch offset and a ½-inch overhang at the rakes and eaves. Changes in these specifications or the type of offset will require modifications of each of the shingling steps. Whenever shingles without a self-sealing adhesive strip are used, each tab should be embedded in a quarter-size spot of roofing cement.

Roofing manufacturers will provide application instructions with the shingles, but the recommended specifications will vary among manufacturers. For example, although a 5-inch shingle exposure is fairly standard, recommendations for overhang at the eave and rake range from ¼ to ¾ inch.

Valley Shingling

Roof valleys are formed where two roofs join at an angle (Fig. 8–20). The joint formed by the two roofs is a potential weak point in the

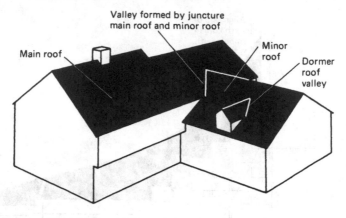

Fig. 8–20. Roof valleys.

structure because it provides a channel for water runoff. In order to prevent leakage, the joint formed by the two roofs must be adequately covered with roofing materials. The three common methods of covering a roof valley are

1. Open valley shingling
2. Woven valley shingling
3. Closed-cut valley shingling

Both strip shingles and individual shingles can be used in open valley shingling. *Only* strip shingles can be used in woven or closed-cut valleys.

Open Valleys—In open valleys, the flashing along the valley joint is left exposed (Fig. 8–21). The shingles ending at the valley are cut to form an edge running in a straight line parallel to the joint. The valley must be covered with flashing before the shingles can be applied. Valley flashing instructions are described in Chapter 7. The procedure for constructing an open valley after the flashing has been applied is as follows:

1. Snap two chalk lines along both sides of the valley. These lines should be 6 inches apart at the ridge, or 3 inches up from the joint on either side of the valley, and must diverge at the rate of ⅛ inch per foot in the direction of the eave.
2. Run the shingle courses from each roof slope to the chalk lines in

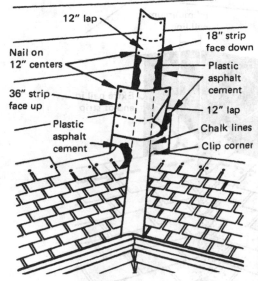

12" lap
Nail on 12" centers
36" strip face up
Plastic asphalt cement
18" strip face down
Plastic asphalt cement
12" lap
Chalk lines
Clip corner

Fig. 8–21. Open roof valley. *(Courtesy Asphalt Roofing Manufacturers Association)*

the valley and cut the last shingle of each course so that it ends parallel to the line.

3. Cut approximately 1 inch off the upper corner of each shingle touching the chalk lines to guide water runoff into the valley. As shown in Fig. 8–21, a diagonal cut should be made.

4. *Cement* the shingles to the surface at the chalk lines. Do *not* nail within 7 inches of the chalk lines because the nail holes will penetrate the flashing strips and may cause leaks. Press the shingles firmly into the asphalt cement before nailing.

Woven Valleys — A woven valley is completely covered with shingles. Alternate shingle courses are laid across the valley and woven together as shown in Fig. 8–22. Because of the additional coverage provided by the shingles, it is not necessary to cover the valley joint with a double thickness of roll roofing. Only a single 36-inch wide strip of mineral surface roll roofing is required, and it is applied according to the instructions given for constructing an open valley. The flashing strip is laid *over* the underlayment (see Chapter 7).

In woven valleys, the roof shingles are first laid to a point approx-

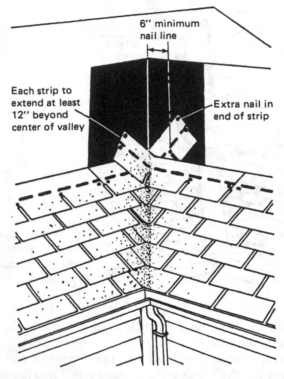

Fig. 8–22. Woven valley. *(Courtesy Asphalt Roofing Manufacturers Association)*

imately three feet from the center of the valley on each roof slope. The remainder of the procedure is as follows:

1. Snap a chalk line down each side of the valley 6 inches from the valley joint.
2. Run the first course of shingles (from the main roof) across the valley and onto the adjoining roof for at least 12 inches. Nail the shingles down, but keep all nails at least 1 inch from the chalk line on each side of the valley.
3. Run the first course of shingles from the adjoining roof across the valley to the main roof and nail it in position. Remember to keep the nails at least 1 inch outside the chalk lines.

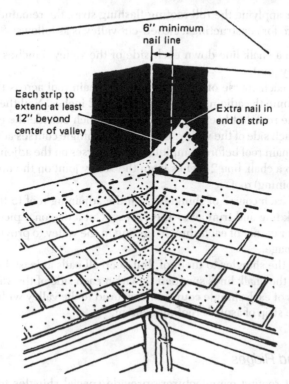

6" minimum
nail line

Each strip to
extend at least
12" beyond
center of valley

Extra nail in
end of strip

Fig. 8–23. Closed-cut valley. *(Courtesy Asphalt Roofing Manufacturers Association)*

4. Run the remaining shingle courses alternately back and forth across the valley, weaving the shingles together and nailing them in position as described in steps 2 and 3 above.

Closed-cut Valleys—In closed-cut valleys, the valley is completely covered with shingles, but the shingle courses are not woven together (Fig. 8–23). A single 36-inch wide flashing strip of mineral surfaced roll roofing is laid down the length of the valley before the shingles are applied. The method used for applying the flashing strip of roll roofing is included in the instructions describing the construction of an open valley.

After applying the roll roofing flashing strip, the remainder of the procedure for constructing a closed-cut valley is as follows:

1. Snap a chalk line down each side of the valley 6 inches from the valley joint.
2. Run each course of shingles from the main roof across the valley and onto the adjoining roof for at least 12 inches. Nail the shingles to the roof, but keep all nails at least 1 inch outside the chalk line on each side of the valley. Bring all the shingle courses across from the main roof before completing the courses on the adjoining roof.
3. Snap a chalk line 2 inches from the valley joint on the unshingled (adjoining) roof.
4. Run each course of shingles from the adjoining roof to the 2-inch chalk line and trim away the excess. Cut a diagonal piece off the upper corner of each shingle ending in the valley to provide better drainage.
5. Nail the shingles to the roof, but keep all nails at least 1 inch outside the chalk lines snapped down on each side of the valley. The ends of each shingle course are secured to the valley with a 3-inch wide strip of asphalt roofing cement.

Hips and Ridges

Most roofing manufacturers provide special shingles to cap the roof ridge and hips. Ridge and hip shingles can also be made from the same ones used to cover the roof.

The ridge and hip shingles for a roof covered with square-butt strip shingles should be at least 9 × 12 inches. Any tabs must be cut off the shingles. The common 12 × 36-inch three-tab shingle strip is large enough to supply three ridge or hip shingles.

All shingling on the flat surfaces of the roof must be completed before the ridge or hip shingles are applied. After the ridge or hip shingles have been cut to the required size and shape, bend each shingle lengthwise down the center so that each side of the ridge or hip will have the same amount of overlap. Avoid splitting or cracking the shingle as you bend it. If the weather is cold, warm the shingle until it is flexible enough to bend without cracking.

Each shingle is secured to the ridge or hip with a nail on each side of the roof juncture. Position the nails 1 inch up from the bottom edge

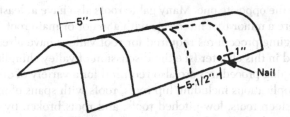

Fig. 8–24. Standard exposure and nail locations for hip and ridge shingles. *(Courtesy Asphalt Roofing Manufacturers Association)*

and 5½ inches back from the exposed edge (Fig. 8–24). These nail positions are for standard shingle exposure.

Ridge shingles are applied by beginning at either end of the roof and working toward the opposite end (Fig. 8–25). Begin hip shingles at the bottom of the hip and work toward the ridge. On hip roofs, the hips should be shingled before the ridge. Allow 5 inches of exposure for both ridge and hip shingles when standard exposures are being used.

Avoid using any kind of metal ridge roll with asphalt shingles. After a period of time the metal may corrode and discolor the shingles.

Special Shingling Procedures

A gable roof with a straight, unbroken surface is usually shingled by starting at the left-hand rake, or the most visible rake, and working

Fig. 8–25. Applying ridge shingles.

across to the opposite one. Many gable roofs also have at least two valleys where a minor roof intersects with a major or main roof. The special shingling procedures required for roof valleys have already been described in this chapter (see the discussion of valley shingling). Special shingling procedures are also required for a variety of other types of roof applications including hip roofs, roofs with spans of more than 30 feet, steep roofs, low-pitched roofs, and roofs broken by dormers, skylights, and other roof openings. These special shingling procedures are described in this section.

Shingling from Roof Center

The recommended method of applying strip shingles to hip roofs, roofs with two equally visible rakes, or roofs longer than 30 feet, is to start at the center and work in both directions toward the two roof rakes.

Begin by snapping a vertical chalk line in the exact center of the roof. Cut off 3 inches from the bottoms of as many shingles as necessary and apply a starter strip running in each direction from the center to the two rakes. The starter strip is applied with the mineral surface coating of each shingle facing down against the roof deck, and should overhang both eaves and rakes of the roof by approximately ½ inch. The amount of overhang must be maintained with each shingle course at the rakes or the roof will have an uneven appearance.

Begin the first course of shingles by aligning the first full shingle with its butt edge even with the starter course edge and its center tab centered on the vertical chalk line (Fig. 8–26). Succeeding courses are laid from the centerline with the required exposure and offset. The starter strip and each shingle course must be trimmed to fit the rake or hip ridge line. The ends of each of the two shingles terminating a course should be secured to the deck with asphalt roofing cement.

Shingling Around Dormers

Shingling around a dormer requires special care because the shingle courses must be laid exactly parallel to the eave line or their pattern will not continue evenly above the dormer roof. The courses can be kept parallel by snapping horizontal and vertical chalk lines on either side of the dormer as guides for shingle alignment (Fig. 8–27).

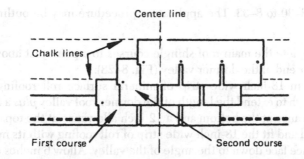

Fig. 8–26. Shingling from roof center. *(Courtesy Asphalt Roofing Manufacturers Association)*

When the roof shingle courses reach the dormer, flashing must be installed along its base. Dormer base flashing serves the same function as the flashing installed along the base of a chimney. Base flashing installation procedures are described in Chapters 6 and 7.

After the base flashing is installed, the shingle courses are continued up the roof to a point slightly above the lower edge of the dormer valley (Fig. 8–28). As each course is laid, step flashing should be installed where the last shingle contacts the dormer wall (Fig. 8–29).

If the dormer has a gable roof, two valleys will be formed at the point it joins the main roof. Flashing must be applied to these valleys before any shingle courses are laid. Mineral surface roll roofing is the most common type of flashing material used on asphalt shingle roofs.

The flashing and shingling of an *open* dormer valley are illustrated

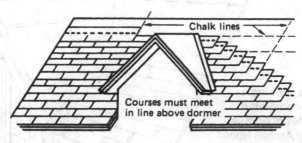

Fig. 8–27. Shingling around dormers. *(Courtesy Johns-Manville Sales Corp.)*

in Figs. 8–30 to 8–33. The application procedure may be outlined as follows:

1. Complete the main roof shingle courses to a point just above the lower end of the dormer valley (Fig. 8–28).
2. Cut an 18-inch wide strip of mineral surface roll roofing long enough to extend the length of the dormer roof valley *plus* a 2-inch allowance at the bottom and a 12-inch allowance at the top.
3. Bend and fit the 18-inch wide strip of roll roofing with its mineral surface face down to the angle of the valley. Allow 6 inches of material to cover the dormer roof deck and 12 inches to cover the main roof deck (Fig. 8–30).
4. Cut the bottom of the flashing strip so that there is a ¼-inch overhang along the eave of the dormer roof deck and a 2-inch projection below the point where the two roofs join.
5. Nail the roll roofing flashing strip to the valley with the mineral surface face down. Locate the nails approximately 1 inch from the edges of the roll roofing strip. Make certain the roll roofing fits tightly into the valley before nailing.
6. Cut and fit a 36-inch wide strip of roll roofing to the valley with the mineral surface side facing up.
7. Center the wider strip in the valley and cut the bottom edge so that it is flush with the edge of the underlying 18-inch wide strip (Fig. 8–31).
8. Nail the 36-inch wide strip to the valley with nails positioned 1

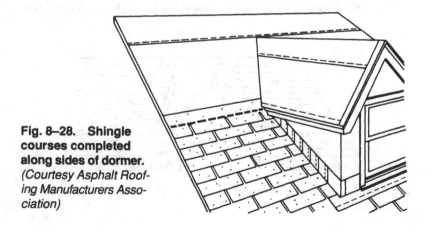

Fig. 8–28. Shingle courses completed along sides of dormer. *(Courtesy Asphalt Roofing Manufacturers Association)*

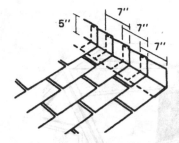

Fig. 8–29. Step flashing details along dormer. (Courtesy Johns-Manville Sales Corp.)

inch in from the edges. Keep all nails at least 7 inches from the valley joint. Additional protection against leakage can be provided by running a strip of asphalt roofing cement along the edges of the roll roofing.

9. Snap two chalk lines along both sides of the valley. These lines should be 3 inches on either side of the center of the valley where the dormer roof ridge intersects with the main roof and must diverge at the rate of ⅛ inch per foot in the direction of the dormer roof eave (Fig. 8–32).

10. Complete all shingle courses on the main roof to the chalk line in

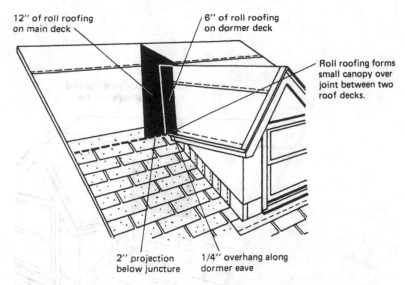

Fig. 8–30. Laying 18-inch wide strip of roll roofing in dormer valley.

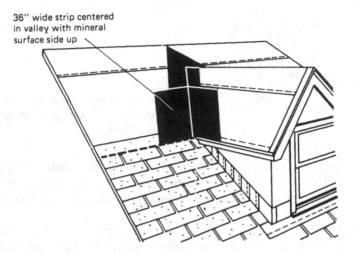

36" wide strip centered
in valley with mineral
surface side up

Fig. 8–31. Laying second or upper strip of roll roofing.

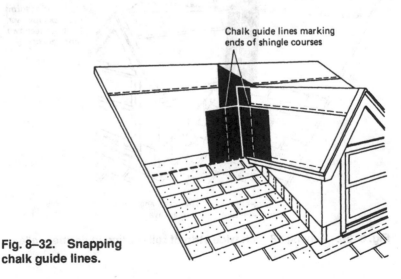

Chalk guide lines marking
ends of shingle courses

**Fig. 8–32. Snapping
chalk guide lines.**

the valley and cut the last shingle to conform to the angle of the line (Fig. 8–33).

11. Embed the end shingles of each course in plastic asphalt cement and press them down firmly onto the valley flashing. Do not nail through the flashing.

12. Repeat steps 1–10 on the other side of the dormer.

13. Shingle the dormer roof by starting at the eave at each rake and running the courses to the chalk lines in the valleys. Cut the ends of the shingles to fit the angle of the valley and embed them in plastic asphalt cement. Do not nail through the flashing. The shingling method is the same as that used for shingling a main roof.

14. Apply shingles to the dormer ridge by starting at the rake and working toward the main roof. Secure the shingles with one nail on each side of the shingle $5\frac{1}{8}$ inches up from its bottom edge (for a 5-inch exposure) and 1 inch up from each side edge. Split the last dormer ridge shingle and nail it to the roof deck with two nails in each half (Fig. 8–34).

Shingling Steep Slopes

The maximum roof slope considered suitable for normal shingle application is 7-in-12 or approximately 60°. On steeper roof slopes, the factory applied self-sealing adhesives on the backs of the shingle tabs show a tendency to pull free. A strong wind will frequently bend these

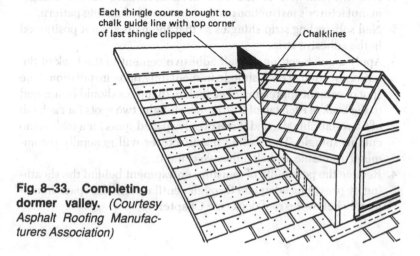

Each shingle course brought to chalk guide line with top corner of last shingle clipped

Chalklines

Fig. 8–33. Completing dormer valley. (Courtesy Asphalt Roofing Manufacturers Association)

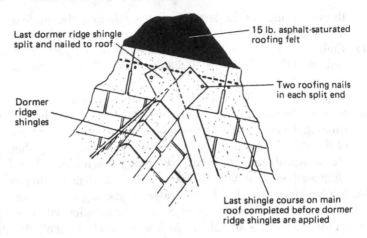

Last dormer ridge shingle
split and nailed to roof

15 lb. asphalt-saturated
roofing felt

Two roofing nails
in each split end

Dormer
ridge
shingles

Last shingle course on main
roof completed before dormer
ridge shingles are applied

Fig. 8–34. Shingling dormer ridge.

loose shingles back, which not only ruins the appearance of the roof, but also exposes the roof to possible leaks.

Mansard roofs and others with slopes exceeding 7-in-12 or 60° require that the shingles be applied with certain modifications of standard shingle application procedures. These modifications are illustrated in Fig. 8–35 and are outlined as follows:

1. Nail the shingles to the roof deck according to the shingle manufacturer's instructions and recommended nailing pattern.
2. Nail self-sealing strip shingles so that the nails are not positioned in the adhesive strip.
3. Apply a quick-setting asphalt adhesive cement to the back of the bottom section of each shingle immediately upon installation. One spot of cement approximately 1 inch in diameter should be applied to the back of each tab of a three-tab shingle; two spots for each tab of a two-tab shingle; and three evenly spaced spots for a tabless (no cutout) shingle. The roofing manufacturer will generally recommend a suitable asphalt adhesive cement.
4. Reduce the possibility of moisture entrapment behind the sheathing by providing adequate through ventilation in the space immediately below the roof deck (see Chapter 4).

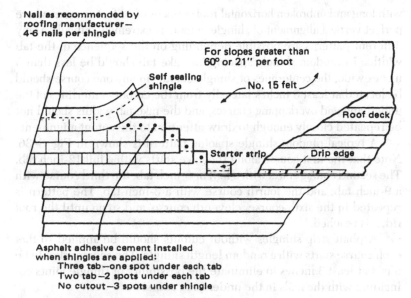

Nail as recommended by
roofing manufacturer—
4-6 nails per shingle

For slopes greater than
60° or 21" per foot

Self sealing
shingle

No. 15 felt

Roof deck

Starter strip

Drip edge

Asphalt adhesive cement installed
when shingles are applied:
Three tab—one spot under each tab
Two tab—2 spots under each tab
No cutout—3 spots under shingle

Fig. 8–35. Shingling steep slopes and mansard roofs. *(Courtesy Asphalt Roofing Manufacturers Association)*

Shingling Low-Pitched Roofs

Low-pitched roofs with slopes of 2-in-12 to just under 4-in-12 should be covered with square tab shingle strips. Because the rate of water runoff is so much slower on a low slope, there should be a double layer of underlayment along the eaves to protect the sheathing of the roof deck. The method for applying underlayment on low slopes is described in greater detail in Chapter 3. An eaves flashing strip of roofing felt should be cemented to the roof edge before any shingles are applied.

If self-sealing shingles are not used, each tab or lower section of the shingle should be cemented down by applying at least three quarter-size spots of roofing cement.

Random Shingle Spacing

Random shingle spacing is an application method in which the shingle courses are started with irregular tab widths. It is used on roofs with long and unbroken horizontal roof surfaces of 30 feet or more where perfect vertical alignment of shingle cutouts is extremely difficult. Many different patterns are possible depending on the sequence of the tab widths. In random shingle spacing, no rake tab should be less than 3 inches wide, the centerlines of shingle cutouts in any one course should be located at least 3 inches laterally from the cutout centerlines of the underlying and overlapping courses, and the rake tab widths should not be repeated closely enough to draw attention to the cutout alignment.

A typical random shingle spacing pattern is shown in Fig. 8–36. Note that the first course along the eave starts with a full 12-inch tab. The second course starts with a half or 6-inch tab, the third course with a 9-inch tab, and the fourth course with a 3-inch tab. The pattern is repeated in the sixth course, eleventh course, and so on until the roof ridge is reached.

Asphalt strip shingles without cutouts should be applied so that each course starts with a random length shingle. The courses should be offset at least 3 inches to eliminate the possibility of shingle joints coinciding with the nails in the underlying course.

Ribbon Courses

A distinctive look can be achieved on asphalt shingle roofs by using ribbon courses at the eaves and every fifth course up the roof. As

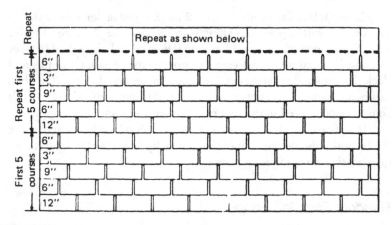

Fig. 8–36. Typical random shingle spacing pattern. *(Courtesy Asphalt Roofing Manufacturers Association)*

shown in Fig. 8–37, each ribbon course results in a triple thickness of shingles that emphasizes the horizontal roof line. The application method is as follows:

1. Cut a 4-inch wide strip from the top of a 12 × 36-inch asphalt strip shingle.
2. Nail the 4 × 36-inch strip to the eave either flush with the flashing, or with the same overhang provided the first course of shingles. Complete this 4-inch wide course the length of the eaves.
3. Nail the 8 × 36-inch strip directly over the 4 × 36-inch strip with the bottom edges of both strips flush (Fig. 8–38). Complete this 8-inch wide course the length of the roof.
4. Lay the first course of 12 × 36-inch shingles along the eaves with its bottom edge flush with the edges of the underlying strips. Offset the cutouts of the first course so that they are not aligned with those of the underlying 8 × 36-inch strip.
5. Lay ribbon courses at every fifth shingle course up the roof using the method described in steps 1–4 above.

Roof Repair and Maintenance

The roof surface is exposed to a variety of potentially harmful weather conditions. Over a period of time the heat of the sun causes asphalt shingles to blister, curl or cup, fade and lose their color, or dry out and become brittle. Temperature changes cause dried out, brittle shingles to split or crack. Strong winds blow away portions of split or cracked shingles. They also lift shingles and loosen the roofing nails, or create bare spots by blowing away the mineral granules. Rain can be blown under loose, curled, or cupped shingles and leak through the sheathing of the roof deck. Rain can also rust galvanized steel flashing, and the sun can dry out and crack the flashing cement.

The service life of an asphalt shingle roof can be extended by proper repair and maintenance. This can be accomplished by periodically checking the roof for signs of damage or deterioration and making the necessary repairs. The checklist should include the following:

1. Roof deck and framing
2. Shingle courses
3. Hip and ridge shingles

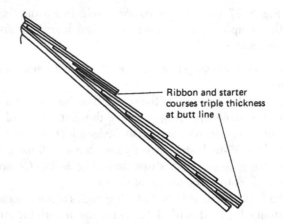

Fig. 8–37. Cross section of roof with ribbon courses. *(Courtesy Asphalt Roofing Manufacturers Association)*

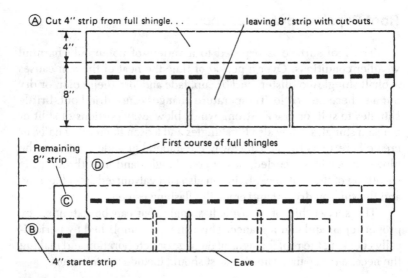

Fig. 8–38. Offsetting cutouts for ribbon courses. *(Courtesy Asphalt Roofing Manufacturers Association)*

4. Flashing
5. Roof openings
6. Gutters and downspouts

Roof Deck and Framing

Go into the attic or attic crawl space and check for leaks. The best time to do this is when it is raining. A minor leak can sometimes be stopped by patching the underside of the sheathing, but most require the removal of some or all of the roofing materials before repairs can be made.

Check the roof framing for adequate support and the condition of both the rafters and roof deck sheathing. Replace warped, split, or broken rafters, and warped or damaged sheathing. If the framing is inadequate for the weight of the roof, it will have to be reinforced. See Chapter 3 for detailed instructions concerning the inspection and repair of roof decks and roof framing.

Shingle Courses

Check the roof surface for damaged or missing shingles, missing portions of shingles, cracked, split, or blistered shingles, cupped or curled shingles, lifted shingles, and loose or missing nails (Figs. 8-39 to 8-41). If these conditions are extensive, it is an indication that the roof is in an advanced stage of deterioration and must be replaced or reroofed. If there are relatively few occurrences of these conditions, only minor repairs are required. Before making any repairs, however, check the condition of the underlayment. If it is dried out and brittle, both the shingles and underlayment should be replaced regardless of the condition of the former.

Replacing Damaged Shingles—Lift the overlapping shingle in the next course to provide access to the nails of the damaged shingle. Carefully remove the nails from the damaged shingle by prying them up with a chisel or pry bar. Lift out the damaged shingle and fill the old nail holes with roofing cement or a sealer. Insert the replacement shingle carefully so that it does not damage the underlayment. Lift the overlapping shingle and nail the new one to the roof with the required number of nails and nailing pattern (Fig. 8-42). *Do not drive the nails*

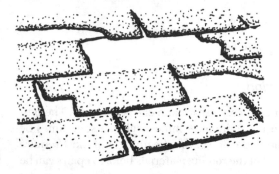

Fig. 8–39. Shingle damage.

into the old nail holes. Cover the nail heads with roofing cement and run a strip of cement around the underside of the edges of both the new and overlapping shingles and press them down firmly.

Repairing Cracked or Loose Shingles—If a portion of a shingle has cracked off but has not blown away, nail down both sections (Fig. 8–43). Cover the heads of the roofing nails with roofing cement and run a strip of cement over the crack between the two shingle sections. If the shingle is only loose, insert dabs of roofing cement under the shingle, press it down firmly, and nail it to the roof.

Replacing Missing Shingles—Follow the procedure described for replacing damaged shingles. Do not use the old nail holes when nailing the replacement shingle to the roof.

Cupped or Curled Shingles—Apply roofing cement under the cupped or curled section, press it down firmly, and nail it to the roof. Cover the nail heads with dabs of roofing cement.

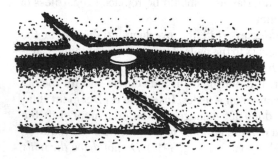

Fig. 8–40. Popped nails.

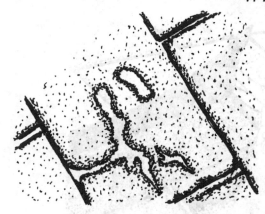

Fig. 8–41. Bare spots.

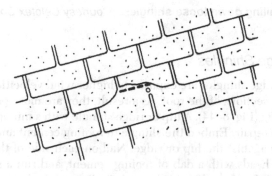

Bend shingles up to expose roofing nails of damaged shingle.

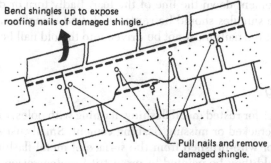

Pull nails and remove damaged shingle.

Fig. 8–42. Removing damaged shingle.

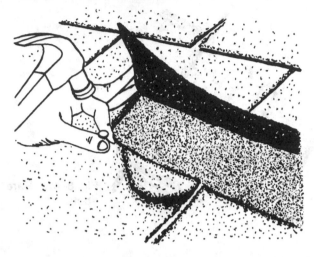

Fig. 8–43. Nailing down loose shingles. *(Courtesy Celotex Corp.)*

Hip and Ridge Shingles

Hip and ridge shingles are exposed to higher wind velocities than those on other sections of the roof. As a result, they are more prone to lifting or tearing (Fig. 8–44). A hip or ridge shingle with a minor tear is not difficult to repair. Embed the shingle in roofing cement and press it down firmly against the hip or ridge. Nail on each side of the tear, cover the nail heads with a dab of roofing cement, and run a strip of roofing cement down the line of the tear. Badly torn or damaged hip and ridge shingles should be replaced. Lifted shingles can be nailed down, but the nails must not be driven into the old nail holes.

Flashing

Check for rusted metal flashing, flashing with holes, torn felt flashing, and cracked or missing flashing cement. Small rust spots can be scraped or wire-brushed from the surface of metal flashing. The surface should then be painted to protect it from corrosion. Where rust has penetrated through the metal, the flashing should be replaced. Small pin holes in the metal can be covered with roofing cement after

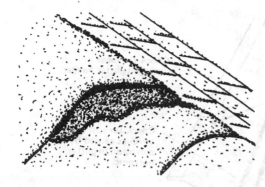

Fig. 8–44. Torn hip and ridge shingles.

the area around the hole has been cleaned with a wire brush. Replace roofing felt used as flashing when it is badly torn or brittle. Cracked or brittle flashing cement should be removed and the joint cleaned and then recemented. Cracks around the embedded edges of chimney flashing should be freed of loose mortar, cleaned with a wire brush, and completely sealed with a suitable sealer. See Chapter 7 for additional information on the inspection, repair, and replacement of flashing.

Gutter Systems

Gutters—Clean leaves and other debris from the gutters. Check metal gutters for rust spots and rusted out areas, and make the necessary repairs (see Chapter 15). Leaks usually develop at seams and end caps of gutters. The points to check for leaks on a metal gutter system are illustrated in Figure 8–45. Scrape away the old sealer, clean the seam, and apply a suitable sealer along all connecting edges and joints.

Gutter Hangers—Check to see if the gutter hangers are properly fastened beneath the shingles. If the gutter hanger straps are nailed to the tops of shingles, check for signs of deterioration and repair as necessary. Deterioration at this point may require replacement of some or all of the starter course shingles.

Downspouts—Check downspouts and downspout fasteners for looseness and repair as necessary. Replace badly rusted metal downspouts.

Splash Blocks—Check the splash block at the bottom of each downspout for accumulations of mineral granules. The heavy loss of mineral

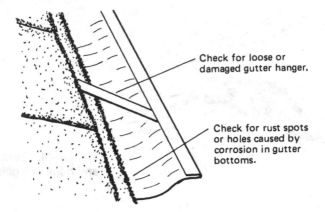

Check for loose or damaged gutter hanger.

Check for rust spots or holes caused by corrosion in gutter bottoms.

Fig. 8–45. Points to check for leaks on metal gutters.

granules from the shingles is usually a sign that the roof is deteriorating. Deposits of mineral granules will also collect in the gutters.

Roof Openings

Inspect the flashing and flanges around skylights to make certain that they are secure and unbroken. The putty or synthetic sealer around the skylight glass should be unbroken and tight. These are all potential sources of water leaks. Check the condition of stack or vent pipe flashing and repair as necessary. Check the roof for possible leaks where television antennas or antenna guy wires have been anchored to the roof (Fig. 8–46). If the television mast has been secured to a vent pipe, strong winds can cause it to loosen the caulking around the vent pipe flashing. Check for these problems and make the necessary repairs. Detailed instructions describing the inspection and repair of flashing and flanges for roof openings are provided in Chapters 5 and 7.

Reroofing with Asphalt Shingles

Asphalt shingles can be used in reroofing to cover any type of roofing material except tile, slate, mineral fiber shingles, and metal roof-

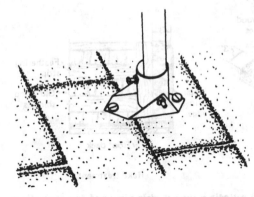

Fig. 8–46. Leaks around television antennas and antenna guy wires.

ing. These roofing materials are too hard and brittle to serve as a suitable nailing base. Asphalt shingles are most commonly applied over an existing asphalt shingle roof or a wood shingle (or shake) roof. Under certain conditions they can also be applied over roll roofing and built-up roofs. *Never reroof over more than one existing roof.* If the existing roof has already been reroofed once, all the roofing materials will have to be removed before the new roofing can be applied.

Reroofing Preparations

Roofing over Asphalt Shingles—Prepare the existing roof so that it provides a smooth, flat nailing surface for the new shingles. Remove loose nails and renail shingles where necessary. Buckled, cupped, or curled shingles as well as surface rolls should be split with a utility knife and nailed flat. Warped sheathing boards should be nailed flat or replaced. See Chapters 2 and 3 for additional details on preparing the roof for reroofing.

Roofing over Wood Shingles—Remove all loose and protruding nails, and renail the shingles. *Do not drive the new nails into the old nail holes.* Split warped or curled wood shingles and nail the sections flat. Replace missing shingles. If the existing shingles are badly curled or turned at their butts, nail beveled wood feathering strips along the butts (Fig. 8–47). If the existing shingles along the rakes and eaves are badly deteriorated, cut them back far enough to install 1 × 4-inch or 1 × 6-inch wood strips with their edges overhanging the eaves and rakes the same distance as the old shingles. Using these strips along

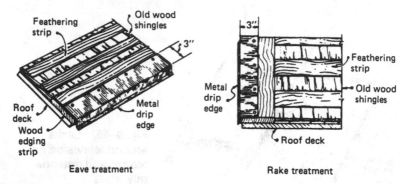

Fig. 8–47. Preparation of existing wood shingle roof for reroofing with asphalt strip shingles. *(Courtesy Asphalt Roofing Manufacturers Association)*

Fig. 8–48. Applying starter course. *(Courtesy Owens-Corning Fiberglas Corp.)*

the eaves and rakes improves wind resistance, provides better alignment of the asphalt shingles, and serves as a suitable nailing base for
the new roofing.

Flashing

Old flashing should be saved if it will last as long as the new roofing. If it is too deteriorated to be saved, it can still be used as a pattern
for new flashing. Use metal drip edges along the rakes and eaves. The
length of the roof along the eaves should also be covered by a strip of
roll roofing. Additional flashing details are discussed in Chapter 7.

Reroofing

After the roof surface has been properly prepared for reroofing,
the procedure used to apply asphalt strip shingles is the same over either an existing asphalt or wood shingle roof.

Starter Course—Use full-length shingles with their tabs removed as
the starter course. Each starter course shingle should be cut equal in
width to the exposure of the old shingles (usually 5 inches). Trim 6
inches off the length of the first shingle and position it at the left-hand
rake with its mineral surface up and its adhesive strip next to the edge
of the roof (Fig. 8–48). Nail it to the roof so that it overlaps the drip
edge ¼ inch to ⅜ inch at the rake and eave. Continue the starter course
with full-length (36-inch) shingles across the bottom edge of the roof to
the opposite rake.

First Course—Cut 2 inches from the top edge of a full-width new shingle, overlap the starter course, and nail it to the roof (Fig. 8–49) with its
cut edge aligned with the butt edge of the old shingle. Complete the
first course by continuing across the roof to the opposite rake.

Second Course—Use full-width shingles with their top edges aligned
with the butt edges of the existing shingles in the next course (Fig.
8–50). Cut the first shingle in the second course to meet the required
offset of the shingling pattern. The full size of the second course shingles will reduce the exposure of the first course. This reduction of exposure occurs only in the first course, which is partially concealed by
the gutter system.

Third and Succeeding Courses—Full-width shingles with their top
edges aligned with the butts of the existing shingles are used in the

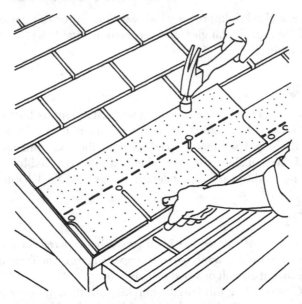

Fig. 8–49. Applying first course. *(Courtesy Owens-Corning Fiberglas Corp.)*

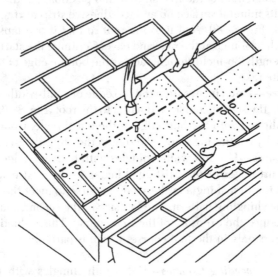

Fig. 8–50. Applying second course. *(Courtesy Owens-Corning Fiberglas Corp.)*

Fig. 8–51. Applying third course. *(Courtesy Owens-Corning Fiberglas Corp.)*

Fig. 8–52. Shingling woven valley. *(Courtesy Owens-Corning Fiberglas Corp.)*

Fig. 8–53. Shingling closed-cut valley. *(Courtesy Owens-Corning Fiberglas Corp.)*

Fig. 8–54. Applying hip and ridge shingles. *(Courtesy Owens-Corning Fiberglas Corp.)*

third course and all succeeding courses (Fig. 8–51). Shingle exposure for the new roofing will coincide with that of the existing roofing.

Valley Details—Woven or closed-cut valleys are recommended for reroofing with asphalt strip shingles (Figs. 8–52 and 8–53). Application details for both types and described in this chapter.

Hips and Ridges—Hips and ridges are completed as in new construction (Fig. 8–54). The hip and ridge shingles will have the same exposure as the shingle courses.

CHAPTER 9

Applying Specialty Shingles

Specialty shingles are designed for the specific purpose of forming a roof that stays flat and secure regardless of weather conditions. When applied to the roof, specialty shingles produce a distinctive woven or hexagonal pattern in each shingle course (Figs. 9–1 and 9–2). Most are individual shingles, but strip types are also available. Some are manufactured with locking devices to interlock adjacent shingles and thereby provide greater resistance to strong winds.

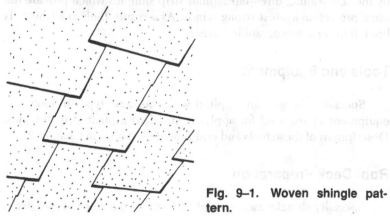

Fig. 9–1. Woven shingle pattern.

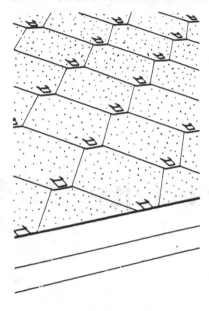

Fig. 9–2. Hexagonal shingle pattern. *(Courtesy Celotex Corp.)*

Specialty shingles are similar in composition to standard asphalt strip shingles. They are made from roofing felt saturated with asphalt and coated with a layer of mineral granules. They are available in a variety of colors, shapes, and sizes, and carry the same UL class C fire rating as standard asphalt strip shingles.

In many areas of the country, specialty shingles are being replaced by the self-sealing, three-tab asphalt strip shingles which provide the same protection against strong winds. As a result, their availability is limited to certain geographical areas.

Tools and Equipment

Specialty shingles are applied with the same types of tools and equipment as are used for applying standard asphalt strip shingles. Descriptions of these tools and equipment are found in Chapter 2.

Roof Deck Preparation

Specialty shingles can be applied to the same types of roof decks as asphalt strip shingles. Roof deck preparation instructions are covered in Chapter 3. Specific instructions for some types of reroofing applications are included in this chapter. Flashing details are described in Chapter 7.

Hex-Type Shingles

Hex-type shingles can be used in new construction or when reroofing over an existing roof. In either case, roofing with hex-type shingles produces a distinctive hexagonal pattern in each shingle course.

Hex Strip Shingles

Hex strip shingles are used primarily in new construction for roof slopes of 4-in-12 or greater. They are not recommended for slopes under 4-in-12. These low-pitched roofs require special roof deck preparation and are covered more effectively with self-sealing square-tab asphalt strip shingles.

Before hex strip shingles can be applied, the roof deck must be properly prepared by applying an underlayment, roofing felt (flashing) along the roof eaves, metal drip edges at the eaves and rakes, and other flashings where required. Either two-tab or three-tab hex strip shingles may be used to finish the roof.

Each two-tab or three-tab hex strip shingle is nailed to the roof with 11- or 12-gauge hot-dipped galvanized steel roofing nails with heads at least ⅜ inch in diameter (Fig. 9–3). The nails must be long enough to penetrate at least ¾ inch into 1-inch thick wood board sheathing or through plywood sheathing. They are driven through the shingles in a line 5¼ inches above the exposed butt edge. On two-tab shingles, the nails are located 1 inch from each end of the strip and ¾ inch from the angled sides of the cutout. Three-tab shingles are secured with nails driven 1 inch from each end of the strip and one nail centered over each cutout.

Each course is applied with the bottom edge of each shingle tab aligned with the top edge of the cutout of the shingle in the underlying course. Free-type tabs should be cemented to the underlying shingle with a spot of roofing cement (Fig. 9–4).

Two-Tab Shingle Application—The construction details of a roof covered with two-tab hex strip shingles are shown in Fig. 9–5. Begin by nailing a starter course of strip shingles along the roof eave with the

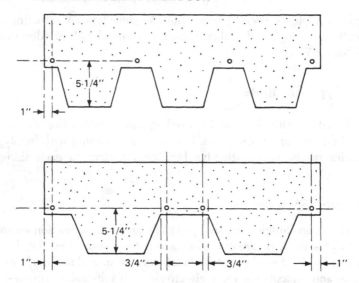

Fig. 9–3. Nail locations for hex-tab strip shingles.

mineral surface side up and the tabs pointing toward the ridge line (Fig. 9–6). Cut one half the tab off the first shingle in the starter course and apply it at the right-hand corner of the roof. Nail it to the sheathing so that it extends about ⅜ inch beyond the rake and roof eave. Nail the rest of the shingles in the starter course with the same ⅜ inch overhang at the eaves as the first one. Make certain that each shingle is in perfect

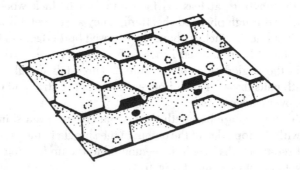

Fig. 9–4. Cement spot locations under hex shingle tabs. *(Courtesy Asphalt Roofing Manufacturers Association)*

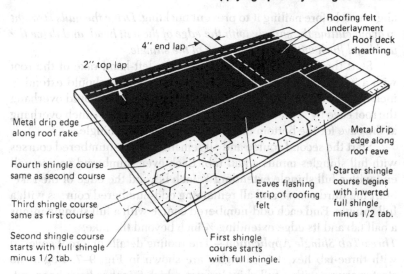

Fig. 9–5. Construction details of roof covered with two-tab hex strip shingles. *(Courtesy Asphalt Roofing Manufacturers Associations)*

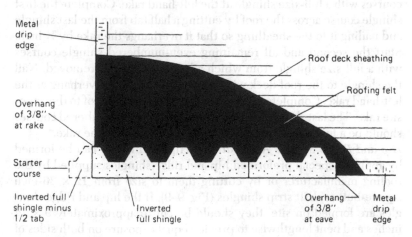

Fig. 9–6. Starter course of two-tab hex strip shingles.

alignment before nailing it to prevent buckling. *Drive the nails straight to avoid cutting the shingle with the edge of the nail head, and drive the nail head flush with the top surface of the shingle.*

Start the first course of shingles at the left-hand rake of the roof with a full shingle. The vertical edge of the shingle should extend ⅜ inch beyond the rake and the bottom edge of its tabs should overhang the roof eave by ⅜ inch. Finish the first course with a ⅜-inch overhang at the eave for the bottom edge of the tabs of each shingle.

Start the second course and all remaining even-numbered courses with full shingles minus a half tab. Each even-numbered course will end with a full shingle extending ⅜ inch beyond the edge of the rake. Start the third course and all remaining odd-numbered courses with a full shingle. End each odd-numbered course with a full shingle minus a half tab and its edge extending ⅜ inch beyond the rake.

Three-Tab Shingle Application—The roofing details of a roof covered with three-tab hex strip shingles are shown in Fig. 9–7. Begin the starter course with a full shingle from which 6 inches have been cut. Start shingling from the left-hand corner of the roof, with the mineral surface up and tabs pointed toward the ridge line. Nail the shingle to the sheathing so that it extends about ⅜ inch beyond the rake and roof eave (Fig. 9–8.). Complete the starter courses by nailing full-size shingles to the sheathing with the same ⅜-inch overhang as the first shingle. Start the first course of shingles and all remaining odd-numbered courses with a full-size shingle at the left-hand rake. Complete the first shingle course across the roof by cutting a half tab from the last shingle and nailing it to the sheathing so that it overhangs the rake by ⅜ inch. Start the second and all remaining even-numbered shingle courses with a full-size shingle from which a half tab has been removed. Nail the shingle to the roof deck with the required ⅜-inch overhang at the left-hand rake. Complete the second course across the roof to the opposite rake. The last shingle in the second and all even-numbered courses should be a full-size shingle extending ⅜ inch beyond the rake.

Hip and Ridge Shingle Application—Hips and ridges can be formed by applying special preformed hip and ridge shingles supplied by the roofing manufacturer or by cutting them to size from 12 × 36-inch square-butt asphalt strip shingles (Fig. 9–9). If the hip and ridge shingles are formed on site, they should be cut to approximately 9 × 12 inches and bent lengthwise to provide equal exposure on both sides of the hip or ridge. Apply the shingles to the hips or ridge with a 5-inch

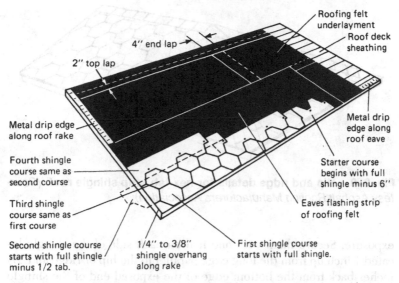

4" end lap

2" top lap

Roofing felt
underlayment

Roof deck
sheathing

Metal drip edge
along roof rake

Metal drip
edge along
roof eave

Fourth shingle
course same as
second course

Third shingle
course same as
first course

Second shingle course
starts with full shingle
minus 1/2 tab.

1/4" to 3/8"
shingle overhang
along rake

First shingle course
starts with full shingle.

Starter course
begins with full
shingle minus 6"

Eaves flashing strip
of roofing felt

Fig. 9–7. Roof covered with three-tab hex strip shingles. *(Courtesy Asphalt Roofing Manufacturers Association)*

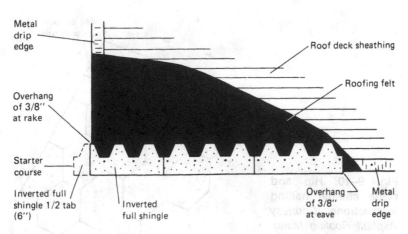

Metal
drip
edge

Roof deck sheathing

Roofing felt

Overhang
of 3/8"
at rake

Starter
course

Inverted full
shingle 1/2 tab
(6")

Inverted
full shingle

Overhang
of 3/8"
at eave

Metal
drip
edge

Fig. 9–8. Starter course of three-tab hex strip shingles.

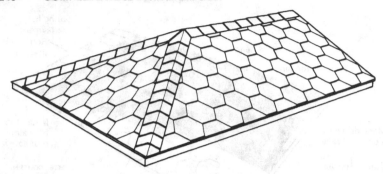

Fig. 9–9. Hip and ridge details for hex-tab strip shingle roof. *(Courtesy Asphalt Roofing Manufacturers Association)*

exposure. Secure them with one nail on each side of the shingle located 1 inch up from the long edge parallel to the hip or ridge and 5½ inches back from the bottom edge of the exposed end of the shingle (Fig. 9–10). The next shingle applied to the hip or ridge will overlap the nails of the underlying shingle. Hips are shingled by beginning at the bottom and working toward the roof ridge. Ridges are shingled by beginning at either end of the roof and working toward the opposite end.

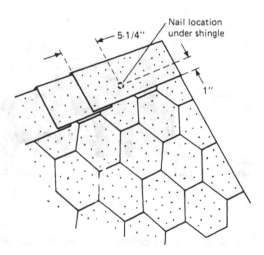

Fig. 9–10. Hip and ridge shingle nailing instructions. *(Courtesy Asphalt Roofing Manufacturers Association)*

Individual Hex Shingles

Individual hex shingles are used primarily for reroofing roofs with slopes of 4-in-12 or more. There are two types of individual hex shingles: the staple-applied shingle and the lock-down shingle. The staple-applied shingle is secured to the roof with 2 nails and one staple fastener. Lock-down shingles have tabs and slots or other devices so that the shingles can be locked together across the roof. Interlocking the shingles in this manner results in better than average wind and weather resistance.

Snap horizontal and vertical chalk lines across the roof to insure proper shingle and course alignment. The top center of each shingle should be aligned with a vertical chalk line. The horizontal chalk lines are used for course alignment.

Staple-Applied Hex Shingles — A staple-applied hex shingle is secured to the roof with one nail driven through each tab and a staple driven through the lower corner (Fig. 9–11). The nails are located 1 inch in from the side edge and 1 inch up from the bottom edge of each tab. Use either 11- or 12-gauge hot galvanized nails with heads at least ⅜ inch in diameter. Minimum nail lengths for different roofing applications are listed in Table 9–1. The nails must be long enough to penetrate at least ¾ inch into solid deck sheathing or through plywood sheathing. The lower corner of the shingle is stapled to the tabs of adjacent shingles in the course immediately below.

The construction details of a wood shingle roof reroofed with individual staple-applied hex shingles are illustrated in Fig. 9–12. The starter course consists of half shingles laid along the roof eaves. The half shingles for the starter course are cut so that the tabs are retained.

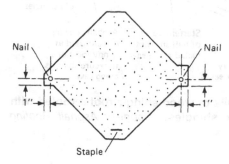

Fig. 9–11. Staple-applied hex shingles. *(Courtesy Asphalt Roofing Manufacturers Association)*

Table 9–1.   Minimum Nail Lengths for Different Roofing Applications

Type of Application	1″ Wood Board	⅜″ Plywood
Over new wood decks	1¼″	Nails should be long
Over old asphalt shingles	1½″	enough to penetrate
Over old wood shingles	1¾″	through plywood deck

(Courtesy Asphalt Roofing Manufacturers Association)

The recommended method of applying staple-applied hex shingles is to start at the center of the roof and work in both directions toward each rake. Begin by nailing the first half shingle of the starter course at the exact center of the roof with about a ⅜ inch overhang at the eave. Drive one nail through each tab and drive two more evenly spaced nails across the bottom of the shingle (Fig. 9–13). Lay the rest of the shingles of the starter course in both directions toward the rakes. Make certain that the bottoms of the shingles are aligned and that there is a

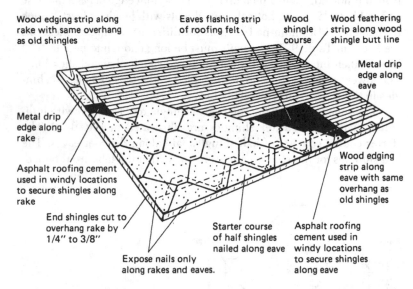

Wood edging strip along rake with same overhang as old shingles

Eaves flashing strip of roofing felt

Wood shingle course

Wood feathering strip along wood shingle butt line

Metal drip edge along eave

Metal drip edge along rake

Asphalt roofing cement used in windy locations to secure shingles along rake

Wood edging strip along eave with same overhang as old shingles

End shingles cut to overhang rake by 1/4″ to 3/8″

Starter course of half shingles nailed along eave

Asphalt roofing cement used in windy locations to secure shingles along eave

Expose nails only along rakes and eaves.

Fig. 9–12.   Construction details of wood shingle roof reroofed with individual staple-applied hex shingles. *(Courtesy Asphalt Roofing Manufacturers Association)*

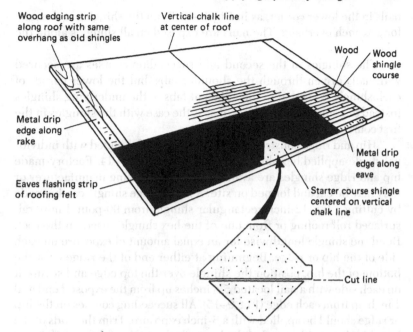

Wood edging strip along roof with same overhang as old shingles

Vertical chalk line at center of roof

Wood

Wood shingle course

Metal drip edge along rake

Metal drip edge along eave

Eaves flashing strip of roofing felt

Starter course shingle centered on vertical chalk line

Cut line

Fig. 9–13. Installing staple-applied shingles with the center starting method.

⅜-inch overhang at the eave. Secure each shingle with four nails, one in each tab and two more along the bottom. Cut the end shingles of the starter course to overhang each rake by ⅜ inch. In areas of the country where strong winds are common, embed the shingles along the eaves and rakes in a 6-inch wide strip of asphalt roofing cement before nailing them to the roof.

Begin the first course by nailing a full-size shingle to the center of the roof with its bottom edge aligned with the edge of the eave flashing strip of roofing felt or extended beyond the eave for a ⅜-inch overhang and centered over the shoulder tabs of the starter course shingles. Apply full-size shingles from the center of the roof toward each rake. Secure each shingle with its tabs butted and aligned with the tabs of adjacent shingles and its top aligned with a vertical chalk line. Nail the shingles to the roof with one nail in each shoulder tab and two exposed nails in the lower corner, as in Fig. 9–12. Trim the shingles at the rake for a ⅜-inch overhang. The nails are exposed on all shingles along the rakes.

The shingles in the second and succeeding courses are secured with nails driven through the shoulder tabs, but the lower corner of each shingle is stapled to the adjacent tabs of the underlying shingles instead of being nailed to the roof as is the case with the shingles in the first course.

Hip and ridge construction details for a roof shingled with individual staple-applied hex shingles are shown in Fig. 9–14. Factory-made hip and ridge shingles are available from the roofing manufacturer or they may be cut and formed on site. A hip or ridge shingle can be made by cutting a 9 × 12-inch rectangular shingle from 90-pound mineral-surfaced roll roofing or from one of the hex shingles used on the roof. Bend the shingle lengthwise for an equal amount of exposure on each side of the hip or ridge. Beginning at either end of the ridge or at the bottom of the hip, position the shingle over the top edge and secure it on each side with a nail located 5½ inches up from the exposed end and 1 inch up from each edge (Fig. 9–15). All succeeding courses on the hip or ridge should be applied with a 5-inch exposure. Trim the ends of the last shingles at either end of the hip to fit the angle of the eave or juncture with the ridge shingle.

Lock-Type Hex Shingles—Lock-type hex shingles are designed with a built-in locking device that ties or locks adjacent shingles to one another. Several different types of lock-type hex shingles are available.

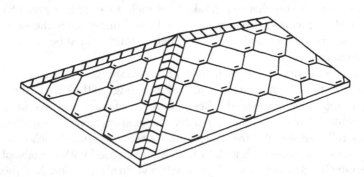

Fig. 9–14. Hip and ridge construction details. *(Courtesy Asphalt Roofing Manufacturers Association)*

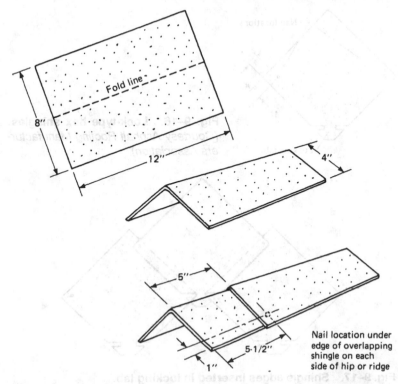

Fig. 9–15. Exposure and nail locations for hip and ridge shingles.

The shingle shown in Fig. 9–16 is secured to the roof with one nail in each shoulder. The lower corner of each shingle is locked by inserting the locking tabs under the sides of shingles in the underlying course (Fig. 9–17). The lock-type shingles shown in Fig. 9–18 are nailed to the roof with three nails instead of two. Two nails are driven through the shingle tab and one through the center of the left shoulder. When the next course of shingles is applied, the shingle tabs in the lower course are inserted through the slots in the lower corner of the shingles in the upper course.

Construction details of a roof covered with lock-type hex shingles are shown in Fig. 9–19. As with staple-applied shingles, the shingles along both the eaves and rakes should be embedded in a 6-inch wide

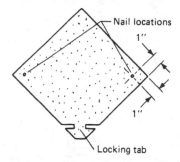

Fig. 9–16. Lock-type hex shingles. *(Courtesy Asphalt Roofing Manufacturers Association)*

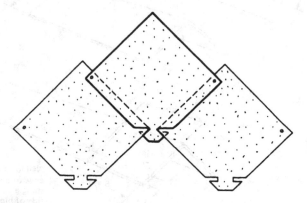

Fig. 9–17. Shingle edges inserted in locking tab.

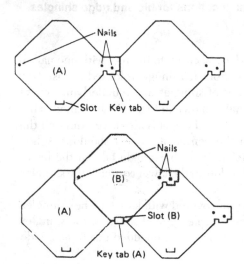

Fig. 9–18. Lock-type hex shingles secured with three nails. *(Courtesy Celotex Corp.)*

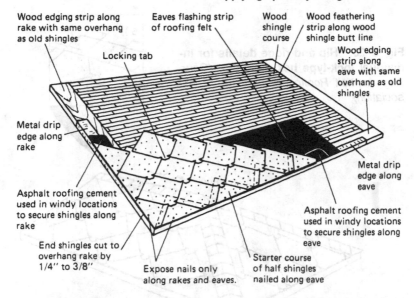

Wood edging strip along rake with same overhang as old shingles

Eaves flashing strip of roofing felt

Wood shingle course

Wood feathering strip along wood shingle butt line

Wood edging strip along eave with same overhang as old shingles

Locking tab

Metal drip edge along rake

Metal drip edge along eave

Asphalt roofing cement used in windy locations to secure shingles along rake

Asphalt roofing cement used in windy locations to secure shingles along eave

End shingles cut to overhang rake by 1/4" to 3/8"

Expose nails only along rakes and eaves.

Starter course of half shingles nailed along eave

Fig. 9–19. Roof covered with lock-type shingles. *(Courtesy Asphalt Roofing Manufacturers Association)*

strip of asphalt roofing cement to provide additional protection against strong winds (Fig. 9–20). The methods used to shingle the hips and ridges are the same as those used when shingling with individual staple-applied hex shingles (Fig. 9–12).

Giant Individual Shingles

Large rectangular 12 × 18-inch individual shingles are used in either new construction or reroofing applications to produce a single coverage roof. Either the Dutch lap or American-application method may be used to apply giant individual shingles.

Dutch Lap Method

The *Dutch lap method,* or *Scotch method* as it is also sometimes called, is a shingling method used to apply giant individual shingles over any kind of old roofing that provides a smooth surface. Although used primarily in reroofing, the Dutch lap method can also be used to

Fig. 9–20. Hip and ridge details for individual lock-type hex shingles. *(Courtesy Asphalt Roofing Manufacturers Association)*

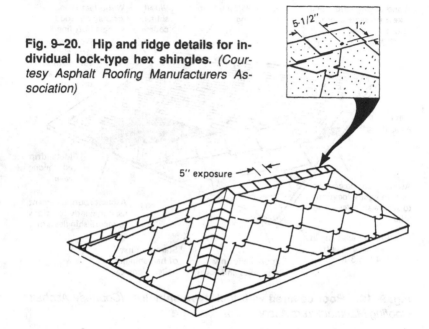

apply shingles to a new roof deck if an adequate underlayment has first been applied.

In new construction, roof deck preparation and the use of underlayment and flashing follow the same methods used with standard asphalt strip shingles (see Chapters 3 and 7).

The Dutch lap application method produces a single coverage roof, and the shingles are laid from either rake (from left to right or from right to left), but *never* from the center of the rock deck toward both rakes. This application method is not recommended for roof slopes of less than 4-in-12.

Both nails and metal fasteners are used in the Dutch lap method. Each shingle is secured with two nails and one fastener. The nails used are 11- or 12-gauge hot galvanized roofing nails with large heads of at least ⅜-inch diameter. In reroofing, nails with a minimum length of 1¾ inches should be used when applying shingles over an existing wood shingle roof. Reroofing over an existing asphalt shingle roof requires nails with a minimum length of 1½ inches. Slightly shorter nails (1¼-

inch minimum length) may be used when applying shingles by the Dutch lap method to a new wood deck.

Some shingle manufacturers provide fasteners in the form of special metal clips. Others recommend using noncorrodible wire staples and a stapling device. If fasteners are unobtainable or undesirable, a spot of asphalt roofing cement applied under the shingle will serve as well.

When applying the shingles from the left rake to the right one, the nails should be located in the upper left-hand and lower right-hand corners of the shingle. The fastener (or spot of asphalt roofing cement) is located in the exposed lower left-hand corner of the shingle. The fastener is used to secure the shingle to the overlapped portion of the adjacent shingle in the same course. *Never* use nails to secure the exposed corner of a shingle to the roof deck *except* along the rake and eave. The locations of the nails and fasteners are reversed when the shingles are applied from the right roof rake to the left one.

The Dutch lap method of applying giant individual shingles to an existing wood shingle roof is illustrated in Fig. 9–21. *A starter course is not required when using the Dutch lap application method.*

Begin by cutting back the old wood shingles approximately 6 inches from the eaves and rakes, and nail a wood edging strip to the wood deck. Nail feathering strips along the bottom edge of each course of wood shingles and metal drip edges to the rakes and eaves (Fig. 9–22). Install an eave flashing strip of roofing felt extending from the eave line up the roof to a point 12 inches inside the interior wall line of the structure (Fig. 9–23).

Cut a shingle 3 inches wide by 12 inches long and nail it to the roof deck, allowing a ⅜-inch overhang at the eaves and rakes. In areas where strong winds are common, embed the strip in roofing cement before nailing it. Cut a second shingle to a length of 15 inches and nail it over the first 3 × 12-inch shingle so that its edges are flush with it. Complete the course by applying full-size shingles with a 3-inch side lap. Trim the last shingle in the course so that it overlaps the opposite rake ¼ to ⅜ inches.

Start the second course with a shingle cut to a 12-inch length. Allow a 2-inch top lap over the first course and a ⅜-inch overhang at the rakes. Complete the second course with full shingles applied with a 3-inch side lap.

The third course begins with a shingle cut to a 9-inch length, and

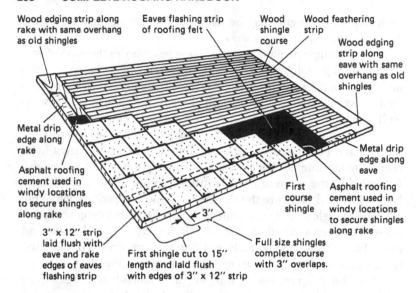

Fig. 9–21. Construction details of wood shingle roof reroofed with giant individual shingles (Dutch lap method). *(Courtesy Asphalt Roofing Manufacturers Association)*

the fourth with a shingle cut to a 6-inch length. The fifth course begins with a full shingle, and the pattern repeats itself. Each course is completed with full-size shingles with a 3-inch side lap and a 2-inch top lap over the shingles in the previous course.

Hip and ridge shingles are made by bending each 12 × 18-inch

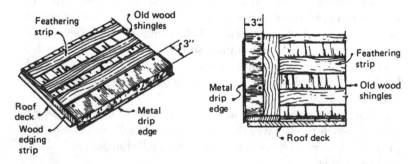

Fig. 9–22. Feathering strips, edging strips, and metal drip edges. *(Courtesy Asphalt Roofing Manufacturers Association)*

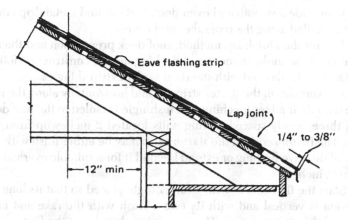

Fig. 9–23. Eave flashing strip. *(Courtesy Asphalt Roofing Manufacturers Association)*

giant individual shingle lengthwise down the middle. This allows a 6-inch overlap of shingle on either side of the hip or ridge. Beginning at the bottom of the hip or at either end of the ridge, apply the shingles with a 5-inch exposure. The nails should be placed 1 inch up from the edge and 5½ inches back from the exposed edge (Fig. 9–15).

American Method

When the American method is used to apply giant individual shingles, the shingles are laid with their long 18-inch dimension perpendicular to the roof eaves and parallel to the rakes. It is primarily this feature that distinguishes the American method from the Dutch lap method.

The American method may be used to apply shingles in either new construction or reroofing applications. When reroofing, no underlayment is required if the existing roof surface is smooth, flat, and in good condition. If the American method is used to apply shingles over an existing wood shingle roof, the same preparations used in the Dutch lap method must be used here. In other words, the wood shingles should be cut back about 6 inches along the roof rakes and eaves and a wood edging strip nailed to the roof deck in their place. Feathering strips should also be nailed to the bottom edge of each course of shingles to provide a smooth and even deck surface, and metal drip edges must be nailed along the roof rakes and eaves.

As with the Dutch lap method, roof deck preparation and the application of the underlayment and flashing in new construction follow the same methods used with standard asphalt strip shingles.

The shingles in the starter strip are laid horizontally along the roof eave with their edges touching. Each shingle is nailed to the roof deck with three evenly spaced roofing nails located 2 inches up from the eave. The bottom edge of the starter strip may be either flush with the edge of the eave flashing or extend beyond it for a suitable overhang of about ¼ inch.

Start the first course with a full shingle placed so that its long dimension is vertical and with its edges flush with the rake and eave edges of the starter course. The remaining shingles in the first course should be separated by a ¾-inch gap and nailed to the roof deck with two roofing nails. Each nail is located 6 inches from the bottom edge of the shingle and 1½ inches in from the sides. The ¾-inch shingle gap and nailing pattern is used in each shingle course.

The way in which the second and succeeding courses are laid will depend on whether the shingle joints break on thirds or on halves. Both patterns are illustrated in Figs. 9–24 and 9–25. In either case, all shingles are laid with the long dimension vertical.

If the shingle courses are to break on thirds, the second course should begin with an 8-inch wide shingle. Provide the same overhang at the rake as was provided for the first course, and allow a 5-inch exposure for the latter (Fig. 9–25). Apply full shingles for the remainder of the second course, and make certain that the ¾-inch gap between shingles is maintained.

Begin the third course with a 4-inch wide shingle, and complete the course with full shingles. The entire pattern is repeated in the fourth course with a full shingle.

If the shingle courses are to break on halves, begin the second course and every *even*-numbered course (fourth, sixth, eighth, and so on) with a 6-inch wide shingle, and the odd-numbered courses with a full shingle. The remaining shingles in every course are full shingles except where they have to be cut at the opposite rake line.

The roof ridge and the hips are covered with shingles bent lengthwise down their center. This allows 6 inches of shingle on either side of the ridge or hip.

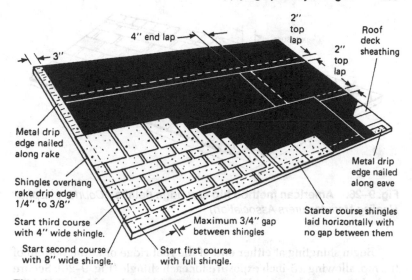

Fig. 9–24. Giant individual shingles applied by the American method with joints breaking on thirds. *(Courtesy Asphalt Roofing Manufacturers Association)*

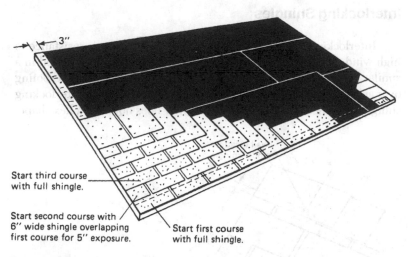

Fig. 9–25. Giant individual shingles applied by the American method with joints breaking on halves. *(Courtesy Asphalt Roofing Manufacturers Association)*

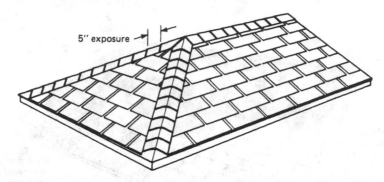

Fig. 9–26. American method hip and ridge details. *(Courtesy Asphalt Roofing Manufacturers Association)*

Begin shingling at either end of the roof ridge or at the bottom of the hip, allowing a 5-inch exposure for each shingle (Fig. 9–26). Secure each shingle to the roof with two nails. Each nail should be located 5½ inches from the exposed end and 1 inch up from the edge (Fig. 9–27).

Interlocking Shingles

Interlocking or lock-down shingles provide the same resistance to high wind damage as the hex lock-down shingles and are applied in a similar manner. They can be used in new construction or for reroofing over an existing roof. The principal difference between an interlocking shingle and a hex shingle is that the former does not have a hex shape.

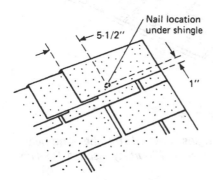

Fig. 9–27. Hip and ridge shingle nail locations.

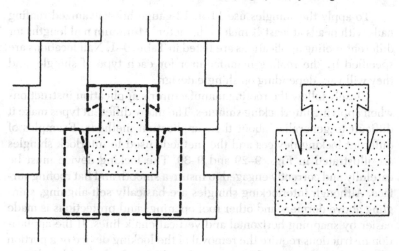

Fig. 9–28. T-lock shingle.

The interlocking shingle produces a woven pattern across the roof instead of a hexagonal one (Fig. 9–1).

Some interlocking shingles are shaped like a T and are called T-lock shingles (Fig. 9–28). Many other shapes are also available, however, and they differ primarily on the design and location of the locking device (Fig. 9–29). Most are individual shingles, but there is also a strip-type shingle available.

Interlocking shingles are usually applied to roofs with slopes of 4-in-12 or more. They can also be applied to roofs with slopes less than 4-in-12 if the roof deck is properly prepared in accordance with the roofing manufacturer's instructions.

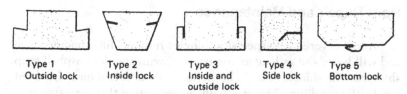

| Type 1
Outside lock | Type 2
Inside lock | Type 3
Inside and
outside lock | Type 4
Side lock | Type 5
Bottom lock |

Fig. 9–29. Typical locking devices for interlocking shingles. *(Courtesy Asphalt Roofing Manufacturers Association)*

To apply the shingles use 11-or 12-gauge hot galvanized roofing nails with heads at least ⅜ inch in diameter. Minimum nail lengths for different roofing applications are listed in Table 9–1. Nail locations are specified by the roofing manufacturer for each type of shingle, and they will vary depending on shingle design.

Closely follow the roofing manufacturer's application instructions when applying interlocking shingles. The many different types make it difficult to generalize about their application methods. Examples of different locking devices and the methods used to interlock shingles are illustrated in Figs. 9–29 and 9–30. The locking device must be carefully and correctly engaged to insure a smooth and flat roofing surface. Although interlocking shingles are basically self-aligning, shingling above dormers and other roof openings and protections is made easier by snapping horizontal and vertical chalk lines. If the application instructions require the removal of the locking device or a portion of it for the starter and first course shingles, embed each shingle in a 6-inch wide strip of roofing cement along the roof eaves and rakes.

Hips and ridges can be finished with preformed shingles available from the manufacturer or rectangular shingles cut to a minimum 9 × 12-inch dimension from 90-pound mineral-surfaced roofing felt or from interlocking shingles if large enough. The hip or ridge shingle should be bent lengthwise to provide an equal 4½-inch exposure on both sides of the hip or ridge. Hips are shingled by starting at the bottom and working toward the point at which the hip meets the roof ridge. Ridges are shingled after the hips are completed by starting at either end of the ridge and working toward the opposite one. Each hip or ridge shingle is laid with a 5-inch exposure and secure with two nails located 5½ inches up from the bottom of the shingle and 1 inch up from each side edge (Fig. 9–27).

Roof Repair and Maintenance

A roof covered with specialty shingles requires little maintenance and will last about as long as one covered with standard asphalt strip shingles. The shingles should be periodically checked for a dried out and brittle condition. This is usually an indication that reroofing is in order. If the underlayment is also dried out and brittle, strip the entire roof and install a new underlayment and shingles.

Repairs to roofs covered with specialty shingles are generally sim-

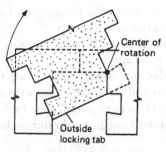

Type 1. Outside lock

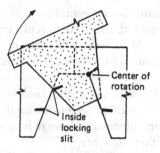

Type 2. Inside lock

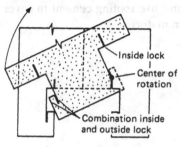

Type 3. Inside and outside lock

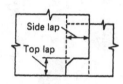

Type 4. Side lock

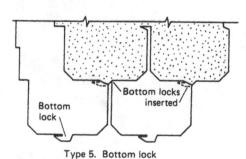

Type 5. Bottom lock

Fig. 9–30. Typical locking methods for interlocking shingles. *(Courtesy Asphalt Roofing Manufacturers Association)*

ilar to those made on roofs covered with standard asphalt strip shingles (see Chapter 8). This is particularly true of the hip and ridge shingles which are easily replaced. Because of the interlocking design of the interlocking type specialty shingle, it is unlikely that any will be missing from the roof. This is not the case with hex strip shingles which share the same problems as standard asphalt strip shingles, and are repaired, removed, or replaced in much the same manner.

Most leaks will occur at or around the roof flashing and can be repaired by following the instructions in Chapter 7.

If the shingles have been embedded in roofing cement along the roof eaves and rakes, check for looseness and use a putty knife to insert fresh roofing cement under the shingles where required. New roofing cement may also have to be inserted under loose hex strip shingle tabs if they are loose.

Nail down any loose nails on the roof deck and cover the exposed nail heads with a spot of roofing cement to prevent leaks.

Nail down any torn shingles and use roofing cement to cover the nail heads and the edges of the torn material.

CHAPTER 10

Roll Roofing

The easiest and least expensive method of covering a roof is by applying layers of smooth or mineral-surface roll roofing. This roofing material is recommended for use on flat or low-pitched roofs with a minimum slope of 1-in-12 or more. Because it is a quick and uncomplicated roofing method, roll roofing is frequently used in rural areas to cover the roofs of sheds, garages, and other types of farm buildings. It is also sometimes used to cover the roofs of houses. Among its disadvantages are its short service life (about 5 to 12 years depending on weather conditions) and its drab, unattractive appearance.

Tools and Materials

Table 10–1 lists some common types of roll roofing with specifications. Most roll roofing is produced in 36-inch wide rolls with each roll containing 36 feet of material. Pattern edge roll roofing is available in longer lengths.

Roll roofing is made by impregnating an organic felt material with an asphalt or coal tar saturant and then coating it with a viscous, weather-resistant asphalt. It is manufactured in 36-inch wide rolls in weights per roofing square of 45 to 120 pounds.

Table 10–1. Roll Roofing Weights and Sizes

Type of Roll Roofing	Shipping Weight Per Roofing Square	Length	Width	Top Lap	Exposure
Smooth	65 lb.	36′	36″	2″	34″
Surface	55 lb.	36′	36″	2″	34″
	45 lb.	36′	36″	2″	34″
Mineral Surface	90 lb.	36′	36″	2″	34″
	90 lb.	36′	36″	3″	33″
	90 lb.	36′	36″	4″	34″
Pattern Edge	105 lb	42′	36″	2″	16″
	105 lb	48′	32″	2″	14″
19″ Selvage Double Coverage	110 lb.	36′	36″	19″	17″
	120 lb.	36′	36″	19″	17″

The principal types of roll roofing are shown in Fig. 10–1. Smooth-surface roll roofing is the least expensive and the easiest to apply. It is used on sheds and other farm buildings where economy and utility outweigh considerations about appearance. The exposed nail application method is generally used when applying smooth-surface roll roofing.

Mineral surface roll roofing is made by embedding colored granules in the surface of the material while the asphalt coating is still hot. The other surface of the roll roofing sheet is identical in composition to smooth-surface roll roofing.

Pattern-edge roll roofing is produced in 36-inch wide rolls with diagonal shaped tabs along the bottom edge. The pattern edge or tabs resemble shingles except that the tabs are not staggered in each course.

Nineteen-inch selvage double coverage roll roofing is a roof covering material produced specifically for double coverage applications. Each roll is 36 inches wide and is divided horizontally into two sections. The lower 17-inch wide section is intended for exposure and is generally covered with an embedded mineral surface. The upper, or 19-inch wide selvage section, has a smooth surface and is covered by the mineral-surfaced lower half of the next overlapping course.

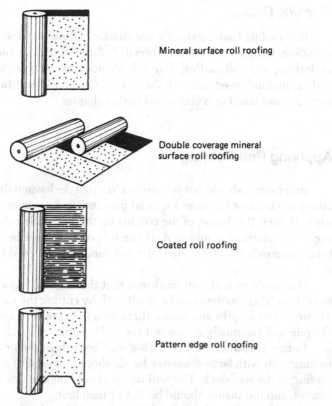

Mineral surface roll roofing

Double coverage mineral
surface roll roofing

Coated roll roofing

Pattern edge roll roofing

Fig. 10–1. Types of roll roofing. *(Courtesy Asphalt Roofing Manufac-turers Association)*

Roof Deck Preparation

No underlayment is required with roll roofing. Before laying the roll roofing, the roof deck should be swept with a broom. Knot holes should be covered with a piece of tin secured to the deck with mastic. The roof sheathing must be completely dry before the roll roofing can be applied.

Edging boards should be applied to the roof eaves and rakes be-fore applying roll roofing over an existing asphalt shingle roof (see the discussion of reroofing in this chapter).

Flashing Details

Roll roofing flashing details are similar to those used with asphalt shingling. Only composition materials (saturated felt) should be used as flashing with roll roofing. For a description of the types of flashing and the methods used to install them, read Chapter 7. Additional information about flashing is contained in this chapter.

Applying Roll Roofing

Roll roofing should not be applied to a roof deck when the temperatures are below 45°F unless special precautions have been taken, because there is the danger of the coating on the roll roofing sheet cracking as the material is unrolled. If the roll roofing must be applied at temperatures below 45°F, then the roll should be warmed before it is used.

One problem with roll roofing is that the material has a tendency to curl up. This problem can be dealt with by cutting the material into 12- to 18-foot lengths and piling them on a flat surface. The weight of the pile will eventually flatten out the roll roofing.

In new construction, 11- or 12-gauge hot galvanized or aluminum roofing nails with large diameter heads should be used to nail the roll roofing to the roof deck. The nail heads should measure ⅜ inch in diameter, and the shanks should be ⅞ to 1 inch long.

End laps and top laps should be cemented with a lap cement or quick setting cement recommended by the manufacturer of the roll roofing material. Always store the cement in a warm place until you are ready to use it. If you must work in temperatures below 45°F, warm the cement container by placing it in a pan of hot water. *Never* heat a can of lap cement over a flame. The cement is highly flammable and could explode.

Roll roofing is most commonly applied with each course running parallel to the roof eaves. However, if smooth surface roll roofing is used, the courses may also be applied so that each course runs parallel to the roof rakes (Figs. 10–2 and 10–3). The horizontal application method offers the best protection, because each course overlaps the underlying one on the downward slope of the roof.

In addition to the direction in which the courses are laid, roofs of

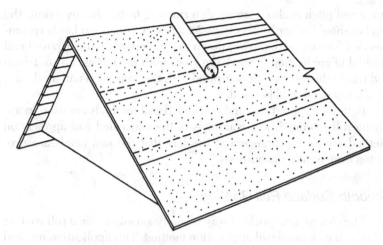

Fig. 10–2. Horizontal application.

this type can also be distinguished by the amount of top lap, the nailing method, and the type of roll roofing material used.

The *top lap* is the amount of material in a course of roll roofing that overlaps the underlying course. The amount of top lap is often recommended by the roll roofing manufacturer (Table 10–1), but the mini-

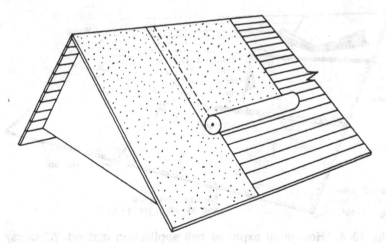

Fig. 10–3. Vertical application.

mum roof pitch is also a strong determining factor. In any event, the top lap should never be less than 2 inches. A 2-inch top lap is recommended for roofs with pitches down to 2 inches when the exposed nail method of application is used. On roofs with pitches as low as 1 inch and where the roll roofing is applied with the concealed nail method, a 3-inch top lap should be used.

Both the exposed nail and concealed nail methods are used in applying roll roofing. The advantages of the concealed nail application method are its appearance and the fact that the nail heads are protected from weather conditions.

Smooth-Surface Roll Roofing

The easiest and quickest way to apply smooth surface roll roofing is to use the exposed nail application method. This application method can be used either when applying the courses parallel to the eaves, or parallel to the rakes. The procedure for applying the roll roofing courses parallel to the eaves is illustrated in Fig. 10–4, and can be outlined as follows:

1. Lay the first course of roll roofing parallel to the eave of the roof so that approximately ¼ to ⅜ inch of material overlaps both the eave and rake.

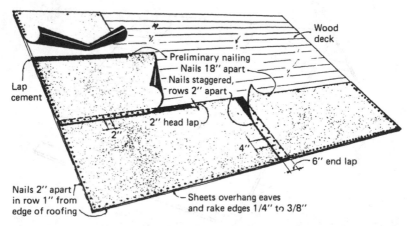

Fig. 10–4. Horizontal exposed nail application method. *(Courtesy Asphalt Roofing Manufacturers Association)*

2. Nail the material along the eave, rake, and end (lapping sections) of the course with nails spaced 2 inches apart in rows 1 inch from all edges. The nails in the rows should be slightly staggered to prevent splitting the wood sheathing boards of the roof deck.

3. Nail the top edge of the first course with nails spaced 18 inches apart and about ½ inch down from the edge. These nails will be covered by the 2-inch lap of the second course.

4. Apply the second course of roll roofing so that it extends ¼ to ⅜ inch over the rakes at each end of the roof, and laps the first course by at least 2 inches. Space the nails 18 inches apart along the top edge of the sheet.

5. Lift the lower edge of the second course and apply lap cement along the top 2 inches of the underlying (first) course. Press down the bottom edge of the second course and nail it to the roof deck, staggering the nails so that they are 2 inches apart and at least ¾ inch above the bottom edge of the material.

6. Apply the succeeding courses in the same manner as the second course, and trim the top of the last course even with the ridge. The top edges of the courses on either side of the roof ridge should butt, not overlap.

7. Cut a 12-inch wide length of roll roofing to cover the ridge.

8. Snap a chalk line 5½ inches down from the ridge on either side of the roof deck. These two chalk lines *must* run exactly parallel to the ridge.

9. Apply a 2-inch wide strip of lap cement along each chalk line. Do not allow the lap cement to extend below either chalk line.

10. Bend the 12-inch wide strip of roll roofing along its center, lay it along the ridge, and embed it in the cement. Nail the roll roofing strip to the ridge with nails staggered 2 inches apart in rows ¾ inch up from the bottom edges of the strip.

Double coverage smooth surface roll roofing is possible with the exposed nail application method by allowing each course to overlap the underlying one by 19 inches. This permits an exposure of 17 inches when using 36-inch wide rolls.

As shown in Fig. 10–5, the exposed nail application method can also be used by lay roll roofing courses parallel to the rake (i.e., perpendicular to the roof eave). The first course or strip of roll roofing should be started at the roof ridge at either rake. This first course is nailed

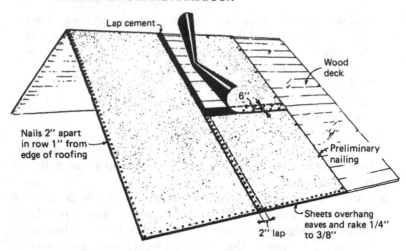

Fig. 10–5. Vertical exposed nail application method. *(Courtesy Asphalt Roofing Manufacturers Association)*

along the rake and eave with nails spaced on 2-inch center and approximately 1 inch in from the edge of the roof. Allow approximately ¼ to ⅜ inch of the lower end of each course to project over the rake at each end of the roof. Nails should be spaced 18 inches apart along the ridge and the lap edge on a line about ½ inch in from the edge of the sheet.

Before applying the second course of roll roofing, coat the 2-inch wide lap edge of the underlying sheet with lap cement. Embed the second course in the lap cement so that it overlaps the underlying sheet by exactly 2 inches, and nail it down. Space the nails on 2-inch centers along the lap edge and eave on a line 1 inch in from the edge. The nail spacing along the ridge will be the same as for the first course. If more than one section of roll roofing is required to complete a course, the end laps should be 6 inches wide, embedded in lap cement, and nailed to the roof deck. The nails should be staggered in two rows and spaced on 4-inch centers in each row. Stagger all end laps so that none of them is parallel in adjacent rows.

The procedures used to lay the first two courses of roll roofing also applies to the remaining courses. Roll roofing strips cut 12-inch wide and bent lengthwise along their centers are used to cap the ridge and hips (Fig. 10–6).

Details of the concealed nail method of applying smooth surface

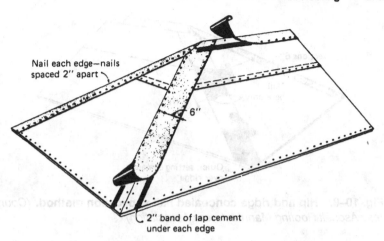

Fig. 10–6. Hip and ridge exposed nail application method. *(Courtesy Asphalt Roofing Manufacturers Association)*

roll roofing are illustrated in Figs. 10–7 and 10–8. The concealed nail application method may be outlined as follows:

1. Nail 9-inch wide edge strips along the rakes and eaves and allow approximately ¼ to ⅜ inch of material to overhang the edges of the roof deck. Use two rows of nails spaced on 4-inch centers in lines located ¾ to 1 inch in from each edge of the strip.

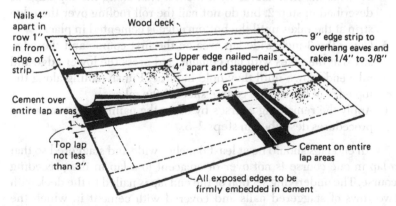

Fig. 10–7. Concealed nail application method. *(Courtesy Asphalt Roofing Manufacturers Association)*

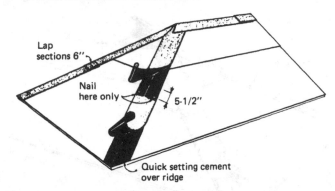

Fig. 10–8. Hip and ridge concealed nail application method. *(Courtesy Asphalt Roofing Manufacturers Association)*

2. Apply the first course of smooth surface roll roofing so that its outer edges are flush with the edges of the edge strips at the rake and eave. Nail along the top edge of the first course with nails spaced 4 inches apart and slightly staggered. The lower nails in the staggered row should never be more than 2 inches below the top edge of the first course.
3. Cover the edge strip with cement and firmly press the lower edge and rake ends of the first course sheet down until they hold in place.
4. Apply the second course of smooth surface roll roofing so that it overlaps the first course by 3 inches. Nail along the top edge as described in step 2, but do not nail the roll roofing over the edge strip at the rakes until the bottom edge is cemented in place.
5. Cover the 3-inch wide lap area on the first course and the edge strips at the rakes with lap cement, and press the lower edge and rake ends down until they hold in place. Finish nailing along the top edge of the second course sheet at the rakes.
6. Apply the remaining courses by using the nailing and cementing procedures described in steps 2–5.

All end laps should be at least 6 inches wide and staggered so that a lap in one course is not over or adjacent to a lap in the preceding course. The underlying section of the end lap is nailed to the deck with two rows of staggered nails and covered with cement in which the overlapping section of the end lap is embedded.

The ridge and hips are covered with 12 × 36-inch sections cut from the roll roofing strip. As shown in Fig. 10–8, each section is nailed and covered with cement over that portion which is to be lapped.

Double Coverage Roll Roofing

Roll roofing can be applied to the roof deck so that it provides double coverage. In other words, there is sufficient overlap between courses to provide a double layer of roll roofing across the surface of the roof deck.

A 19-inch selvage double coverage roll roofing is specially manufactured for this type of roofing application. Some types of double coverage roll roofing are applied with a cold asphalt adhesive, whereas others require the use of hot asphalt. Hot asphalt applications will also differ according to how the roll roofing is saturated and coated. Because of these differences, the manufacturer's instructions should be carefully followed when applying this type of roofing material.

The application of 19-inch selvage double coverage roll roofing to a new roof deck is illustrated in Fig. 10–9. The procedure may be outlined as follows:

1. Nail a drip edge along the roof eaves and rakes. The drip edge should extend 3 inches onto the roof deck and the nails should be placed about 10 inches apart along its inner edge (Fig. 10–10).

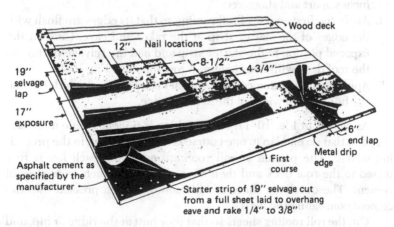

Fig. 10–9. Application of 19-inch selvage double coverage roll roofing. *(Courtesy Asphalt Roofing Manufacturers Association)*

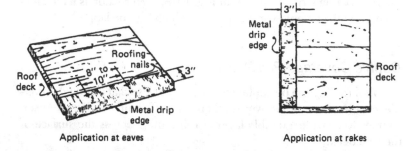

Application at eaves   Application at rakes

Fig. 10–10. Metal drip edge application. *(Courtesy Asphalt Roofing Manufacturers Association)*

2. Cut the 19-inch wide selvage portion from a length of roll roofing to use as a starter strip.
3. Nail the starter strip to the roof deck with about ¼ to ⅜ inch of material projecting over the edges of the rake and eave. The nails along the eave and rake should be spaced 6 inches apart in rows 1 inch in from the edge of the roof deck. The nails along the top edge of the starter strip are spaced 12 inches apart in rows 4 inches down from the top edge. A third (middle) row of nails is located exactly halfway between the top and bottom nail rows. As shown in Fig. 10–9, the nails in the top and middle rows are placed 12 inches apart and staggered.
4. Apply the first course of roll roofing so that its edges are flush with the edges of the starter strip at the rakes and eave. Cement the exposed portion to the starter strip and nail the selvage portion to the roof deck with two staggered rows of nails (Fig. 10–9).
5. Apply the remaining courses so that each laps the underlying one the full 19-inch width of the selvage edge.

As shown in Fig. 10–11, each end lap is 6 inches wide and staggered so that no end lap in one course is adjacent to one in the preceding course. The portion of roll roofing underlying each lap is first nailed to the roof deck and then covered with a 6-inch wide strip of cement. The overlying sheet of roll roofing is then pressed into the cement concealing the nails.

Cut the roll roofing sheets so that they butt at the ridge or hip, and nail them to the roof deck. Both the ridge and hip can be covered with

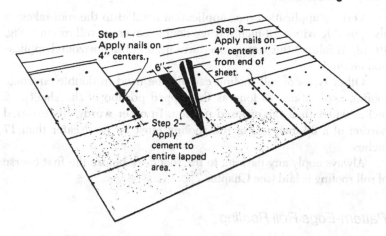

Fig. 10–11. Staggering end laps. *(Courtesy Asphalt Roofing Manufacturers Association)*

12 × 36-inch cut strips of roll roofing bent lengthwise along their centers. Allow a 17-inch exposure on both the ridge and hip by overlapping each section a distance equal to the length of the selvage portion (Fig. 10–12). When covering the hips, begin at the bottom with a starter piece and then lay the first section of roll roofing. Begin at either end of the ridge and work toward the other one.

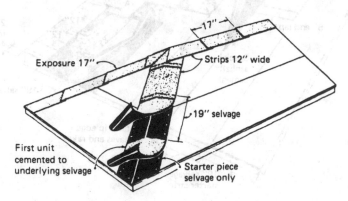

Fig. 10–12. Ridge and hip application method. *(Courtesy Asphalt Roofing Manufacturers Association)*

Vertical application (i.e., application parallel to the roof rakes) is also possible with a 19-inch selvage double coverage roll roofing (Fig. 10–13). It follows the same procedures used in the horizontal application method.

Other types of roll roofing can also be used in double coverage roofing applications as long as the lapped portion of the sheet is 2 inches wider than the exposed portion. In other words, the exposed portion of a 36-inch roll of roll roofing must be no greater than 17 inches.

Always apply any flashing to the roof deck before the first course of roll roofing is laid (see Chapter 7).

Pattern-Edge Roll Roofing

Pattern-edge roll roofing is applied parallel to the roof eaves. The concealed nail application method is used in new construction (Fig. 10–14). Both the concealed nail and exposed nail application methods are used for reroofing over existing materials (see the discussion of reroofing).

In new construction, 9-inch wide strips of roll roofing are nailed

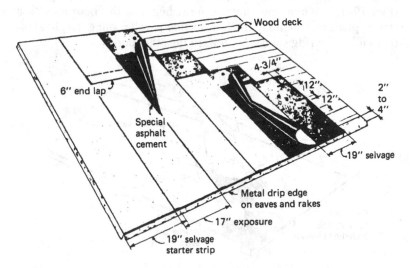

Fig. 10–13. Vertical application of 19-inch selvage double coverage roll roofing. *(Courtesy Asphalt Roofing Manufacturers Association)*

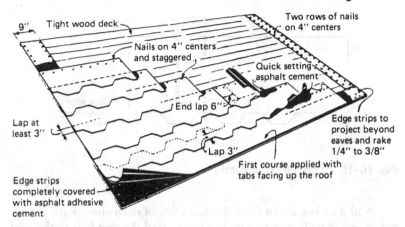

Fig. 10–14. Pattern-edge roll roofing concealed nail application method in new construction. *(Courtesy Asphalt Roofing Manufacturers Association)*

along the eaves and rakes of the roof deck with two rows of nails spaced 4 inches apart. Each edge strip projects approximately ¼ to ⅜ inch beyond the edge of the eave or rake.

When using the concealed nail application method, the first or starter course is laid with its pattern edge facing up the roof toward the ridge. The straight edge is allowed to extend ¼ to ⅜ inch beyond the eaves. Nail the first or starter course along the pattern edge with staggered rows of nails spaced 4 inches apart, but none more than 2 inches below the top of the sheet or within 18 inches of the rake. Thoroughly coat the 9-inch wide edge strip with asphalt roofing cement, press the first or starter sheet down firmly, and nail to the edge of the rake. *Do not nail along the bottom of the sheet.* The bottom of each course will be secured with the roofing cement.

Apply the second and succeeding courses with their pattern edge facing down the roof toward the eaves. Cut each overlying sheet so that its tabs are centered over the cutouts of the course below. Overlap each course so that there is a minimum effective top lap of 3 inches. All end laps should be at least 6 inches wide. Form the end lap by cutting diagonally from the center of the cutout on the pattern edge over to a point that will give a 6-inch width and then vertically to the top of the sheet (Fig. 10–15).

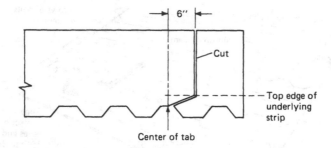

Fig. 10–15. End lap construction.

Nail 2 inches down from the top edge of each sheet with a staggered row of nails spaced 4 inches apart. Nail the end laps to the roof deck with staggered nails spaced 4 inches apart in two rows 1 inch and 5 inches from the end of the sheet. Stagger the end laps so that they are offset between adjacent courses.

Lift the lower edge of each course and apply lap cement in a continuous layer over the full width of the top lap and over 5½ inches of each 6-inch width of end lap. Apply pressure uniformly to the overlying sheet over the entire cemented area.

Hip and ridge shingles are formed by cutting a tab from the pattern-edge roofing sheet. The cuts are made at the center of each cutout in either side of the tab (Fig. 10–16). Apply the shingles to the hip or ridge with an overlap of at least one third of the length of the overlying unit. As shown in Fig. 10–17, this will provide an exposure of about 11½ inches. Firmly press the first shingle into lap cement applied to either side of the hip or ridge and nail the shingle at the end that will be covered by the overlapping unit. As successive shingles are applied, nail and then coat each portion of the shingle that is to be lapped with lap cement before applying the next one.

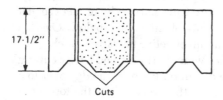

Cuts

Fig. 10–16. Cutting hip and ridge shingles.

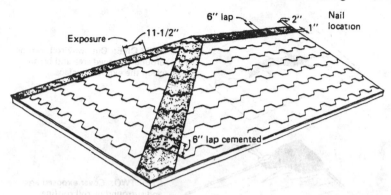

Fig. 10–17. Hip and ridge construction details. *(Courtesy Asphalt Roofing Manufacturers Association)*

After the roof has been completed, check all top and end laps for a good bond. Press down any that may have pulled loose and apply additional lap cement where necessary.

Always apply any flashing to the roof before the first course of roll roofing is laid (see Chapter 7).

Roll Roofing Repairs

Small holes, rips, or loose seams that are not too extensive can be repaired by applying roofing cement and pressing the roll roofing back in place.

Larger holes or rips in the roll roofing can be patched. Thoroughly clean the surface around the damaged area and allow it time to dry. Cut a piece of roll roofing large enough to cover it. Apply roofing cement to the damaged area, press the patch firmly in place, and nail it around the edges (Fig. 10–18). Cover the edges of the patch with cement to prevent it from being lifted by the wind. This will also provide a watertight seal.

Loose, broken, or rusting roofing nails should be removed and replaced with new ones. Do not use the old nail holes when renailing. Cover the old holes with a dab of roofing cement to prevent the possi-

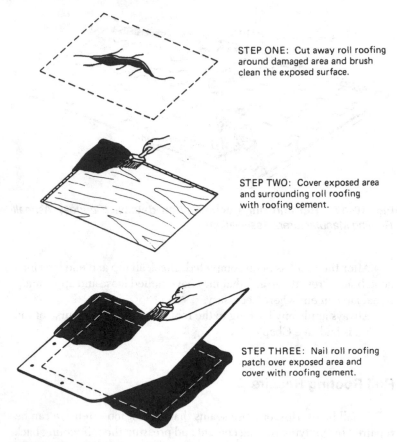

STEP ONE: Cut away roll roofing around damaged area and brush clean the exposed surface.

STEP TWO: Cover exposed area and surrounding roll roofing with roofing cement.

STEP THREE: Nail roll roofing patch over exposed area and cover with roofing cement.

Fig. 10–18. Repairing roll roofing.

bility of water leaking through the roll roofing and damaging the sheathing.

Reroofing

Pattern-edge roll roofing can be used for reroofing over an existing asphalt shingle roof with a slope of 4-in-12 or more. The concealed nail or exposed nail application method can be used when reroofing.

Pattern-edge roll roofing is always applied with each course parallel to the roof eave. Except for the starter strip, each course must be laid with the diagonal-shaped tabs on the bottom edge and facing the roof eave.

The exposed nail method of applying pattern-edge roll roofing to an existing asphalt shingle roof is illustrated in Fig. 10–19 and may be outlined as follows:

1. Cut back the existing asphalt shingles and nail a 1 × 6-inch board along the rakes and eaves. Allow the boards to extend ⅜ to ½ inch beyond the edges of the roof deck and nail them in place.
2. Apply the first or starter course of pattern-edge roll roofing so that its tabs face in the direction of the roof ridge. The lower or straight edge and the ends of the first course must extend ¼ to ⅜ inch over the rakes and eave.
3. Nail along the bottom edge of the first course with a staggered row of nails spaced 2 inches apart and no more than 2 inches up from the top of the sheet. Nail each sheet along the rake edge.

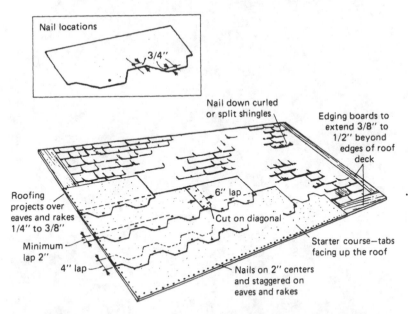

Fig. 10–19. Pattern-edge roll roofing exposed nail application method in reroofing. *(Courtesy Asphalt Roofing Manufacturers Association)*

4. End laps are cut as shown in Fig. 10–15. The application method is the same as the one described for pattern-edge roll roofing.
5. Apply the remaining courses with the tabs facing the eave and with a minimum horizontal or top lap of 2 inches. Succeeding courses are applied with a nail centered at each tab and at each cutout.

The ridges and hips are formed with 12-inch wide strips bent lengthwise so that 6 inches extends down either side (Fig. 10–6). The ridge and hip sections are laid with a 11½-inch exposure with a 6-inch lap. Nail each ridge or hip strip along its outer edge. Cover the lapped area with a 2-inch band of lap cement before applying the next ridge or hip section.

CHAPTER 11

Built-Up Roofing

Flat roofs on commercial structures are often covered with a roof membrane consisting of three, four, or five layers of roofing felt, each coated with hot- or cold-mopped asphalt, or with a cap sheet. This type of roof is commonly called a *built-up roof* (BUR). It is also customary to refer to a built-up roof as a 10-year, 15-year, or 20-year roof depending on the number of layers or plies in the roof membrane and the method of application.

A built-up roof is generally applied by a roofing contractor who specializes in this type of work. Depending on the job specifications, a reputable contractor will provide a warranty that guarantees the roof membrane for terms of 10, 15, or 20 years. These are limited guarantees subject to the provisions that specified materials be used and that these materials be applied in accordance with the manufacturer's instructions. Furthermore, they must be applied by a local roofing contractor who operates as a field representative of the manufacturer. In general, the roofing contractor is responsible for making certain repairs to the roof membrane during the term of the warranty.

The membrane of a built-up roof is subject to alligatoring, splitting, ridging, and other forms of deterioration caused by its exposure to weather conditions. This deterioration of the roof membrane eventually produces areas on the surface of the roof where leaks can occur. It should be clearly understood from the outset that a built-up roof membrane is as susceptible to leaks as any other type of roof covering material. The condition of the roof should be inspected at regular intervals. If signs of deterioration are detected, the roofing contractor who did the work should be notified. If the roof is not under warranty, however, you will probably have to make the necessary repairs at your own expense. Never sign a roofing contract without first reading the small print. Make sure you understand all the provisions contained in the contract.

Equipment, Tools, and Materials

Many different types of materials are produced for use in built-up roofing. Because of this great variety, care must be taken not to use chemically incompatible combinations. These can sometimes occur when the products of different manufacturers are mixed. Mixing materials will also invalidate any warranty that applies to the roof. The manufacturer of a roofing product will usually recommend other products with which it can be satisfactorily used.

The tools used in built-up roofing will depend on the requirements of the specific job. They will range from simple hand tools used for mopping and spreading the roof coating onto the surface to sophisticated spraying equipment (Fig. 11–1). Most of the tools and equipment used in built-up roofing are available through local building supply dealers or rental stores. Some manufacturers of roofing materials will even loan spray equipment free with complete operating instructions in return for a minimum purchase of their product. Unfortunately, the minimum purchase is well beyond the means of the average small businessman.

All of the roof covering materials taken together to form a built-up roof, exclusive of the roof deck itself and the flashings, are referred to as the *roof membrane*. Unlike a shingle roof, the roof covering materials of a built-up roof are bonded together to form a single inseparable cover or membrane.

The roofing materials of a built-up roof may be bonded together either by cold or hot applied coatings. These coatings are applied by mopping, brushing, or spraying. When a cold adhesive is used, the ap-

Fig. 11–1. Asphalt coating being applied over roof deck insulation.

plication method is referred to as *cold roofing* or *cold-process roofing,* and the completed roof is sometimes called a *cold roof.* Cold-process roofing is most commonly used to maintain or repair an existing built-up roof. Although this roofing method can also be used in new construction, the roof will not last as long as a hot mopped roof.

In *hot-process roofing,* or *hot roofing,* the bonding agent is heated to a high temperature in special kettles before it is applied to the roof. The completed roof is sometimes called a *hot roof.* Because the hot roof process requires special equipment and more workers than the cold roof process, it is generally applied by a professional roofing contractor specializing in this type of work.

Roof Coatings

The roof membrane of a built-up roof consists of one or more plies of roofing felt separated by hot or cold applied roof coatings.

In cold-process roofing, a fibrated cut-back asphalt cement is used as the interply adhesive. Fibrated asphalt cut-back coatings or emulsions are used as top coats.

Because hot-process roofing terminology is not standardized, there has been some confusion and misunderstanding between roofing contractors and property owners about job specifications and the types of materials used. Either roofing asphalt or coal-tar pitch can be used for the top coat and interply moppings in hot-process roofing. Although these two bituminous materials are virtually identical in appearance *after* they have been applied to the roof, there are important differences because one is a petroleum by-product and the other is derived from coal. Unfortunately, the terms *roofing asphalt* and *coal-tar pitch* are often incorrectly used as synonyms for a hot-process roof coating. To confuse the issue further, the roof coating on a hot-process roof is also sometimes called *tar, pitch,* or *roofing bitumen.*

Roofing asphalt or *asphalt* may be obtained from one of several varieties of naturally occurring bitumen, but it is much more commonly produced as a petroleum by-product. Asphalt obtained from a natural source is sometimes called *bitumen, pitch,* or *natural asphalt.*

The three basic types of asphalt produced from petroleum are (1) straight-run asphalt, (2) air-blown asphalt, and (3) cracked asphalt. Only air-blown asphalt is used as a roofing material. It reacts with air at approximately 400°–600° F, and has proved less susceptible to temperature change than straight-run asphalt.

Tar is a dark brown or black substance formed from the destructive distillation of such organic materials as wood, coal, and petroleum. *Coal tar* is obtained as a by-product in the manufacture of coke and coke oven gas from soft coal. *Coal tar* (also called *coal-tar pitch* or *roofing pitch*) is one of a number of different coal tar by-products. It is used primarily as a roof covering material.

Strictly speaking, *pitch* is the residue remaining after the distillation of tar. Pitch and tar are chemically identical, but the term *pitch* is used to refer to wood (southern pine) by-products or to pitch obtained from natural deposits. In its natural form, pitch is also called *asphalt.*

There are certain differences between roofing asphalt and coal-tar pitch that will affect their use. For example, the higher melting point of roofing asphalt makes it a more suitable roof coating than coal-tar pitch on steeper slopes. Because coal-tar pitch melts more quickly, it should be protected from the sun with a layer of gravel or slag. One advantage of using coal-tar pitch is that it is a self-sealing roof coating.

For this reason, it can be effectively used to repair or recoat an existing roof.

Coal tar chemicals used as roof covering materials are now being largely replaced by petroleum by-products, such as roofing asphalts.

Roofing Asphalts

Roofing asphalts are available in several grades or classes for field application to all types of built-up roof assemblies. The proper grade should be selected for each job, taking into consideration the slope of the roof, weather conditions, and the type of roof assembly.

Roofing asphalt should be heated slowly and the temperature kept relatively stable. This can be accomplished by maintaining a constant fluid level in the kettle and by adding fresh asphalt in small chunks. Kettles must be equipped with a thermostatic device to control asphalt temperatures. For large jobs, asphalt pumps and circulating systems will permit proper mopping temperatures with little heat loss in transporting the hot asphalt from the kettle to the roof surface.

The temperature of the roofing asphalt must be kept within a specific range while being heated in the kettle. The temperature range will depend on the type and grade of roofing asphalt. For example, a *dead level* asphalt has a softening point of 145°F, whereas a *high melt* or *steep asphalt* will begin to soften at 190°F. The kettle temperature will generally be maintained at approximately 400–450°F.

Once again, the temperature maintained in the kettle is very important. Excessive kettle temperatures result in degrading the roofing asphalt, and temperatures below specified minimums result in lack of adhesion. Not only is temperature control important to the quality of the roofing asphalt, it also reduces the amount of smoke pollution.

The recommended temperature range for the roofing asphalt will be indicated by the manufacturer. The type and grade of roofing asphalt selected for the job will depend on the slope of the roof deck and the process used to apply the built-up roof membrane.

Roofing asphalt should be added to the kettle in small chunks. Large chunks should not be used. Asphalt is brittle and pieces of a large chunk will sometimes split or break off while it is being handled. If this should happen while it is being added to hot asphalt already in the kettle, the split-off piece may splash hot asphalt onto the worker's face or hands.

Hot roofing asphalt should not be applied too thickly, or alligatoring will probably occur within a year. This will result in the development of cracks that will eventually penetrate the roof membrane.

Mineral Surfacing Materials

The surfacing materials used on gravel surfaced built-up roofs are (1) opaque gravel, (2) crushed rock, and (3) crushed blast furnace slag. Each of these surfacing aggregate materials must comply with the provisions and requirements of ASTM Designation D1863–64 (Table 11–1).

The size of the surfacing aggregate particles should range from ¼ to ⅝ inch. Furthermore, the aggregate must be dry, hard, and clean before being applied to the roof. Wet aggregate will cause a hot coating, such as asphalt or tar, to foam and form bubbles on its surface. These surface bubbles remain after the hot coating has cooled and hardened.

The surfacing aggregate is embedded in a poured flood coat of hot asphalt. Approximately 300 pounds of slag and 400 pounds of opaque gravel or crushed rock should be used per 100 square feet of roof surface.

Roof Deck Preparation

The Built-Up Roof Committee of the Asphalt Roofing Manufacturers Association recommends a minimum of ¼ inch per foot of roof slope to assure adequate drainage. Roofs with slopes less than ¼ inch per foot are highly susceptible to water accumulation. Roof decks with

Table 11–1. Sieve Requirements According to ASTM Specification D1863–64

Sieve Size	Percentage Total Passing
¾″	100%
½″	90–100%
⅜″	0–70%
No. 4	0–15%
No. 8	0–15%

(Courtesy American Society for Testing and Materials)

slopes less than ¾ inch per foot should be designed with interior roof drains and raised gravel stops or parapet walls.

Before a built-up roof can be applied to a deck, the surface of the deck must be thoroughly cleaned. All dirt, debris, and other materials that might prevent the adhesive from adhering to the surface must be removed.

The deck surface must also be dry before the bonding agents and other roofing materials are applied or they will not adhere properly. Any snow, ice, or water on the deck must be removed and sufficient time allowed for the surface to dry before actual roofing begins.

The deck surface should be as smooth as possible. Any warped boards must be removed and replaced. Raised or loose boards or panels must be securely nailed down to underlying purlins, rafters, or beams to prevent deck movement after the roof membrane has been applied to the surface.

If the roof membrane is applied over insulation, the insulation must be completely dry. The insulation must also be suitable as a base for built-up roofs. Urethane, polystyrene, or any other roof insulation material with a soft or spongy composition is generally not recommended as a base for a built-up roof, *unless* the insulation material is sandwiched between asphalt-saturated facer sheets (Fig. 11–2).

Built-Up Roof Flashing Details

Most roofs require flashing to protect the deck from leaks at those points on the surface subject to heavy water runoff or water accumulation, or where structural joints are formed. A built-up roof is no excep-

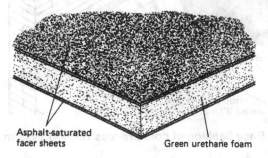

Asphalt-saturated
facer sheets   Green urethane foam

Fig. 11–2. Built-up roof insulation.

tion to this rule, but the flashing details and application methods used are in many ways different from those used on pitched roofs. For that reason, they have been included in this chapter instead of Chapter 7.

Metal flashing should always be embedded in plastic asphalt cement before it is applied to the surface of the roof membrane. The top surface of the metal flashing should be coated with a suitable primer to protect it from chemical reactions resulting from contact with bituminous roofing materials.

The principal flashing applications used in built-up roofing are

1. Base flashing
2. Edge flashing
3. Metal gravel stop
4. Vent and stack flashing
5. Projection flashing
6. Interior roof drain flashing
7. Expansion joint flashing

Base Flashing

Base flashing is most commonly used where the roof deck abuts a parapet wall or a vertical wall of the structure. Examples of base flashing applications are illustrated in Figs. 11–3 and 11–4.

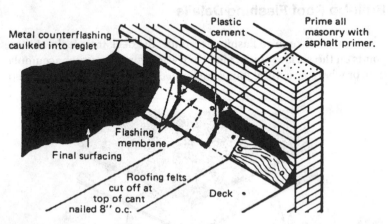

Fig. 11–3. Base flashing of cold-process roofing system. *(Courtesy Flintkote Co.)*

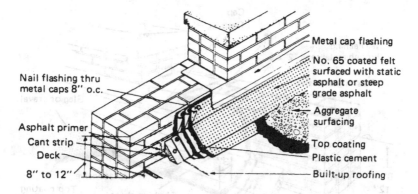

Metal cap flashing

No. 65 coated felt
surfaced with static
asphalt or steep
grade asphalt

Nail flashing thru
metal caps 8″ o.c.

Aggregate
surfacing

Asphalt primer

Cant strip

Top coating

Deck

Plastic cement

8″ to 12″

Built-up roofing

Fig. 11–4. Base flashing of hot-process roofing system. *(Courtesy Flintkote Co.)*

Before the base flashing is applied, a fiberboard *cant strip* with a minimum 4-inch face is installed in the angle formed by the vertical wall and the roof deck. The purpose of the cant strip is to soften the angle formed between the flat and vertical surfaces and to provide support for the flashing. If roof insulation is applied directly to the roof deck, a treated wood nailer the same thickness as the insulation should be installed the length of the joint with one edge abutting the surface of the vertical wall (Fig. 11–5). The cant strip is then nailed to the wood nailer. Plastic asphalt cement is applied to the wood nailer before the cant strip is installed to provide additional waterproofing (Fig. 11–6).

After the cant strip has been installed, lay the roofing felts of the built-up roof so that they extend up the face and are cut off evenly at the top of the cant. Nail the roofing felts into the top edge of the cant 8 inches on center.

Coat those portions of the masonry wall surface to be covered by the base flashing with an asphalt primer. Install one ply of No. 15 perforated asphalt roofing felt set in plastic asphalt cement extending from the bottom of the cant to approximately 6 inches onto the vertical wall surface above the top of the cant.

Install a second ply of No. 15 perforated asphalt roofing felt in the same manner, but allow it to extend 2 inches beyond the top and bottom of the first ply.

A final ply of roofing felt is applied over the first two to complete

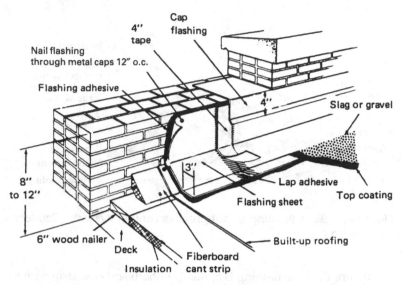

Fig. 11–5. Construction details of insulated built-up roof. *(Courtesy Celotex Corp.)*

the flashing membrane. It should be wide enough to extend 2 inches beyond the bottom edge of the second ply. Like the other two plies, it is embedded in plastic asphalt cement, but it is also nailed to the vertical masonry wall with large head roofing nails through tin caps. The nails should be located 8 inches on center along the top edge of the final ply.

After the base flashing membrane has been completed, cover the roof deck surface *and* the flashing with a top coating. Remove 1½ inches of mortar from the first horizontal joint above the base flashing.

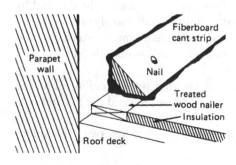

Fig. 11–6. Cant strip nailed to wood nailer. *(Courtesy Celotex Corp.)*

Caulk the lip or bent edge of a metal counter or cap flashing into the reglet formed by the removal of the mortar. The metal counter flashing should be large enough to extend downward a minimum of 3 inches over the top edge of the finished flashing (Figs. 11–3 and 11–4).

Edge Flashing

Raised edge flashing should be installed along the eaves and rakes when a building roof has no parapet wall along its edges. Its purpose is to provide a gentle rise in the roofing at the perimeter of the building in order to raise the connection between the bituminous roofing materials and the metal edge trim above any standing water that might accumulate on the flat roof. It also prevents this water from running down the outside face of the building. When raised edge flashing is used, interior drains should be installed on the roof to remove any accumulated water within 24 hours.

Basically a raised edge flashing consists of a tapered edge strip, a wood nailer, and the flashing membrane. A metal fascia strip is sometimes nailed over the edge flashing. A typical raised edge flashing construction is illustrated in Fig. 11–7.

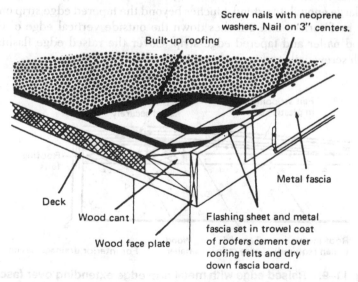

Fig. 11–7. Typical raised edge flashing construction details. *(Courtesy Celotex Corp.)*

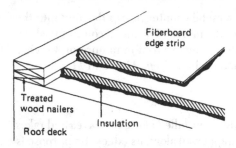

Fig. 11–8. Tapered edge strip cemented to insulation board. *(Courtesy Celotex Corp.)*

The tapered edge or cant strip is either nailed to the wood nailer or secured to the insulation sheet with plastic asphalt cement *before* the roof membrane is applied to the deck. The various plies of roofing felt in the roof membrane should extend across the face of the cant to at least its top edge. If the tapered edge strip is cemented to the insulation and abuts the wood nailer (Fig. 11–8), then extend the roofing felt to the outer edge of the wood nailer and down its outside vertical edge at least 2 inches where it should be nailed into the nailing strip 6 inches on center. Cover the tapered edge strip with a flashing sheet or roll roofing embedded in plastic asphalt cement. A cross section of this type of construction is shown in Fig. 11–9. The flashing sheet should be large enough to extend 4 inches beyond the tapered edge strip onto the roof surface and 2 inches down the outside vertical edge of the wood nailer and tapered edge strip. Cover the raised edge flashing with screw nails and neoprene washers on 3-inch centers.

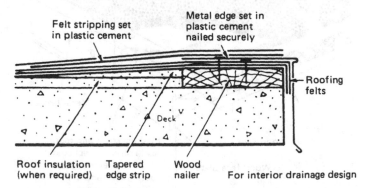

Fig. 11–9. Raised edge with metal drip edge extending over fascia. *(Courtesy Flintkote Co.)*

Metal Gravel Stop

When the top of the roof is covered with gravel, crushed stone, or slag, a gravel stop should be installed along the edges. A gravel stop can also function as a raised flashing edge (Fig. 11–10).

The various layers of roofing felt are applied first, and the flange is stagger-nailed 3 inches on center in lines ¾ inch from the edge of the flange and ¾ inch from the lip. The bottom of the flange should be covered with plastic asphalt cement before it is nailed down.

Two plies of No. 15 asphalt roofing felt, one 8 inches wide and one 12 inches wide, are applied over the flange. Coat the metal surface of the flange with a suitable primer. Apply a coating of bitumen or roofing asphalt and the 8-inch felt stripping. Cover the first ply of felt stripping with a bitumen or asphalt coating and apply the 12-inch felt stripping. Apply the top coating to the surface of the roof deck and allow it to cover the 2-ply felt stripping up to the raised edge of the gravel stop. Add the mineral-surfacing aggregate to the top coat while it is still hot.

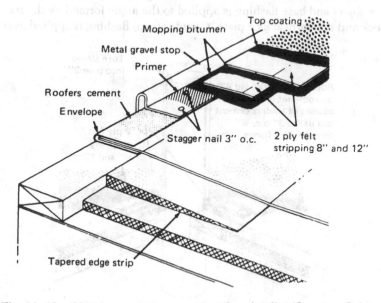

Fig. 11–10. **Metal gravel stop construction details.** (Courtesy Celotex Corp.)

Vent and Stack Flashing

Vents, stacks, and similar types of circular pipes projecting through the roof deck are usually equipped with a metal flange which extends approximately 4 inches onto the surface of the deck (Fig. 11-11). The flange should be set in plastic asphalt cement over the roofing felt and nailed 3 inches on center at a distance approximately ¾ inch from the edge of the pipe. Cut two collars of roofing felt to fit over the pipe. They should be large enough to overlap the metal flange on all sides by 8 and 12 inches respectively. Embed both roofing felt collars in plastic asphalt cement before applying them around the pipe. By increasing the size of the collars around the pipe, a slight incline is created away from it for better drainage. The final coating is applied to the roof after the flashing around the pipe is completed.

Projection Flashing

Sometimes objects other than circular pipes will project through the roof deck. Usually a vertical wood face plate is constructed around the object and base flashing is applied to the angle formed by the roof deck and the vertical face plate. Counter (cap) flashing is applied over

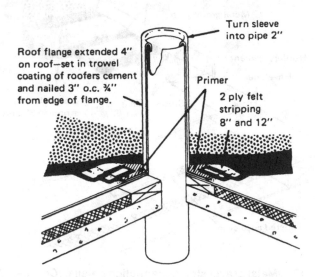

Fig. 11-11. Typical vent pipe flashing. *(Courtesy Celotex Corp.)*

each of the face plates. A typical example of projection flashing is the application of wood face plates and metal flashing to an I-beam projecting through the roof deck (Fig. 11–12).

Interior Roof Drain Flashing

A flat roof with raised edge flashing will have one or more interior roof drains installed on its surface. The metal flashing flange furnished with the drain should be cemented to the roofing felts with plastic asphalt cement. The metal flange should be covered with two layers of felt stripping embedded in plastic asphalt cement. The top layer of felt should extend 9 inches beyond the flange on all sides, and the bottom layer should extend 6 inches. After the flashing is completed, cover the roof and the flashing with the final coating (Fig. 11–13).

Expansion Joint Flashing

Expansion joints are used on large flat roofs to prevent the buildup of destructive stresses caused by the expansion and contraction of structural elements.

Expansion joint flashing consists of base flashing applied to the

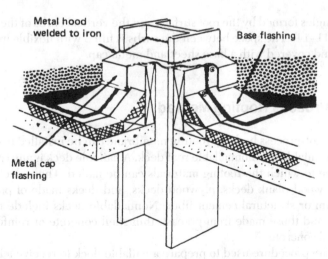

Fig. 11–12. Projection flashing details. *(Courtesy Celotex Corp.)*

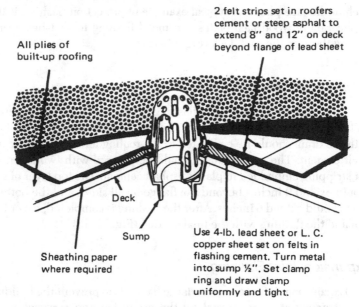

Fig. 11–13. Interior roof drain flashing details. *(Courtesy Celotex Corp.)*

two angles formed by the roof surface and the vertical curbs of the joint (Fig. 11–14). The space between the curbs is filled with flexible insulation and covered with a base sheet and metal cap.

Hot-Process Roofing Method

A hot-process built-up roofing membrane can be installed over either a nailable or a nonnailable roof deck. A nailable deck is any type of surface to which the roofing materials can be nailed. These decks include wood plank decks, plywood decks, and decks made of poured gypsum or structural cement fiber. Nonnailable decks include metal decks and those made from precast, thin shell concrete or reinforced poured concrete.

The procedure used to prepare a nailable deck to receive a built-up roof membrane differs to some extent from the procedure used for

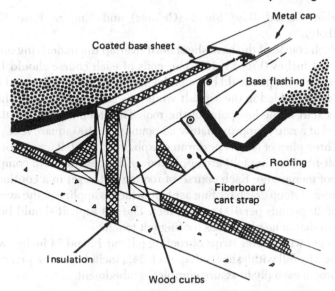

Metal cap

Coated base sheet

Base flashing

Roofing

Fiberboard
cant strap

Insulation

Wood curbs

Fig. 11–14. Expansion joint construction details. *(Courtesy Celotex Corp.)*

a nonnailable deck; however, in either case, the purpose of the preparation is to make certain the roof membrane will adhere properly to the surface of the deck.

Roofing Nonnailable Decks

The first step in preparing a nonnailable deck for a hot roof membrane is to coat it with a suitable asphalt primer. The asphalt primer should be spread over the roof surface at the rate of ½ to ¾ gallon (two to three quarts) per 100 square feet. The primer coating should be kept back at least 4 inches from all slab joints on precast concrete decks. Always solid mop poured concrete decks.

After the asphalt primer has *completely* dried, a base for the roof membrane is added to the surface of the deck. This generally consists of a single ply of a coated base sheet embedded in steep asphalt. These coated base sheets are made of organic felt, saturated and coated on both sides. They are sold under a variety of trade names, including

Vaporbar Coated Base Sheet (Celotex) and Empire Base Sheet (Flintkote).

Each course of the base sheet *must* overlap the underlying one by at least 4 inches (Fig. 11–15). The ends of each course should lap 6 inches, and all laps should be sealed with hot asphalt. The coated base sheet is embedded in the asphalt while it is still hot. Mopping should always start at the low point of the roof. The steep asphalt should be applied at a rate of approximately 23 pounds per 100 square feet.

Three plies of No. 15 perforated asphalt roofing felt (or perforated asphalt-fiber roofing felt) are applied over the base sheet to complete the roof membrane. Each course of roofing felt is set in a continuous mopping of steep grade roofing asphalt, which is applied at the average rate of 20 pounds per 100 square feet. The hot asphalt should be applied so that in no place does roofing felt touch.

Begin with starter strips of roofing felt cut 12 and 24 inches wide, followed by full width sheets. Lap each 24¾ inches over the preceding ply. Broom each ply to assure complete embedment.

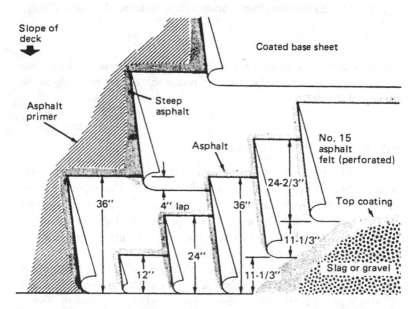

Fig. 11–15. Nonnailable deck with asphalt felt roofing membrane. *(Courtesy Celotex Corp.)*

Pour a uniform top coating of steep grade roofing asphalt over the entire surface at the rate of approximately 60 pounds per 100 square feet. While the top coat is still hot, embed not less than 400 pounds of gravel or 300 pounds of slag per 100 square feet.

Roofing Nailable Decks

A nailable roof deck should first be covered with overlapping courses of 36-inch wide 20-pound rosin sheathing paper. Each course should be nailed to the deck with an overlap of 2 inches. End laps should be at least 6 inches (Fig. 11–16).

Cover the entire surface with overlapping courses of a suitable base sheet, such as 30-pound asphalt-saturated roofing felt. The purpose of the base sheet is to prevent the hot asphalt from penetrating the roof deck sheathing and entering the rafter spaces. Lap each sheet 4 inches over the underlying sheet and lap ends 6 inches. Nail along the lap at 9-inch maximum intervals and stagger nails through the cen-

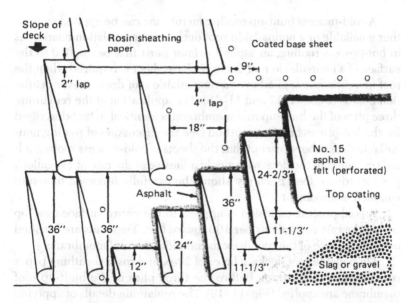

Fig. 11–16. Nailable deck with asphalt felt roofing membrane. *(Courtesy Celotex Corp.)*

ter of each sheet at 18-inch intervals. Use ⅞-inch barbed roofing nails on wood plank decks and ¾-inch annular threaded-shank nails on plywood decks.

Complete the roof membrane by applying three plies of No. 15 roofing felt, each course set in a continuous mopping of hot steep grade roofing asphalt at the average rate of 20 pounds per 100 square feet so that in no place does felt touch felt.

Use starting strips of roofing felt cut 12 and 24 inches wide, followed by full width sheets. Lap each ply 24¾ inches over the preceding one. Broom each ply to assure complete embedment.

Over the entire roof surface pour a uniform top coating of hot steep grade roofing asphalt. While the top coat is still hot, embed not less than 400 pounds of gravel or 300 pounds of slag per 100 square feet.

Cold-Process Roofing Method

A cold-process built-up roofing membrane can be applied over either a nailable or a nonnailable roof deck or over insulation boards. As in hot-process roofing, an asphalt primer must first be applied to the surface of a nonnailable roof deck. No base sheet is required when the roof membrane is applied over a nonnailable roof deck or over insulation boards (Figs. 11–17 and 11–18). The application of the remaining three plies of the built-up roof membrane is identical to that described for the hot-process roofing method (see the discussion of roofing nonnailable decks above) except that the sheets of cold-process roofing felt are cemented to the deck with a cold adhesive at the rate of 1½ gallons per 100 square feet. Each ply should be carefully broomed to assure complete embedment.

A cold-process emulsion is applied to the entire surface as a top coat at the rate of 5 gallons per 100 square feet. Two coats are brushed onto the surface of the roof deck to complete the roof membrane.

On nailable roof decks, a layer of 20-pound rosin sheathing paper is nailed to the sheathing before the three plies of the built-up roof membrane are applied (Fig. 11–19). The remaining details of applying a cold-process roof to a nailable roof deck are described in the preceding sections.

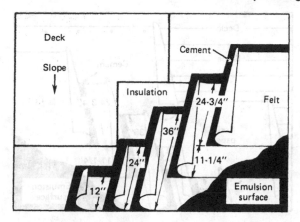

Fig. 11–17. Cold-process roof applied over insulated deck. *(Courtesy Flintkote Co.)*

Repairing Built-Up Roofs

A properly maintained built-up roof will have a service life of from 10 to 20 years. Sunlight is an important factor in the deterioration of this type of roof because the ultraviolet rays oxidize and shrink the coatings. The sun also bakes out the roofing oils, which causes a pliable

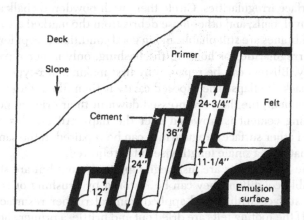

Fig. 11–18. Cold-process roof applied over nonnailable deck. *(Courtesy Flintkote Co.)*

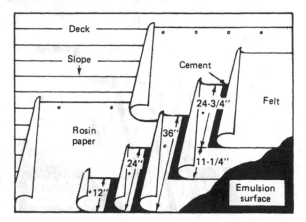

Fig. 11–19. Cold-process roof applied over nailable deck. *(Courtesy Flintkote Co.)*

roof to turn hard and brittle. As the coatings shrink and pull back from the edges of the roof, the underlying roofing felt is exposed and begins to rot. Foot traffic, vibration, and shaking also contribute to roof deterioration.

Carefully inspect the roof for blisters, breaks, large cracks, and other surface irregularities. Circle them with powdered chalk and remove gravel, nails, and other loose debris from the marked area. If the roof membranes are still pliable and in good condition except for minor surface irregularities or holes in the flashing, only minor repairs are necessary. Blisters can be repaired by first making an x-type cut and folding back the flaps. The exposed cavity is then filled with a plastic roofing cement, the flaps are pressed down in their original position, and roofing cement is troweled over the flaps. All buckles, ridges, folds, and other surface irregularities can be repaired in the same way except that only a single lengthwise cut is required.

If the roofing felts are just beginning to dry out but are still in a soft, pliable condition, they can be restored by brushing or spraying the roof surface with a cold-applied mineral rubber resurfacer and sealant. If the roofing felts are dried out and brittle, a primer should be applied before the resurfacing top coat. The primer is a penetrating oil that restores the roofing felts to a pliable condition and seals cracks.

Remove all gravel and debris from the roof surface before applying a primer or resurfacing top coat.

Cold-process resurfacing coatings can be applied without having to hire professional roofers to do the work. However, this work should only be done with the advice of a knowledgeable expert on the condition of the roof and the materials and methods to use. The manufacturers of these top coatings will often supply all the necessary equipment to apply them if the order is large enough.

Reroofing

Reroofing over an existing built-up roof is not generally recommended because defects in the old roofing material are easily transferred to the new covering, especially if the defects are severe or of long duration.

CHAPTER 12

Wood Shingle and Shake Roofing

Western red cedar is the commercial wood most commonly used in the production of roofing shingles and shakes (Fig. 12–1). It has an extremely fine and even grain, considerable strength, and a low expansion and contraction rate. It also sheds water well and provides a high degree of thermal insulation for the structure. As a roofing material, western red cedar shingles and shakes are twice as resistant to heat transmission as asphalt shingles, and three times as resistant as built-up roofing.

Tools and Equipment

Most of the tools and equipment used in other types of roofing are also used when applying wood shingles or shakes. These tools and various types of equipment are described in Chapter 2. Many roofers also use a shingler's or roofer's hatchet similar to those illustrated in Fig. 12–2. This is a multiple-purpose tool that is used to align, nail, or split the shingles or shakes. Each hatchet is equipped with an adjustable weather exposure gauge for cutting or fitting the roofing pieces.

Fig. 12–1. Rough-textured western red cedar shake roof. *(Courtesy Red Cedar Shingle & Handsplit Shake Bureau)*

Roof Deck Preparation

Wood shingles will absorb potentially damaging moisture unless there is adequate air circulation beneath them. For this reason, they are normally applied over spaced sheathing with no underlayment (Figs. 12–3 and 12–4). Closed sheathing is often used along the eaves and rakes if the undersides of the rafters are left open.

If the local building code requires closed (solid) sheathing for a wood shingle roof, the roof deck must first be covered with an underlayment of roofing felt and then built up with 1 × 4-inch wood furring strips to provide adequate air circulation beneath the shingles (Fig. 12–5). Leave ⅛-inch gaps in the furring strips to provide for roof drainage.

The 1 × 4's used in spaced sheathing or to build up a closed roof deck should be spaced on center the same distance as the shingle or shake exposure selected for the roof.

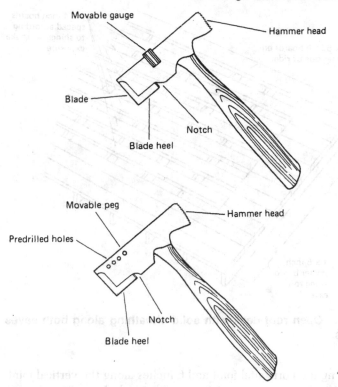

Fig. 12-2. Shingler's hatchets.

Because wood shakes are produced with rough, irregular surfaces, enough air can circulate around them to prevent moisture accumulation and absorption. As a result, they can be applied over either spaced (open) or closed sheathing. Closed sheathing is recommended for areas where heavy snows or strong winds are common.

Underlayment

Closed Roof Decks— Sweep the surface clean and cover any knotholes with small pieces of tin. Use mastic tape to secure the tin to the sheathing. Cover the roof deck sheathing with an underlayment of 36-inch wide 15-pound asphalt-saturated roofing felt. Lap the roofing felt 4

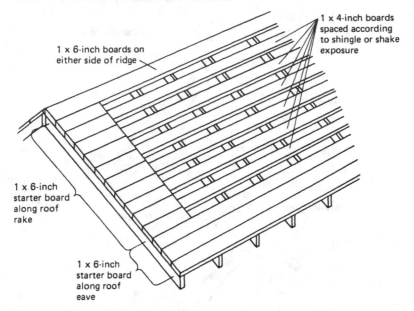

1 x 4-inch boards
spaced according
to shingle or shake
exposure

1 x 6-inch boards on
either side of ridge

1 x 6-inch
starter board
along roof
rake

1 x 6-inch
starter board
along roof
eave

Fig. 12–3. Open roof deck with solid sheathing along both eaves and rakes.

inches along the horizontal joint and 6 inches along the vertical joint. Nail 1 × 4-inch wood furring strips over the underlayment and space them according to the shingle or shake exposure.

Open Roof Decks—An underlayment is not used over spaced sheathing if wood shingles are applied to the roof. If shakes are applied, lay a 36-inch wide strip of 30-pound asphalt-saturated roofing felt along the roof edge and 18-inch wide strips between each shake course (Fig. 12–6).

Flashing

Metal flashing is preferred for wood shingle or shake roofs (Figs. 12–7 to 12–9). Copper flashing is often recommended, but other types of less expensive, rust- and corrosion-resistant metals can also be used just as effectively. Flashing application procedures are described in Chapter 7.

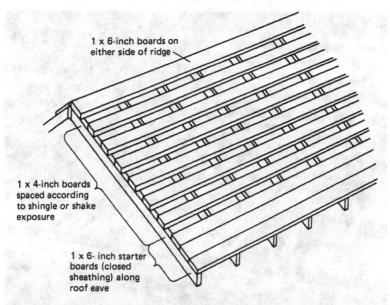

1 x 6-inch boards on either side of ridge

1 x 4-inch boards spaced according to shingle or shake exposure

1 x 6-inch starter boards (closed sheathing) along roof eave

Fig. 12–4. Open roof deck with solid sheathing along eaves.

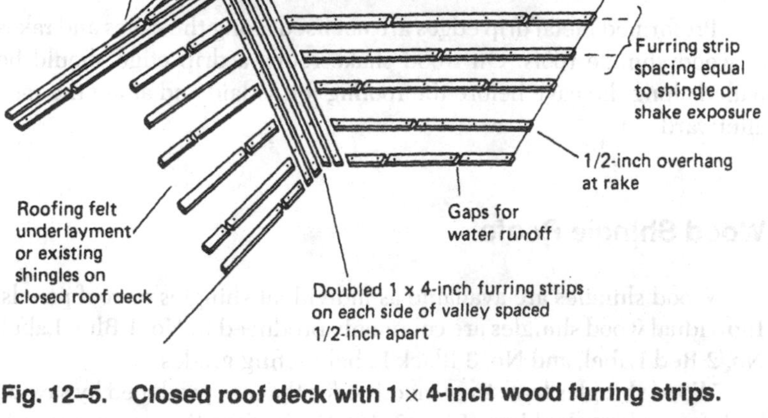

Doubled 1 x 4-inch furring strips along ridges spaced 1/2-inch apart

Furring strip spacing equal to shingle or shake exposure

1/2-inch overhang at rake

Gaps for water runoff

Roofing felt underlayment or existing shingles on closed roof deck

Doubled 1 x 4-inch furring strips on each side of valley spaced 1/2-inch apart

Fig. 12–5. Closed roof deck with 1 × 4-inch wood furring strips.

Fig. 12–6. Construction details of shake roof with roofing felt inter-ply between each shake course. *(Courtesy Red Cedar Shingle & Hand-split Shake Bureau)*

Preformed metal drip edges are not used along the eaves and rakes of wood shingle roofs. On wood shake roofs, a drip edge should be nailed along the eave before the roofing felt is laid and along the rake afterward.

Wood Shingle Roofs

Wood shingles are available as individual shingles or roof panels. Individual wood shingles are commonly produced in No. 1 Blue Label, No. 2 Red Label, and No. 3 Black Label roofing grades.

Wood shingles have two smooth sides that are produced by sawing both faces of a cedar block (Fig. 12–10). Each of the three roofing grade

Fig. 12–7. Step flashing along side of chimney. *(Courtesy Red Cedar Shingle & Handsplit Shake Bureau)*

shingles is cut in 16-, 18-, and 24-inch lengths with thicknesses at the butt of .40-, .45-, and .50-inch respectively.

Shingle grades are determined by the Cedar Shake and Shingle Bureau. No. 1 Blue Label is the premium grade of wood shingles used in roofing. These shingles are cut entirely from the heartwood, and are 100 percent clear and 100 percent edge-grained. A flatgrain and a limited amount of sapwood are found in the No. 2 Red Label grade of wood shingles. No. 3 Black Label is the lowest grade of wood shingle used in roofing. It is strictly a utility grade shingle used primarily for inexpensive applications on secondary buildings.

To accurately estimate the number of shingles or roof panels required to cover a roof, the roof pitch or slope must be determined first (see Chapter 2). After the roof pitch or slope has been determined, find

Fig. 12–8. Flashing details along bottom of chimney. *(Courtesy Red Cedar Shingle & Handsplit Shake Bureau)*

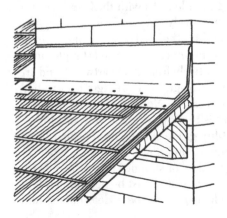

Fig. 12–9. Base and counter flashing used along chimney and vertical masonry walls. *(Courtesy Red Cedar Shingle & Handsplit Shake Bureau)*

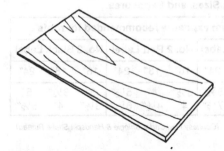

Fig. 12–10. Tapered, smooth-faced wood shingle.

the best maximum exposure to use. Exposure is the amount of uncovered shingle surface exposed to the weather. A well-constructed shingle roof is usually three shingle layers thick through each exposed section of shingle (Fig. 12–11). Once the roof pitch or slope and shingle exposure are known, the wood shingle type and size can be selected by using the information listed in Table 12–1.

Applying Wood Shingles

Proper weather exposure is important and depends largely on the pitch or slope of the roof. On roof slopes of 4-in-12 and steeper, the standard exposures for No. 1 grade shingles are 5 inches for 16-inch shingles, 5½ inches for 18-inch shingles, and 7½ inches for 24-inch shingles. On roof slopes less than 4-in12, but not less than 3-in-12, reduced exposures of 3¾ inches, 4¼ inches, and 5¾ inches, respectively,

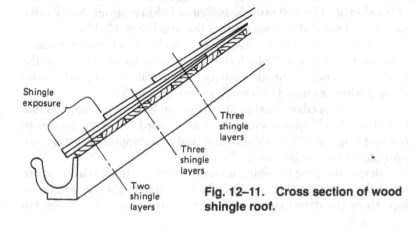

Shingle
exposure

Three
shingle
layers

Three
shingle
layers

Two
shingle
layers

Fig. 12–11. Cross section of wood shingle roof.

Table 12–1.   Wood Shingle Types, Sizes, and Exposures.

Pitch or Slope	Maximum exposure recommended for roofs								
	No. 1 Blue Label			No. 2 Red Label			No. 3 Black Label		
	16″	18″	24″	16″	18″	24″	16″	18″	24″
3-in-12 to 4-in-12	3¾″	4¼″	5¾″	3½″	4″	5½″	3″	3½″	5″
4-in-12 and steeper	5″	5½″	7½″	4″	4½″	6½″	3½″	4″	5½″

(Courtesy Red Cedar Shingle & Handsplit Shake Bureau)

are required. Reduced exposures are also recommended for No. 2 and No. 3 shingles on all 4-in-12 or steeper roof slopes (Table 12–1).

Use only rust-resistant nails when applying wood shingles. Rusting nails lose their holding power and create unsightly stains down the roof surface. Galvanized steel, aluminum, or copper roofing nails may be used to apply wood shingles. Use 3d nails for 16-inch and 18-inch shingles and 4d nails for 24-inch shingles *in new construction.* A threaded, ring-shank nail is sometimes recommended for plywood roof sheathing less than ½ inch thick.

Only two nails should be used in each shingle. The *exact* location of the nails will depend on the amount of shingle exposure that is dictated by roof pitch, the type of coverage, and the shingle manufacturer's recommendations. Nails should be placed no more than ¾ inch from each edge of the shingle. Make certain that the nails are also located at least ¾ inch to 1½ inches above the butt line of the next course so that they will be completely covered by those shingles (Fig. 12–12). Drive the head of the roofing nail flush against the surface of the shingle. This will provide sufficient holding power. *Never* drive the nail so hard that its head crushes the wood (Fig. 12–13).

Lay the shingles with a ⅛-inch to ¼-inch space between them to allow for expansion. A 1½-inch side lap is recommended between the joints in successive shingle courses. Never allow two joints to be aligned when separated by a single course of shingles.

The bottom edge of a shingle course can be kept in a straight line by butting the shingles against a 1 × 4-inch board nailed temporarily to the roof (Fig. 12–14). Thatch, weave, and other application styles are possible by varying the shingle butt line.

Begin shingling by laying a double or triple thick starter course along the roof eave (Fig. 12–15). The best quality shingle will be on top. Allow the starter course to extend approximately 1½ inches be-

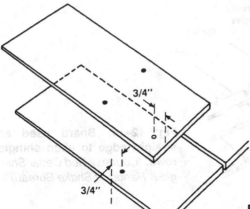

Fig. 12–12. Shingle nail locations.

yond the eave or the end of the roof sheathing to insure proper drainage into the gutters (Fig. 12–16). The starter course and all other shingle courses should also extend 1½ inches beyond the rake (gable) edge at each end of the roof.

Lay successive shingle courses up the roof to the ridge line with each course overlapped to provide the required exposure. Remember to maintain a separation of at least 1½ inches between joints in adjacent courses. Do not allow direct alignment between joints in alternate courses (Fig. 12–17).

Laying a 36-inch wide strip of roofing felt along the roof eaves will

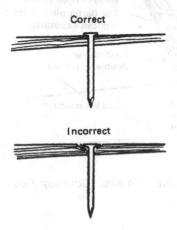

Correct

Incorrect

Fig. 12–13. Correct and incorrect methods of driving nails. *(Courtesy Red Cedar Shingle & Handsplit Shake Bureau)*

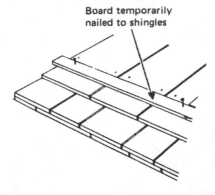

Board temporarily
nailed to shingles

**Fig. 12–14. Board used as
straight edge to align shingle
rows.** *(Courtesy Red Cedar Shingle & Handsplit Shake Bureau)*

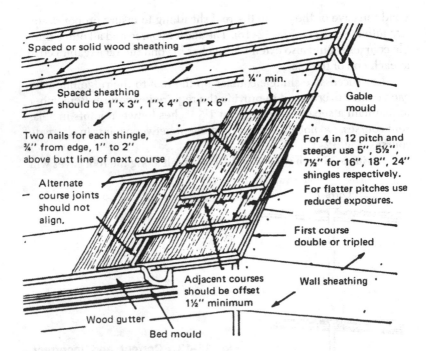

Spaced or solid wood sheathing

Spaced sheathing
should be 1"x 3", 1"x 4" or 1"x 6"

Two nails for each shingle,
¾" from edge, 1" to 2"
above butt line of next course

Alternate
course joints
should not
align.

¼" min.

Gable
mould

For 4 in 12 pitch and
steeper use 5", 5½",
7½" for 16", 18", 24"
shingles respectively.
For flatter pitches use
reduced exposures.

First course
double or tripled

Adjacent courses
should be offset
1½" minimum

Wall sheathing

Wood gutter

Bed mould

Fig. 12–15. Wood shingle roof application details. *(Courtesy Red
Cedar Shingle & Handsplit Shake Bureau)*

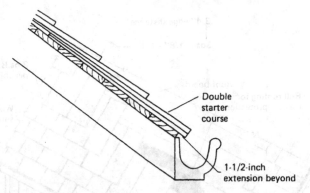

Double
starter
course

1-1/2-inch
extension beyond

Fig. 12–16. Shingle extension beyond sheathing along roof eave.
(Courtesy Red Cedar Shingle & Handsplit Shake Bureau)

provide some protection against ice dam water damage during the cold winter months. Apply the roofing felt up the roof deck far enough to extend beyond the inside surface of the exterior walls (Fig. 12–18).

Hip, Ridge, and Rake Details for Wood Shingle Roofs

The hips and ridges should be of the Boston-type construction with covered (protected) nailing. Hip and ridge details are illustrated in Fig. 12–19. Either the 6d or 10d nail is recommended for applying

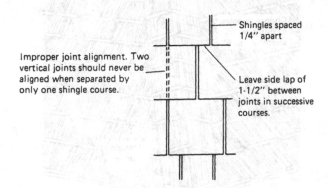

Shingles spaced
1/4" apart

Improper joint alignment. Two
vertical joints should never be
aligned when separated by
only one shingle course.

Leave side lap of
1-1/2" between
joints in successive
courses.

Fig. 12–17. Shingle joint alignment, side lap, and spacing. *(Courtesy Red Cedar Shingle & Handsplit Shake Bureau)*

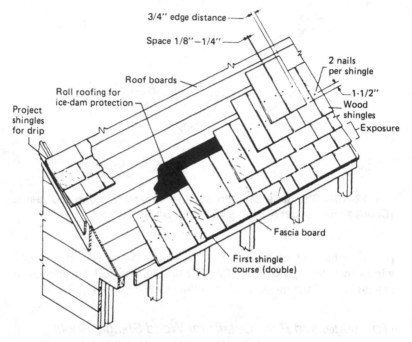

Fig. 12–18. Construction details of wood shingle roof.

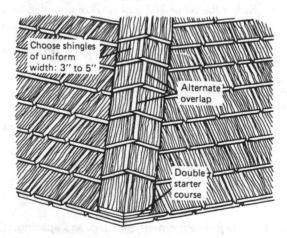

Fig. 12–19. Shingle roof hip and ridge construction details. *(Courtesy Red Cedar Shingle & Handsplit Shake Bureau)*

hip and ridge shingles. The nail must be long enough to penetrate the underlying sheathing.

Factory assembled hip and ridge units are available from shingle manufacturers. Hip and ridge shingles can also be cut to size at the site, but extreme care must be used to maintain a uniform width.

The same amount of exposure should be used when applying hip and ridge shingles as was used for the shingles on the rest of the roof. Overlap the hip or ridge shingles so that the laps alternate (Fig. 12–19). Cut and plane the overlapping shingles to form an even edge with the lapped ones. Ridge flashing is often installed under the shingles of a Boston-type ridge to provide additional protection to the joint (Fig. 12–20).

Hip shingling should begin at the eave with a double starter course (Fig. 12–19). Two 6d or 10d nails are used with each shingle and placed where they will be covered by the next hip shingle. Continue up the hip with *single* hip shingles overlapped to provide the required exposure. Cut the edges of the last shingle to provide a snug fit with the ridge or another hip.

The ridge is shingled by starting at one end of the roof and working toward the other or at both ends and working toward the center. In either case, the first shingles are laid doubled just as they are when shingling a hip (Fig. 12–19). The two ridge shingles are laid with the joints alternately facing opposite directions and secured to the roof deck with 10d nails. The nails should be placed where the next over-

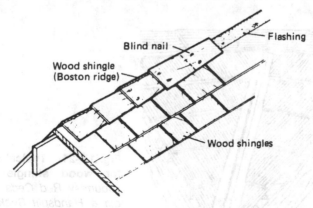

Blind nail

Flashing

Wood shingle
(Boston ridge)

Wood shingles

Fig. 12–20. Flashing installed under wood shingles of Boston-type ridge.

lapping shingle will cover them. Continue the ridge by laying *single* overlapping shingles along the roof ridge line. The amount of overlap will depend on the exposure required. The nails must be placed where they will be covered by the overlap of the next shingle. At the opposite end of the roof, nail a second shingle over the last shingle in the ridge course. The last shingle will be laid with its tip or tail facing outward from the end of the roof. The shingle applied over this one will be positioned with its butt end facing outward.

The shingles ending at each rake must be cut and planed to form an even, straight edge. Allow the shingles to extend (overhang) approximately 1½ inches beyond the edge of the roof at the rakes.

Valley Details for Wood Shingle Roofs

Cover the valley with a suitable metal flashing before applying the shingles. Leading wood shingle manufacturers recommend 26-gauge, center-crimped, painted galvanized steel or aluminum for valley flashing (Fig. 12–21).

The width of the flashing will depend on the pitch of the roof. The flashing should extend a minimum of 7 inches from both sides of the valley center line on roofs with one-half pitch or steeper. Roofs with less than one-half pitch should have valley flashing that extends at least 10 inches from both sides of the valley center line (Fig. 12–22).

Shingles that extend into the valley should be cut to the proper miter and their edges should form a straight line parallel to the valley

Fig. 12–21. Typical valley for wood shingle roof. *(Courtesy Red Cedar Shingle & Handsplit Shake Bureau)*

On roofs flatter
than half pitch,
valley sheets should
extend at least 10"
from valley center.

On half pitch and
steeper, valley
sheets should extend
at least 7" from
valley center.

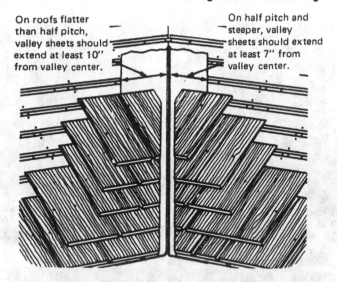

Fig. 12–22. Wood shingle roof valley flashing details. *(Courtesy Red Cedar Shingle & Handsplit Shake Bureau)*

center line. Never break joints into the valley, lay shingles with their grain parallel with the center line of the valley, or nail through the metal flashing.

Swept Eave Shingling

Some roofs have eaves that curve or sweep upward (Fig. 12–23). Wood shingles can be applied to this type of roof if the curve is no greater than one inch per foot. The amount of exposure is determined both by roof pitch and the type of wood shingle used. When in doubt, always follow the shingle manufacturer's recommendations.

Fig. 12–24 illustrates the construction details of a typical swept eave roof. The shingles must be soaked in water overnight and installed the next day. Soaking the shingles makes them more pliable and insures that they will conform to the curve of the roof without splitting. Begin by laying a double course of shingles along the eave with approximately 1½-inch overhang. The remaining courses are applied in the conventional manner.

Fig. 12–23.   Swept eave roof. *(Courtesy Masonite Corp.)*

Shingling Convex, Concave, and Apex Roof Junctures

Convex Juncture—The convex juncture is found on gambrel roofs (Fig. 12–25). The section of the roof above the juncture usually has a normal pitch. At the juncture line, however, the pitch of the lower section of roof becomes much steeper. Construction details of a typical convex juncture are shown in Fig. 12–26.

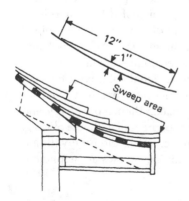

Fig. 12–24.   Construction details of swept eave roof. *(Courtesy Red Cedar Shingle & Handsplit Shake Bureau)*

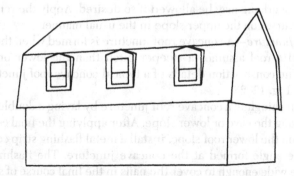

Fig. 12–25. Gambrel roof with convex juncture.

When shingling a convex juncture, the lower portion or slope of the roof is shingled first. Shingling begins at the eaves with a double starter course. After the final shingle course has been nailed in place, a metal flashing strip is installed along the juncture. The flashing strip should be about 8 inches wide. It must be wide enough to cover the nails in the final course of shingles on the lower slope. Bend the metal flashing strip so that it conforms to the angle formed at the juncture, allowing approximately 4 inches to overlap each slope. Bend the metal carefully to avoid fracturing it.

After you have installed the flashing strip, apply a double starter course along the edge of the upper roof slope. Approximately 1 to 1½

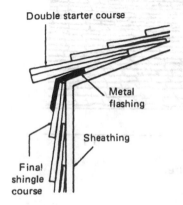

Double starter course

Metal flashing

Sheathing

Final shingle course

Fig. 12–26. Construction details of typical convex roof juncture. *(Courtesy Red Cedar Shingle & Handsplit Shake Bureau)* inches of overhang may be allowed if so desired. Apply the remaining shingle courses on the upper slope in the usual manner.

Concave Juncture—A concave roof juncture is formed when the upper portion of a roof assumes a steeper pitch than the lower one (Fig. 12–27). The construction details of a typical concave roof juncture are shown in Fig. 12–28.

Begin shingling a concave roof juncture by laying a double starter course along the eave or lower slope. After applying the final course of shingles on the lower roof slope, install a metal flashing strip conforming to the angle formed at the concave juncture. The flashing strip should be wide enough to cover the nails in the final course of shingles on the lower slope.

Shingling of the upper slope should begin with a double starter course applied along its lower edge. The remaining courses are applied in the usual manner.

Apex Juncture—Some roofs have angles which form apex junctures (Fig. 12–29). Note that the apex juncture shown is formed by a vertical wall and roof slope.

Shingling must be completed to the juncture. Before the juncture

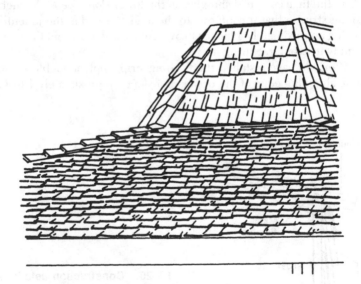

Fig. 12–27. Roof with concave juncture.

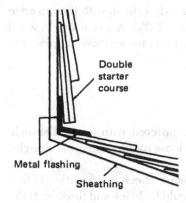

**Double
starter
course**

Metal flashing

Sheathing

**Fig. 12–28. Construction details
of typical concave roof juncture.**
*(Courtesy Red Cedar Shingle & Hand-
split Shake Bureau)*

itself is covered, apply a 12-inch wide metal flashing strip over the
juncture. Bend the flashing strip to conform to the angle formed by the
juncture, but allow 8 inches of the strip to cover the roof slope and 4
inches to cover the top of the wall.

Complete the roof slope by laying a course of shingles over the
metal flashing. Allow the shingle tips of this course to extend slightly
beyond the juncture line. Complete the wall by laying shingles up to
the juncture line. Trim the tip from each shingle in the last wall course
to provide a snug fit with the overlapping roof shingles.

Apply a moulding strip along the top edge of the last course of wall

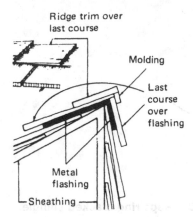

**Ridge trim over
last course**

Molding

**Last
course
over
flashing**

**Metal
flashing**

Sheathing

**Fig. 12–29. Construction details
of typical apex roof juncture.**
*(Courtesy Red Cedar Shingle &
Handsplit Shake Bureau)*

shingles. The top edge of the moulding strip is flush with the top edge of the last course of roof shingles (Fig. 12–29). Apply a single wood strip along the roof parallel with the juncture and overlapping the moulding strip.

Wood Shingle Repairs

The most frequent problems encountered with a wood shingle roof are cracked or splintered shingles, loose roofing nails, and shingles that have been lifted by the wind.

If the roof deck has an underlayment, check its condition before attempting repairs. The roofing felt should be black and flexible. If the roofing felt is crumbly, dry, and hard, the existing roof should be removed and the roof deck reroofed.

If both sections of a cracked shingle remain, nail them to the roof and cover both the nail heads and joint with a suitable roofing cement (Fig. 12–30). The nail holes should be drilled before nailing to prevent splitting the shingle. Where the damage is more extensive and a replacement shingle is not available, a temporary repair can be made by inserting a piece of galvanized steel or aluminum under the damaged shingle and nailing through the shingle and metal with two nails (Fig. 12–31). The piece of galvanized steel should be wide enough to extend

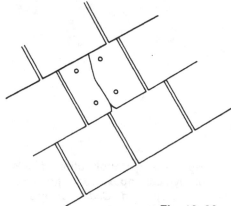

Fig. 12–30. Repairing cracked shingle.

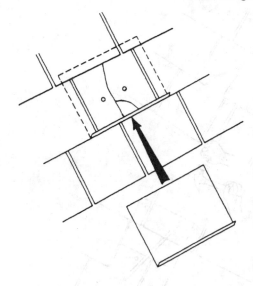

Fig. 12–31. Temporary shingle repair.

2 inches beyond both edges of the shingle and one inch under the butt line of the overlying shingle in the next course.

Replace badly curled shingles, those with extensive splintering or other signs of deterioration, and cracked or split shingles with missing pieces. Individual shingles can be removed by first splitting them and removing the pieces, and then cutting the roofing nails with a shingle ripper or hacksaw blade (Fig. 12–32). It may be necessary to lift the overlying shingle slightly to provide enough room to insert the shingle ripper or hacksaw. After the shingle has been removed, make certain that the old nails are cut off flush with the roof deck sheathing or underlayment. Take care when cutting them not to damage the sheathing or underlayment.

Trim the replacement shingle to the required width, slide the new shingle into place, and tap it gently with a hammer and wood block to align its butt with the other shingles in the course (Fig. 12–33). Nail the shingle to the roof deck and cover the nail heads with roofing cement to protect them from corrosion or leakage through the nail holes.

Lifted shingles can often be nailed back down in place. Use two to four nails and cover the nail heads with roofing cement. If the shingle is bowed, split it down the center the remove about ¼ inch of wood from the inside edge of one section to form a joint for roofing cement.

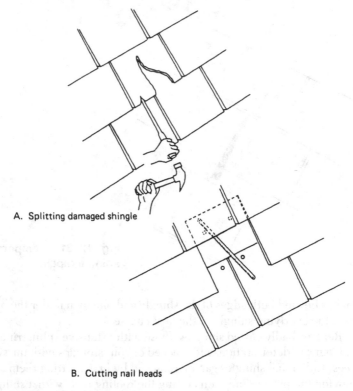

A. Splitting damaged shingle

B. Cutting nail heads

Fig. 12–32. Removing shingles.

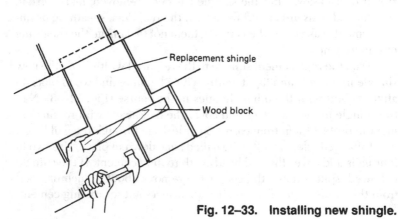

Replacement shingle

Wood block

Fig. 12–33. Installing new shingle.

Nail the two sections on either side of the joint. Cover the nails and fill the joint with roofing cement (Fig. 12–34).

Wood Shake Roofs

Types of Shakes

Wood shingles have a relatively smooth surface, whereas shakes have at least one highly textured, natural-grained split surface. The difference results from the methods used to produce them. Wood shingles are sawed. Shakes are split (rived) from cedar blocks (Fig. 12–35).

Wood shakes are available as individual cedar shakes or roof panels. There are three types of wood shakes:

1. Straight-split
2. Tapersplit
3. Handsplit

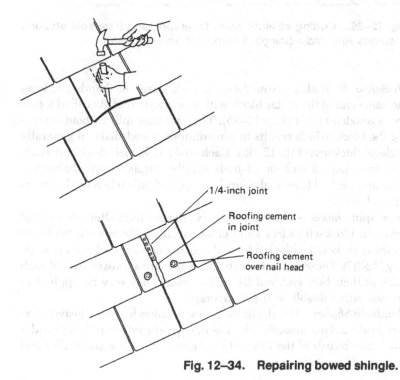

1/4-inch joint

Roofing cement in joint

Roofing cement over nail head

Fig. 12–34. Repairing bowed shingle.

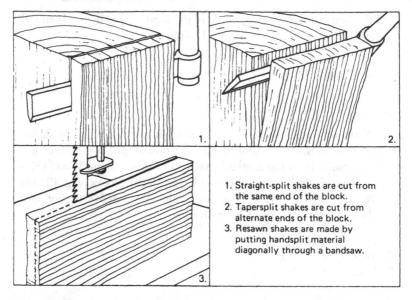

1. Straight-split shakes are cut from the same end of the block.
2. Tapersplit shakes are cut from alternate ends of the block.
3. Resawn shakes are made by putting handsplit material diagonally through a bandsaw.

Fig. 12–35. Cutting straight-split, tapersplit, and handsplit shakes. *(Courtesy Red Cedar Shingle & Handsplit Shake Bureau)*

Straight-split Shakes — Straight-split shakes are cut (handsplit) from the same end of the cedar block with a heavy steel blade called a froe and a wooden mallet (Fig. 12–35). They are handsplit without reversing the block, which results in a medium-textured shake of generally uniform thickness (Fig. 12–36). Each shake is ⅜ inch thick and available in either 18-inch or 24-inch lengths. Straight-split shakes are sometimes called barn shakes and are applied only when triple coverage is desired.

Tapersplit Shakes — Tapersplit shakes are split from alternate ends of the cedar block with a froe and mallet. To obtain the tapered thickness for these shakes, the block is turned end for end after each split is made (Fig. 12–37). These medium-textured shakes are approximately ½ inch thick at their butt end and 24 inches long. They may be applied to provide either double or triple coverage.

Handsplit Shakes — Handsplit (or resawn) shakes have one heavily textured side and one smooth side. They are produced by splitting a cedar block into boards of the desired thickness with a froe and mallet and

Fig. 12–36. Straight-split shake.

then passing the board through a thin bandsaw to form two tapered shakes (Fig. 12–38). These shakes are thicker and heavier through the butt than other shakes, ranging in thickness from ½ to ¾ inch. They are produced in 18-, 24-, and 32-inch lengths, and may be applied to provide either double or triple coverage.

A typical wood shake roof panel is shown in Fig. 12–39. It consists of sixteen 24-inch cedar shakes bonded to ½ inch thick exterior sheathing grade plywood. Each panel is 8 feet long. Only 15 panels are required to cover 100 square feet of roofing area at a 10-inch exposure. Each of the 24-inch shakes on the roof panel has a handsplit face 10 inches to 12 inches from the butt. They are sawn the rest of the way to form a uniform extra heavy tip.

Fig. 12–37. Tapersplit shake.

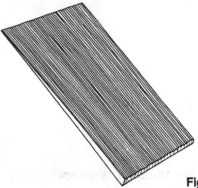

Fig. 12–38. Handsplit-resawn shake.

Applying Wood Shakes

Begin roofing by laying a 36-inch wide strip of 30-pound roofing felt over the sheathing boards along the roof eave. Lay a double or triple layer starter course at the eave line. The bottom course or courses can be either 15-inch or 18-inch shakes. The 15-inch shake is designed specifically for this purpose.

Secure each handsplit shake to the sheathing with two galvanized steel, aluminum, or copper roofing nails. Drive the nails into the shake about 1 inch from each edge, and 1 or 2 inches above the butt line of the overlapping shakes of the next course. The nails must be long enough to penetrate at least ½ inch into the sheathing boards. The 2-inch length of the 6d nail is usually long enough, but longer nails may have to be used for special shake thicknesses or exposures. The nails should be driven in flush with the shake surface. Avoid driving the nails so hard that they damage the wood.

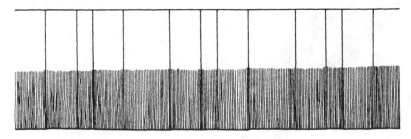

Fig. 12–39. Typical wood shake roof panel.

After the starter course has been laid, lay an 18-inch wide strip of 30-pound asphalt-saturated roofing felt over the top portion of the shakes in each course. Lay the roofing felt interply strip with its bottom edge located above the butt at a distance equal to twice the exposure length of the shakes. This allows the top edge of the roofing felt to extend onto the sheathing.

Fig. 12–6 illustrates the construction details of a shake roof with a roofing felt interlay inserted between each shake course. The 24-inch shakes used in this example are laid with a 10-inch exposure. This exposure requires that the bottom edge of the roofing felt be applied 20 inches above the shake butts (twice the exposure distance). As a result, 4 inches of the roofing felt cover the tops of the shakes and the remaining 14 inches extend onto the sheathing boards. Note that the use of a roofing felt interply is not necessary when tapersplit or straight-split shakes are applied in snow-free areas at weather exposures less than one third the shake length.

The space between shakes should be approximately ¼ to ⅜ inch to allow for expansion when moisture is absorbed. These spaces or joints should be offset by at least 1 inch adjacent courses. In 3-ply roof construction, the joints in alternate rows should not be in direct alignment.

Straight-split shakes differ from tapersplit shakes by having the same thickness throughout their length. When applying straight-split shakes, the smooth end should be laid uppermost. Doing so will produce a tighter and more weather-resistant roof.

Ridges are constructed in much the same way as hips. On unbroken ridges that terminate in a gable at each end, however, begin by laying a double starter course at each gable and then work from both gables to the center of the ridge. When the center is reached, splice the two courses with a small saddle of shake butts.

Hip, Ridge, and Rake Details for Wood Shake Roofs

Prefabricated shakes are available from manufacturers for use on hips and ridges. Shakes can also be cut to size at the site for this purpose. In either case, an 8-inch wide strip of 15-pound roofing felt must first be placed over the ridge and hip crowns before any shakes are laid.

Select shakes approximately 6 inches wide if the hips and ridges are to be fabricated at the site. Nail wood guide strips to the roof on both sides of the hips and ridges. Each guide must be positioned exactly 5 inches from the hip and ridge center lines.

Use nails long enough to penetrate at least ½ inch into the underlying sheathing boards. Two 8d nails on each side of the shake should be adequate for this purpose.

Nail the first hip shake in place with its long edge resting against the wood guide strip and its bottom edge flush with the edge of the roof. Cut back the edge of the shake projecting over the center of the hip to form a bevel. Apply the shake on the opposite side and cut back the projecting edge to fit the bevel cut of the first shake. Double the starting course and then proceed up the hip by laying a single shingle on either side of the hip line. The shakes in the successive courses are applied alternately in reverse order (Fig. 12–40). Use the same exposure for the hip shakes as was given for the shakes on the roof.

Valley Details for Wood Shake Roofs

Either an open or a closed valley can be used on a wood shake roof, but the former is the more common of the two. The construction details of a typical open valley are shown in Fig. 12–41.

Fig. 12–40. Shake hip roof construction details. *(Courtesy Red Cedar Shingle & Handsplit Shake Bureau)*

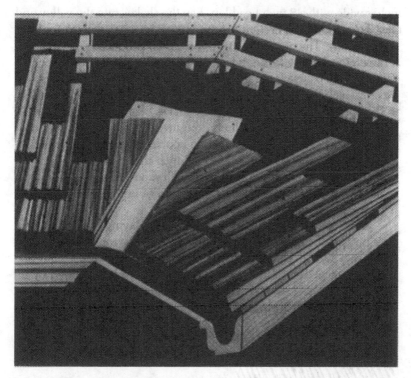

Fig. 12–41. Construction details of typical open valley. *(Courtesy Red Cedar Shingle & Handsplit Shake Bureau)*

Both types of valleys are lined with roofing felt and metal flashing. Construction of an open valley begins by laying a strip of 15-pound roofing felt over the sheathing the entire length of the valley. The roofing felt is overlaid with metal flashing at least 20 inches wide with a 4- to 6-inch head lap. Galvanized metal flashing should be painted on both sides with a good grade of metal paint to provide protection against corrosion. Copper or aluminum flashing do not require painting. Shakes adjoining the valley must be trimmed parallel to the center line.

Begin the construction of a closed valley by nailing a 1 × 6-inch wood strip into the saddle. Cover the wood strip with 15-pound roofing felt and then lay metal flashing over it. The metal flashing should be 20 inches wide with a 2 inch head lap. Lay the shakes into the valley and trim their edges to form a joint extending along the valley center line.

Shake Roof Variations

Distinctive shake roof designs can be created by using special application methods. The one shown in Fig. 12–42 produces a perspective or dimensional effect. It is obtained by gradually reducing the exposure of each shake course from the eaves to the ridge line. It requires the use of 24-inch shakes on the lower half of the roof and 18-inch shakes on the upper half. Application of the shakes begins at the eaves with 24-inch shakes laid with a 10-inch exposure. The exposure is gradually reduced to 8½ inches as successive courses are laid over the bottom half of the roof. The courses of 18-inch shakes used to cover the upper half of the roof begin with an 8-inch exposure that is gradually reduced to 5 inches. The gradual reduction of exposure is illustrated in Fig. 12–43.

Fig. 12–42. Shake roof with perspective or dimensional effect. *(Courtesy Red Cedar Shingle & Handsplit Shake Bureau)*

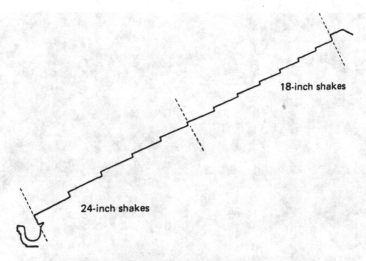

18-inch shakes

24-inch shakes

Fig. 12–43. Cross section of roof with reduced exposure. *(Courtesy Red Cedar Shingle & Handsplit Shake Bureau)*

A textured appearance can be created by laying the shakes with their butts placed slightly above or below the horizontal line of each course (Fig. 12–44). This application method produces an irregular and random roof pattern. Still greater irregularity can be obtained by interspersing longer shakes at random points along each course. A more rugged appearance is possible by mixing rough and relatively smooth surface shakes.

The gable line of a shake roof can be tilted upward slightly by inserting a strip of cedar bevel siding the full length of each gable end (Fig. 12–45). The bevel siding strip is installed with its thick edge flush with the edge of the roof deck sheathing. Tilting the gable line results in an inward pitch of the roof surface. This inward pitch accentuates the gable line and channels moisture away from the gable edge.

Wood Shake Repairs

The procedures used to repair or replace wood shingles also apply to wood shakes (see the discussion of wood shingle repairs). When removing wood shakes, however, special care should be taken to avoid damaging the roofing felt overlapping each shake course.

Fig. 12–44. Shake roof with textured appearance. *(Courtesy Red Cedar Shingle & Handsplit Shake Bureau)*

Shingle and Shake Panel Roofs

Panelized shingles and shakes are available from several manufacturers for use in roofing and reroofing. These roof panels eliminate the need to handle the many small pieces normally found in traditional shingle and shake roofing applications. As a result, the work progresses much more quickly and with less waste, clutter, and clean-up. The

Fig. 12–45. Shake roof with tilted gable line. *(Courtesy Red Cedar Shingle & Handsplit Shake Bureau)*

panels are easy to handle, usually self-aligning, and do not require workers with special skills or training to apply them.

Cedar roof panels are available for use in new construction and reroofing. Cedar panels commonly consist of up to 16 shingles or shakes bonded to an 8-foot long plywood base (Fig. 12–39). Construction details of typical cedar panels are shown in Figs. 12–46 and 12–47. Each panel combines both the shingles or shakes and the sheathing in a single unit. As a result, they may be applied over both open and closed (solid) roof decks.

Applying Cedar Panels

Cedar shingle or shake roof panels are normally used on roofs with slopes of 4-in-12 or more. They should be nailed to the sheathing (closed roof decks) or directly to the rafters (open roof decks) with 8d galvanized box head nails.

Roofing felt is not required with cedar shingle roof panels or shake panels having shakes with less than an 8-inch exposure. Cedar shake roof panels with exposures greater than 8 inches require an 18-inch wide strip of 30-pound roofing felt applied as an interply between the panel courses.

To apply the roof panels, begin by snapping a chalk line or nailing a 1 × 6-inch straight edge near the ends of the rafters. Position the chalk line or straight edge (starter strip) so that the starter roof panel will carry water runoff to the center of the gutter. Make sure the chalk line or straight edge is exactly parallel to the ridge line because it will be used to align the starter panels (Fig. 12–48).

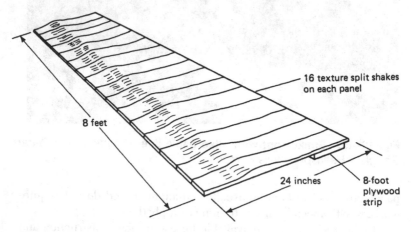

16 texture split shakes on each panel

8 feet

24 inches

8-foot plywood strip

Fig. 12–46.   Construction details of Shakertown roof panel.

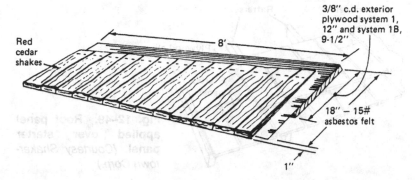

Fig. 12–47. Construction details of Foremost-McKesson roof panel. (*Courtesy Foremost-McKesson Building Products, Inc.*)

Starting at the gable edge, nail the starter roof panels at each rafter with two nails. Drive the first nail into the panel 1 inch down from the top of the plywood strip and the second nail 2 inches up from the bottom of the strip. Make certain that the ends of the starter panels break on rafter centers. Double the starter panels by laying full size shingle panels directly over them (Fig. 12–49). Although the panels are designed to center on 16-inch or 24-inch rafter spacing, they may meet between them. When this happens, make certain that the panels on adjacent courses do not end between the same two rafters. Plywood chips may be used between panel joints to provide a stronger roof underlayment.

Cut each panel at the rakes so that its outside edge is flush with the outside edge of the rafter. Remove the last shingle and apply the

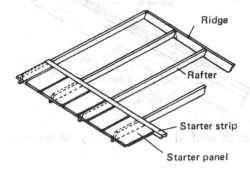

Fig. 12–48. Straight edge (starter strip) used to align starter panel. (*Courtesy Shakertown Corp.*)

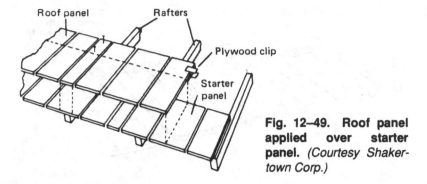

Fig. 12–49. Roof panel applied over starter panel. *(Courtesy Shakertown Corp.)*

rake moulding over the plywood edge of the sheathing so that the top edge of the moulding is flush with the top edge of the sheathing (Fig. 12–50). Apply a shingle or shingles to obtain the desired overlap and lay the next course of panels with a minimum 1½-inch offset. Vertical joints in alternate courses should be varied enough to avoid a straight line.

Use a saber saw to cut the panels around vents and other roof projections. Install the metal flashing between the top and bottom panels

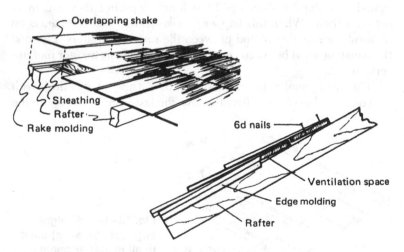

Fig. 12–50. Cross section of roof at rake. *(Courtesy Shakertown Corp.)*

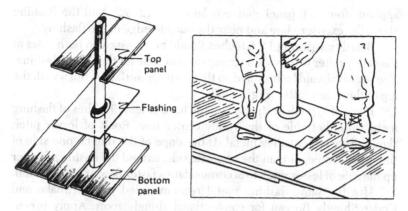

Fig. 12–51. Vent pipe details.

with a small portion extending beyond the bottom edge of the two panels to minimize ice-dam damage (Fig. 12–51).

A saber saw should also be used to cut the panels around the chimney if the flashing is installed prior to roofing. Flashing methods are described in Chapter 7. If the panels are applied first, remove some of the shingles around the chimney and weave individual shingles into the flashing.

Flashing at the juncture of a vertical wall with the roof should be

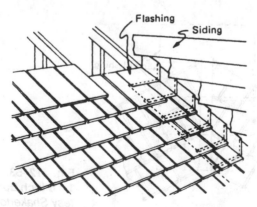

Fig. 12–52. Vertical wall flashing. *(Courtesy Shakertown Corp.)*

applied after each panel course is laid (Fig. 12–52). Nail the flashing above the exposure line and near the outside edge of the flashing.

Boards measuring 1 × 6 inches should be recessed on both sides of the valley rafter to provide a strong support base for the valley flashing. The boards should be installed so that their top surface is flush with the top of the valley rafter (Fig. 12–53).

Roofs of ½ pitch or steeper should have at least 9 inches of flashing metal on either side of the valley center line. Roofs of lesser pitch should have wider valley metal. If the slope of the roof on one side of the valley is steeper than the opposite side, extend the flashing further up the side of lesser pitch to accommodate a faster flow of water runoff.

Use the same flashing metal recommended by the Shake and Cedar Shingle Bureau for conventional shingle roofs. Apply precut flashing sheets after each panel course is laid (Fig. 12–54). Nail the flashing near its outside edge and well above the panel exposure line. The service life of the flashing can be extended by painting the sheets on both sides with a suitable metal paint before they are installed.

Ridge units are available from the panel manufacturer for use on both hips and ridges. To construct a hip, cut the panels at the proper angle to fit the hip line and nail them to both sides of the hip. Then, cover the length of the hip crown with a narrow strip of roofing felt and apply the shingle ridge units over the felt with the same exposure as the rest of the roof (Fig. 12–55). The ridge is constructed in a similar manner.

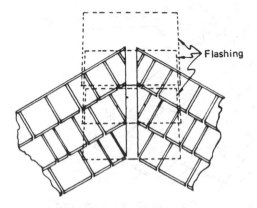

Fig. 12–53. Recessed 1 × 6-inch boards. *(Courtesy Shakertown Corp.)*

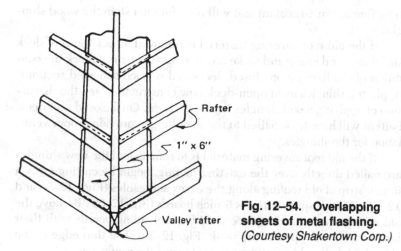

Fig. 12–54. **Overlapping sheets of metal flashing.** (Courtesy Shakertown Corp.)

Reroofing

Wood shingles can be applied over any roofing material except tile, slate, and fiber-cement shingles, but their life will be significantly shorter than if old roofing is removed. The shingles may be nailed directly over the old roofing material, to battens, or directly to the roof sheathing after the old roof covering has been removed.

Use 5d galvanized steel or aluminum roofing nails if the old roof covering is not removed. The nails must be long enough to penetrate the old shingles and the sheathing underneath. If the old roofing is removed, shorter nails may be used. The nails recommended for

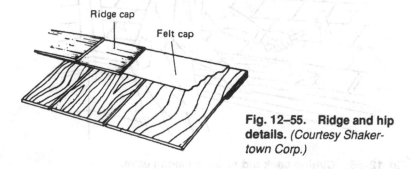

Fig. 12–55. **Ridge and hip details.** (Courtesy Shakertown Corp.)

reroofing are rust-resistant and will therefore not stain the wood shingles.

If the old roof covering material is removed, check the roof deck sheathing for damage and make any necessary repairs. Check the condition of the flashing on closed decks and replace damaged sections. Apply the shingles as in open-deck new construction (see the discussion of applying wood shingles in this chapter). On a closed roof deck, battens will have to be nailed to the sheathing to provide proper ventilation for the shingles.

If the old roof covering material is to remain and the new shingles are nailed directly over the existing roofing, begin by cutting away a 6-inch strip of old roofing along the eaves and gables (Figs. 12–56 and 12–57). Fill this space with 1 × 6-inch boards (Fig. 12–58). Remove the old ridge covering and replace it with bevel siding boards with their butt edges overlapped at the peak (Fig. 12–59). The thin edge of the ridge boards should point downward toward the roof eaves.

Cover the old valley flashing with 1 × 6-inch boards nailed on each side of the valley center line (Fig. 12–60). Cover the boards with new metal flashing. Nail a drip edge along each roof eave and rake, and begin shingling (Fig. 12–61). The shingling procedure will be the same

Fig. 12–56. Cutting back old shingles along eave.

Fig. 12–57. Cutting back old shingles about six inches from gable edges. *(Courtesy Red Cedar Shingle & Handsplit Shake Bureau)*

Fig. 12–58. Nailing 1 × 6-inch boards along gable edges and eaves.

Fig. 12–59. Nailing bevel siding along ridge.

as that used in new construction (see the discussion of applying wood shingles in this chapter).

In wet climates wood shingles should not be nailed directly to the old roof covering material or the roof deck sheathing. Unless the shingles are properly ventilated they will absorb excessive amounts of moisture. This can be avoided by nailing 1 × 2-inch or 1 × 3-inch bat-

Fig. 12–60. Nailing 1 × 6-inch boards in valley.

Fig. 12–61. Nailing red cedar shingles over old roofing.

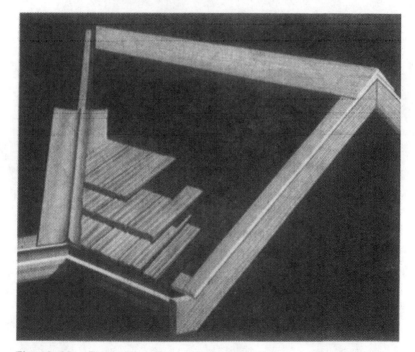

Fig. 12–62. Reroofing with wood shakes. *(Courtesy Red Cedar Shingle & Handsplit Shake Bureau)*

tens to the roof before applying the shingles. The 1 × 3-inch battens are used for shingle exposures of 5½ inches or more.

Allow the ends of the battens to extend about ½ to ¾ inch beyond the edge of the roof rake. Provide drainage gaps at regular intervals along the battens and leave the space between parallel battens open at the rake to allow ventilating air to enter and circulate beneath the shingles. Nail a metal drip edge along the ends of the battens parallel to the rake.

Shakes can be applied over the same types of roofing materials as wood shingles. The roof preparation procedures for reroofing with either wood shingles or shakes are also essentially the same. After the roof deck or existing roof has been properly prepared, apply the shakes and a roofing felt interply between each shake course as in new construction (Fig. 12–62).

CHAPTER 13

Slate Roofing

Slate is one of the most expensive roofing materials, but it more than compensates for its initial high installation cost by providing an attractive roof that will last four or five times as long as one covered with asphalt shingles (Fig. 13–1). If the owner of a house or building is interested in low first cost, a slate roof is certainly not the answer to his roofing problem. On the other hand, a slate roof is definitely the answer if a strong roof with a long service life is desired. The better grades of slate have been known to last 100 years. Lower quality grades are expected to last at least 50 years or more.

Roofing Slate

Slate is a stone that requires no admixture of materials, heat treatment, or special processes to convert it to a roofing material. It is simply extracted from the ground in blocks, which are then split and trimmed to the desired size and thickness.

In addition to being a particularly tough and durable roofing material, slate is completely fireproof and waterproof. It is also highly resistant to climatic changes and will not disintegrate as other roofing

Fig. 13–1. Rough-textured slate roof. *(Courtesy Evergreen Slate Co., Inc.)*

materials have a tendency to do. However, it may become brittle with age and will crack or split when struck by a falling branch or other hard object.

A principal disadvantage of using slate is its weight (900 to 1,000 pounds per roofing square), which requires reinforced framing to support it. Slate is also difficult to install. For this reason, slating is usually done by trained and experienced professional roofers.

Slate is available in a variety of colors, such as gray (neutral), green, blue, purple, black, or red; or in mottled colors, such as gray and green, blue and gray, blue and black, or green and purple. Some slate colors are permanent and nonfading; whereas others (so-called weathering slate) will fade to softer tones after extended exposure to the weather. Ribbon slate has one or more dark stripes crossing the unexposed portion. The useful life of ribbon slate is 60 to 75 years, which makes it cheaper than "one-hundred-year" slate. A slate roof may be

Table 13–1. Standard Slate Sizes and Weights

Sizes		
Length	Standard Widths	Exposure with standard 3″ head lap
24″	16″–14″–12″	10½″
22″	14″–12″–11″	9½″
20″	14″–12″–11″–10″	8½″
18″	14″–12″–11″–10″–9″	7½″
16″	14″–12″–11″–10″–9″–8″	6½″
14″	12″–11″–10″–9″–8″–7″	5½″
12″	12″–10″–9″–8″–7″–6″	4½″
10″	10″–9″–8″–7″–6″	3½″

To relieve uniformity of shadow line, architects have continued to specify one length in its random widths.

Weights	
Standard Smooth Texture	700–800 lbs. per Sq.
Standard Rough Texture	800–900 lbs. ″ ″
¼″ ″ ″	900 lbs. ″ ″
⅜″ ″ ″	1200 lbs.″ ″
½″ ″ ″	1800 lbs. ″ ″
¾″ ″ ″	2500 lbs. ″ ″

(Courtesy Evergreen Slate Co., Inc.)

composed of slates of a single color, or a harmonious pattern of two or more colors.

Slates are cut to at least 30 or 40 different sizes and a number of different thicknesses and weights. Some of these different slate sizes and weights are listed in Table 13–1. Slate is also cut with either a smooth or rough textured surface (Fig. 13–2). Many of the different slate sizes are identified by special names. The most commonly used slate sizes are called "large ladies" (16 × 8 inches), "countess" (20 × 10 inches), and "duchess" (24 × 12 inches).

The standard thickness of smooth and rough textured slate is commonly ¼ to ⅜ inch, but it is also available in textural thicknesses of ½ and ¾ inch. A graduated roof appearance is possible by laying ¾-inch thick slate at the eaves and graduating to ¼-inch slate at the ridge. Several thicknesses of slate may be intermingled in each course for additional roughness and texture.

A common method of applying slate to a sloping roof is to use slate

Smooth surface   Rough surface

Fig. 13–2. Smooth and rough textured slates. *(Courtesy Evergreen Slate Co., Inc.)*

of one uniform standard length and width with all slates laid to a line (Fig. 13–3). If desired, this pattern may be varied by laying two or more sizes of slate.

A textured slate roof can be formed by using a textured or rough grade of slate (Fig. 13–4). Another method is to use slates of varying thickness, size, and color laid so that the bottom line of each course is slightly uneven (Fig. 13–1).

Slates of any thickness laid tile fashion may be used as a surfacing material for flat or low-pitched roofs instead of slag or gravel. In most cases, however, the minimum recommended slope for a slate roof is 4-in-12 or greater.

Slate can also be used on mansard roofs. Because of the short rafters on this type of roof, a small size slate with a 5- or 6-inch exposure is used. Some slate companies produce a special slate for mansard roofs.

Slate Roofing Nails

Always use nails that will resist rust and corrosion. Solid copper, copper weld, or zinc-coated nails are recommended for slate roofs.

Fig. 13–3. Uniform courses of smooth surface slate. *(Courtesy Evergreen Slate Co., Inc.)*

Special slating nails are designed with large thick heads that will fit into countersunk holes in the slate.

Used 3d (1¼-inch) nails with slates of standard thickness and 18 inches or less in length, and 4d (1½-inch) nails with slates 20 inches or longer. The general rule of thumb is to add 1 inch to twice the thickness of the slate to determine the length of nail that should be used.

Fig. 13–4. Textured slate courses. *(Courtesy Evergreen Slate Co., Inc.)*

For example, slate ¼ inch thick should be nailed down with nails at least 1½ inches long (¼ × 2 plus 1 = 1½ inches).

A 6d (2-inch) nail is recommended for nailing slates to hips and ridges, because greater penetration and holding power is required at these points in the roof structure.

Nail holes are machine-punched in each slate. Extra holes can be made by using a hammer and center punch to mark their location and then drilling them with an electric drill. When making these extra holes, make certain that the slate is placed on a flat and smooth surface.

Slates are generally produced with two nail holes machine-punched in the upper end or head of each slate. The advantage of nailing the slate at its upper end is the additional protection provided the nail heads by two layers of slate. In areas of the country where strong winds are common, greater roof strength can be obtained by nailing each slate at the middle. Longer nails are required, particularly if open battens are used to support the slate, but the leverage effect of strong winds is greatly reduced when the center nailing method is used.

Tools and Equipment

Two special tools used in slating are illustrated in Figs. 13–5 and 13–6. The slate nail ripper is inserted under the slate until a nail is

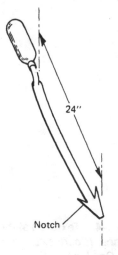

24″

Notch

Fig. 13–5. Slate nail ripper.

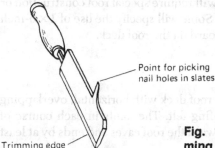

Point for picking
nail holes in slates

Fig. 13–6. Slating tool for trimming slate and picking nail holes.

Trimming edge

hooked by one of the notches. When the ripper is pulled out, it cuts the nail off level with the surface. If a slate nail ripper is unavailable, a hacksaw blade can be used to cut the nails. Some roofers still use the cutting tool shown in Fig. 13–6 for trimming slate and picking nail holes. Slate can also be trimmed with a cold chisel and hammer. The slate is first scored along a straight edge with the chisel and the unwanted section is knocked off. In addition to nailing the slates to the roof, a claw hammer can be used to mark extra nail holes with a center punch. The holes are then drilled with an electric drill. Another useful tool for working with slate is a putty knife, which can be used to coat the bottom of a replacement slate with asphalt roofing cement or clear butyl cement.

Ladders, scaffolds, safety harnesses, and similar types of equipment are described in Chapter 2. The work safety measures described in the second chapter also apply to slating and should be carefully read and followed.

Roof Construction Details

Slate producers generally consider the weight of a slate roof to be relatively insignificant when compared with the combined weights of rafters, sheathing, wind pressure, and the water saturation of certain other types of roofing materials, and do not recommend any special roof construction details.

Regardless of what a slate company states in its sales literature, the local building code should be consulted before applying a slate roof. Many of these local codes will require special roof construction or reinforcement of existing roofs. Some will specify the use of 1 × 6-inch tongue-and-groove sheathing board on the roof deck.

Roof Deck Preparation

Cover the sheathing of the roof deck with horizontal, overlapping layers of 30-pound asphalt roofing felt. The joints in each course of roofing felt should be lapped toward the roof eaves and ends by at least 4 inches.

Lap the roofing felt 4 inches over all hips and ridges, and 2 inches over the flashing metal of any valleys or built-in gutters. Secure the roofing felt along laps and at ends with roofing nails to hold it in place.

Flashing

Rust- and corrosion-resistant copper flashing should be used when roofing with slate. Application procedures are described in Chapter 7. Other types of metal flashing can be used if they have the same rust- and corrosion-resistant characteristics as copper flashing. Copper flashing is generally preferred, however, because the flashing should have the same long service life as the slate.

Slating

The application of a slate roof is called *slating* or *slate roofing.* This type of roofing generally requires experienced workers.

For years the popular method of slating called for the delivery of a mixed assortment of slate to the job where it was sorted and graded into a variety of different sizes and thicknesses. The courses along the roof eave contain the heaviest and longest slates. Medium-sized slates are used in the center courses, and the smallest slates are used along the roof ridge. This slating method is sometimes called *random* (or *texture*) *slating.*

Standard slating (or *standard commercial slating*) is a method that uses slates that have already been graded at the quarries before shipment to the job site. The slates are graded by length, width, and thickness. Standard slating is a modern application method used primarily in commercial roof construction. It is also used on less expensive house roofs.

Special precautions must be taken when working on a slate roof, because slate is a brittle roofing material and will crack or break if too much weight is placed on it. Consequently, a footing scaffold or a similar device should be used to distribute the worker's weight more evenly when working on a slate roof.

When laying the slate, allow a 2-inch projection at the eaves and 1 inch at all gable ends. The slate should be laid in horizontal courses with the standard 3-inch head lap, and each course should break joints with the preceding one. Slates at the eaves or cornice line are doubled and canted ¼ inch by nailing a wood cant strip along the edge of the roof.

Slates overlapping flashing should be laid so that the nails do not penetrate the metal. Any exposed nail heads must be covered with a suitable roofing cement.

All hip and ridge slates must be laid in roofing cement spread thickly over the unexposed surfaces of the underlying courses of slate. These slates should be nailed securely in place and pointed with roofing cement.

Slate should be laid so that all joints are broken and there is a minimum lap of 3 inches over the slate in the course below it. It is customary to provide a 3-inch lap for sloping roofs with a rise of 8 to 12 inches per foot. A 4-inch lap is used if the rise is 4 to 6 inches per foot.

The amount of exposure for each slate is equal to one half the length of the slate minus the lap dimension:

$$\text{Slate Exposure} = \frac{\text{Length of Slate} - \text{Lap}}{2}$$

A typical method for laying slate can be outlined as follows:

1. Inspect and reinforce the roof framing and sheathing as required.
2. Cover the roof sheathing with a suitable 30-pound asphalt roofing felt (see the discussion of roof deck preparation).
3. Nail a wood cant strip along the eave or cornice line (Fig. 13–7). The cant strip along the eave will tip the bottom end of the slate upward slightly. A similar strip (usually lath) nailed along both sides of the roof ridge provides additional support for the slates, but it is not always used.

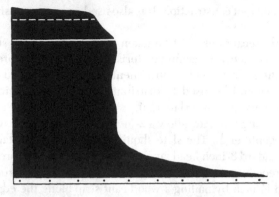

Fig. 13–7. Wood cant strip nailed along roof eave.

4. Lay a starter course of slate along the roof eave with each slate laid lengthwise approximately ¹⁄₁₆ inch apart and projecting 1 inch beyond the edge of the roof (Fig. 13–8). Begin and end the starter course with the slates extending 1 inch beyond the rakes. Nail the slates through the machine-punched nail holes. Do not drive the nails too tightly against the slate because you may crack or break

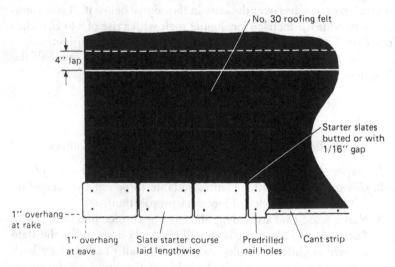

Fig. 13–8. Slate starter course.

it. The nail head should be driven in just far enough to barely touch the surface of the slate. Slates hang on the nail, rather than being fastened down to the sheathing. Use a counter punch and hammer to make a third hole in the upper-right corner of each slate.

5. Begin the first course of slate at the left roof rake with a full slate that extends 1 inch beyond the rake line and 1 inch beyond the eave line. Lay the slates of the first course vertically (at a right angle to the starter course). Complete the first course across the roof with the same projection (1 inch) at the right roof rake (Fig. 13–9). Leave approximately a $\frac{1}{16}$-inch space between each slate. The bottom edge of each slate in the first course should be even with the bottom edge of each slate in the starter course.

6. Begin the second course with a half slate and allow the same projection beyond the left roof rake as was provided for the first course. Complete the second course to the right roof rake and cut the slate (if necessary) to maintain the required projection beyond the rake (Fig. 13–10).

The remaining slate courses are laid according to the instructions outlined in steps 3–7 above. The following points are important to remember:

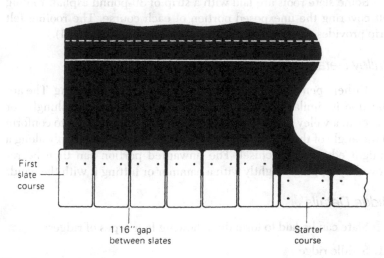

First slate course

1 16" gap between slates

Starter course

Fig. 13–9. First slate course.

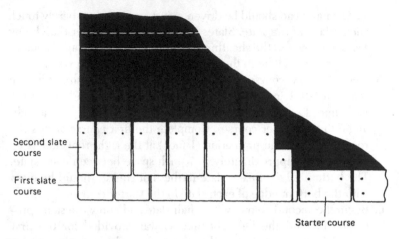

Second slate course

First slate course

Starter course

Fig. 13–10. Second slate course.

1. Maintain the same projection beyond the rakes on gable roofs for each slate course. The roof line at the rake must be even.
2. Lay the slate so that the vertical joints on adjoining courses break and are at least 3 inches apart.

Some slate roofs are laid with a strip of 30-pound asphalt roofing felt covering the unexposed portion of each course. The roofing felt strip provides a cushion for the next slate course (Fig. 13–11).

Valley Details

Either open or closed valleys can be used in slate roofing. The application is similar to the one used when applying wood shingles or shakes to a valley (see Chapter 12). Each slate must be cut to conform to the angle of the valley. Slate can be cut by scoring it deeply along a straight edge with a chisel. The unwanted portion can then be removed by tapping it lightly with a hammer or hitting it with the hand.

Ridge Details

Slate can be laid to form the following four types of ridges:

1. Saddle ridge
2. Strip saddle ridge

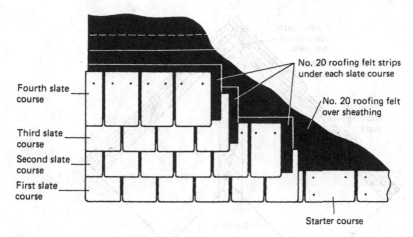

Fourth slate course

Third slate course

Second slate course

First slate course

No. 20 roofing felt strips under each slate course

No. 20 roofing felt over sheathing

Starter course

Fig. 13–11. Overlapped roofing felt.

3. Comb ridge
4. Coxcomb ridge

Most slate roofing instructions will specify a saddle ridge or a strip saddle ridge. Sectional views of both types of ridges are shown in Fig. 13–12. An overlapping plaster lath or beveled strip is sometimes used along both sides of the ridge line for additional reinforcement, but it may be omitted. Each ridge slate is first laid in elastic cement spread thickly over the unexposed surface of the undercourse of slate and then nailed securely in position with two roofing nails. The nails are covered with a spot of slater's elastic roofing cement. The cement is available in colors to match as nearly as possible the general color of the slate. Point the joint formed by the slate at the ridge line with elastic cement.

The principal difference between a saddle ridge and a strip saddle ridge is that the slates on the latter form butt joints, whereas on a saddle ridge they overlap approximately 3 inches (Figs. 13–13 and 13–14).

A sectional view of a comb ridge is shown in Fig. 13–15. In this type of ridge application, the combing slate projects about ⅛ inch beyond the ridge line. The combing slate can be laid with the grain vertical or horizontal (Figs. 13–16 and 13–17).

A coxcomb ridge is similar in detail to a comb ridge except that the combing slates are laid to alternately project on either side of the ridge.

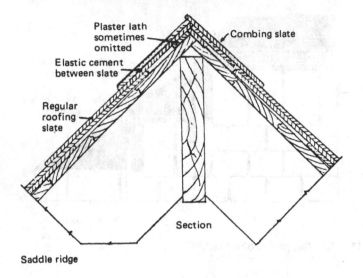

Plaster lath sometimes omitted

Combing slate

Elastic cement between slate

Regular roofing slate

Section

Saddle ridge

Point with elastic cement

Plaster lath sometimes omitted

Elastic cement

Section

Strip saddle ridge

Fig. 13–12. Sectional views of saddle and strip saddle ridges. *(Courtesy National Slate Association)*

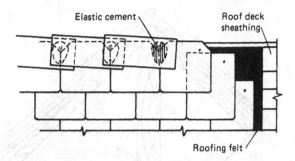

Fig. 13–13. Saddle ridge construction details. *(Courtesy National Slate Association)*

The slate course applied along the ridge should break joints with the preceding one. In other words, never allow a joint in the ridge course to line up with a joint in the preceding (underlying) course on the roof deck.

The ridge course should be given the same exposure as the other roof courses. This may mean cutting the slate used in the ridge course or using smaller slate of a proper size.

The projection of the ridge course at the roof rake must be exactly the same as the other courses on the roof.

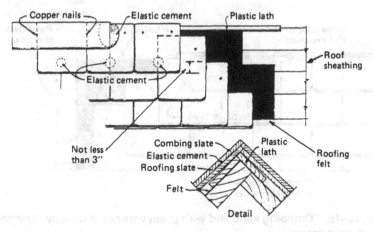

Fig. 13–14. Strip saddle ridge construction details. *(Courtesy National Slate Association)*

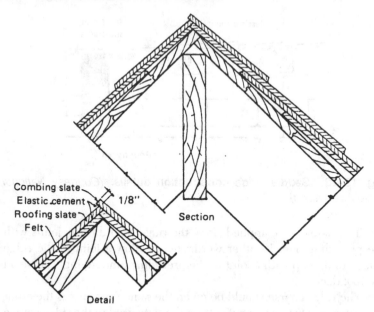

Combing slate
Elastic cement
Roofing slate
Felt

1/8"

Section

Detail

Fig. 13–15. Sectional view of comb ridge. *(Courtesy National Slate Association)*

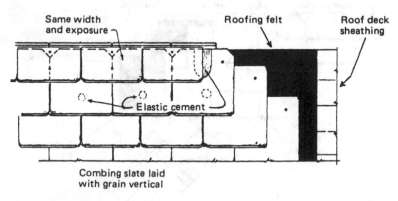

Same width and exposure

Roofing felt

Roof deck sheathing

Elastic cement

Combing slate laid with grain vertical

Fig. 13–16. Combing slate laid with grain vertical. *(Courtesy National Slate Association)*

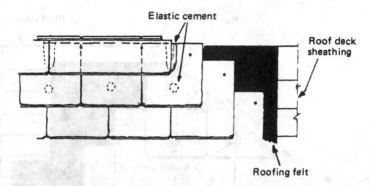

Fig. 13–17. Combing slate laid with grain horizontal. *(Courtesy National Slate Association)*

Hip Roofs

There are four principal types of hip construction used in slate roofing:

1. Saddle hip
2. Mitered hip
3. Boston hip
4. Fantail hip

Beveled strips, or one or two strips of plaster lath, are sometimes nailed along the hip line to provide additional reinforcement and a slightly elevated nailing base.

Both the mitered and fantail hips are constructed by extending each roof course to the roof rake and trimming the last slate to conform to the angle of the hip (Figs. 13–18 and 13–19). Fantail hip construction differs by using slates with their slips trimmed at a right angle to the hip line. Point with elastic roofing cement along the line at which the hip slates meet and secure them with two or three roofing nails placed in the unexposed portion of each slate. The nails should be embedded in a spot or circle of roofing cement for additional holding power.

The saddle hip is formed by laying overlapping slates up the hip ridge after each course of slates on the roof has been extended to the rake and trimmed to conform to the angle of the hip (Fig. 13–20). The

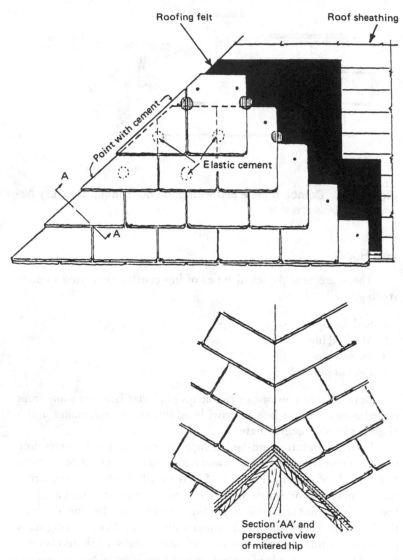

Fig. 13–18. **Mitered hip construction.** *(Courtesy National Slate Association)*

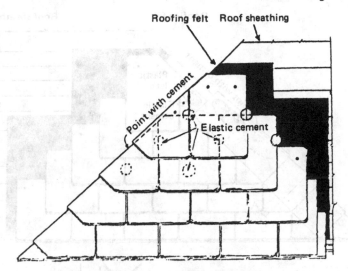

Roofing felt Roof sheathing

Fig. 13-19. Fantail hip construction. *(Courtesy National Slate Association)*

nailing procedure and the use of roofing cement is the same as that described for constructing mitered and fantail hips.

The Boston hip is constructed by applying overlapping slates, as is the case with a saddle hip, but a narrower slate (marked *B* in Fig. 13-21) is used to complete each course of roof slates. The nailing and cement application procedure is the same as that described for constructing mitered and fantail hips.

Hip slates should always be laid from the bottom of the hip toward the ridge line of the roof if the hip construction requires that the slates cap the hip. The nail holes in each hip slate are made with a counter punch after the amount of exposure has been marked on the slate with a piece of chalk or a pencil. The hip exposure should match that of the main slate courses used on the roof.

Repairing Slate Roofs

Individual slates will sometimes break or crack (Fig. 13-22). When this occurs, the broken or cracked slate should be repaired or

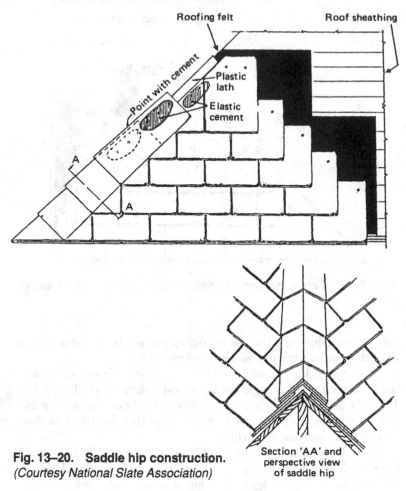

Fig. 13–20. Saddle hip construction.
(Courtesy National Slate Association)

Section 'AA' and
perspective view
of saddle hip

replaced immediately to avoid the possibility of roof leaks. A cracked slate can be repaired by filling the crack with asphalt roofing cement, putty, or a suitable synthetic sealer (Fig. 13–23). These materials can also be used to reattach a loose slate. A slate too damaged to repair should be replaced with a new one.

Do not attempt to remove a damaged slate by breaking it into smaller pieces with a hammer because you will run the risk of breaking the slates in the lower course. The nails should be cut first; the dam-

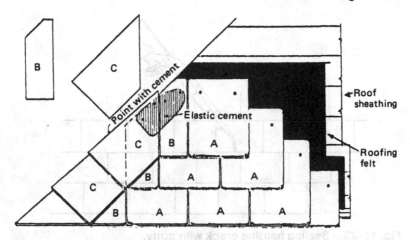

Fig. 13–21. Boston hip construction. *(Courtesy National Slate Association)*

aged slate can then be pulled out. A shingle ripper or a long chisel blade without its handle should be inserted under the damaged slate. Several hammer blows against the nail cutter or chisel blade should be enough to cut the nails and free slate (Figs. 13–24 and 13–25).

After you have removed the damaged slate, cut a 5-inch wide strip of copper long enough to extend 6 inches under the slate course above the damaged one and still bend around the bottom edge of the replacement slate (Fig. 13–26). Nail the copper strip in position with two roofing nails and cover the new nails and the tops of the old nails with

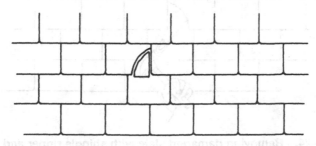

Fig. 13–22. Cracked slate.

Fig. 13–23. Sealing hairline crack with putty.

roofing cement. Place the slate on a flat surface and mark the locations of the new nail holes with a center punch. Makes certain the new holes are *not* aligned with any existing ones on the roof. Use an electric drill to make new holes in the slate. Coat the bottom of the replacement slate with asphalt roofing cement. The cement can be applied with a putty knife. Insert the new slate, press it down, and bend the metal strip up around its bottom edge to give it support (Fig. 13–27). Trim off any excess metal from the copper strip.

The individual slates will generally outlast the nails used to fasten

Fig. 13–24. Removing damaged slate with shingle ripper and hammer.

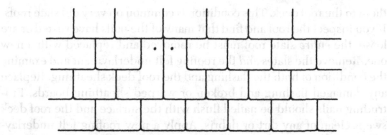

Fig. 13–25. Slate removed.

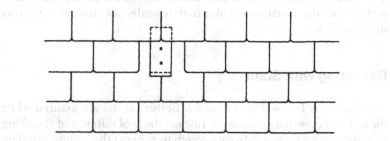

Fig. 13–26. Nailing metal strip.

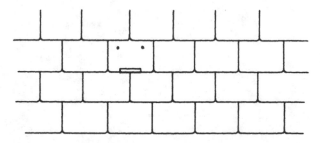

Fig. 13–27. Bending metal strip around bottom of replacement slate.

them to the roof deck. This condition is common on very old slate roofs. If you inspect the roof and find that many of the nails have rusted or are loose, the *entire* slate roof must be removed and replaced with a new one. Remove the slates *and* the roofing felt underlayment and examine the condition of both the flashing and the roof deck sheathing. Replace any damaged flashing and broken or warped sheathing boards. Protruding nails should be nailed flush with the surface and the roof deck swept clean of any dirt or debris. Apply a new roofing felt underlayment and lay the new courses of slate.

When removing slate from an existing roof, try to save slates that are still in good condition. New slates should be the same size and thickness as the old ones and should closely match them in color. Do not be concerned about a difference in color tone between the old and new slates because this adds to the attractiveness of a slate roof. Make certain that the variation in color tone is uniformly distributed across the roof deck.

Reroofing with Slate

Slate may be used for reroofing. Better results are assured when the old roof covering is removed. Inspect the roof rafters and sheathing to make certain they are strong enough to support the weight of a slate roof. It may be necessary to reinforce them. Consult the local building code before applying a slate roof. The procedures used in reroofing with slate are generally the same as those used in new construction.

CHAPTER 14

Tile Roofing

Tile has been used as a roofing material for centuries. Some of the earliest and finest examples of tile roofs date from the time of ancient Greece and Rome where they were used on both public buildings and private residences. Its popularity as a roofing material is still widespread in regions with warm climates, such as the Mediterranean area, Latin America, Florida, the southwestern United States, and California. Elsewhere it has been replaced to a great extent by cheaper roofing materials, particularly asphalt-based products.

Tools and Equipment

Tile roofing is applied with the same basic tools and equipment used in the application of other roofing materials. These roofing tools and equipment are described in Chapter 2. Roofing tiles should be cut with a circular saw equipped with a masonry (carborundum) blade. *Safety goggles must be worn when cutting tile with a circular saw.* Flying particles of clay or concrete tile may cause serious injury to unprotected eyes.

Roof Deck Preparation

Always consult the local building code before laying tile. Because of the extra weight of a tile roof, special roof construction may be necessary. Inspect the roof framing and reinforce the roof with braces and extra rafters if the framework needs strengthening.

Either a closed or open roof deck can be used to support a tile roof. A closed roof deck is generally constructed of 1 × 6-inch tongue-and-groove sheathing board or plywood sheathing panels of suitable thickness with 1 × 2-inch battens mounted between the roof eaves and ridge. The use of battens is always recommended and is required for roofs with slopes of 7-in-12 or more. The battens should be made from strips of douglas fir, redwood, or cedar. Both redwood and cedar battens offer the best resistance to moisture. The battens should be nailed to the sheathing with nails placed approximately every 2 feet along the strip (Fig. 14–1). Adequate means of drainage must be provided at the

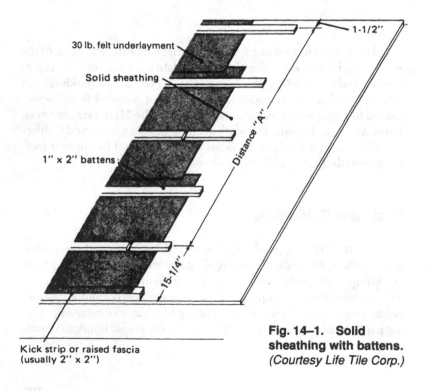

30 lb. felt underlayment

Solid sheathing

1" × 2" battens

1-1/2"

Distance "A"

15-1/4"

Kick strip or raised fascia
(usually 2" × 2")

Fig. 14–1. Solid sheathing with battens.
(Courtesy Life Tile Corp.)

battens to prevent the accumulation of moisture on the roof. This can be done by allowing ½-inch gaps every 4 feet, (Fig. 14–2), or by slightly raising the battens off the roof deck with shims cut from asphalt shingles.

As shown in Fig. 14–1, a 2 × 2-inch kick strip is nailed along the edge of the sheathing directly above the fascia board. The kick strip slightly elevates the bottom edge of the first course of tile to give it the same cant as succeeding tile courses. In new construction, a raised fascia board is sometimes used instead of a kick strip to raise the first tile course. In either case, a 28-gauge galvanized metal strip should be nailed along the eave to support the underlayment over the raised fascia board or kick strip (Fig. 14–3). This will prevent the accumulation of water along the edge of the roof.

An open roof deck is constructed with spaced sheathing boards. The spacing of the sheathing boards will depend on the size of the tile. There are two basic types of open roof decks used to support a flat tile roof: (1) spaced sheathing and (2) combination solid and spaced sheathing.

A typical open roof deck with spaced sheathing is shown in Fig. 14–4. The spacing of the 1 × 6-inch battens depends on the size of the tile. Either a kick strip nailed directly to the top of the fascia board or a raised fascia board may also be used on an open roof deck in new construction to elevate the bottom edge of the first course of tile.

Sometimes a transition from solid to spaced sheathing is used to support a tile roof (Fig. 14–5). The solid sheathing may consist of either

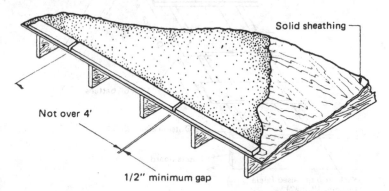

Fig. 14–2. Drainage gaps in battens. *(Courtesy Monier Co.)*

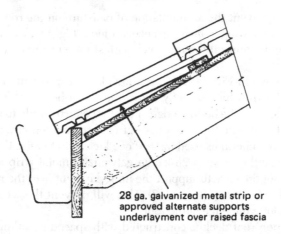

28 ga. galvanized metal strip or
approved alternate supports
underlayment over raised fascia

Fig. 14–3. Underlayment along roof eave supported by galvanized metal strip to prevent ponding of water. *(Courtesy Monier Co.)*

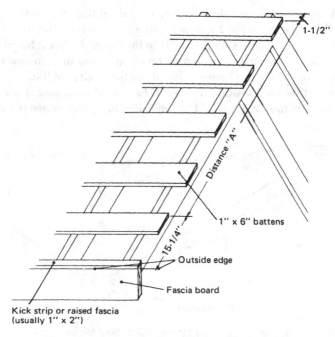

1-1/2"

Distance "A"

1" x 6" battens

15-1/4"

Outside edge

Fascia board

Kick strip or raised fascia
(usually 1" x 2")

Fig. 14–4. Spaced sheathing. *(Courtesy Life Tile Corp.)*

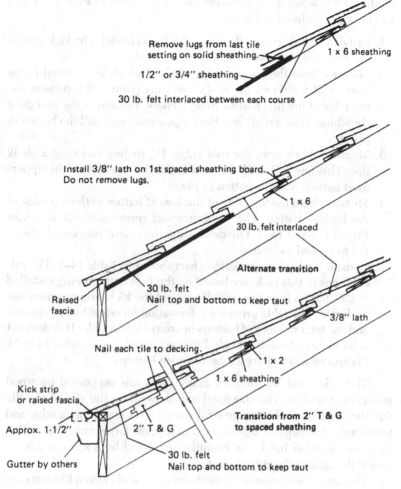

Remove lugs from last tile setting on solid sheathing.

1 x 6 sheathing

1/2" or 3/4" sheathing

30 lb. felt interlaced between each course

Install 3/8" lath on 1st spaced sheathing board. Do not remove lugs.

1 x 6

30 lb. felt interlaced

Alternate transition

30 lb. felt

Nail top and bottom to keep taut

Raised fascia

3/8" lath

Nail each tile to decking.

1 x 2

1 x 6 sheathing

Kick strip or raised fascia

Approx. 1-1/2"

2" T & G

Transition from 2" T & G to spaced sheathing

Gutter by others

30 lb. felt

Nail top and bottom to keep taut

Fig. 14–5. Overhang transition from solid to spaced sheathing. *(Courtesy Life Tile Corp.)*

½-inch or ¾-inch thick plywood or 2-inch tongue-and-groove sheathing board.

The spaced battens must be mounted exactly parallel to the roof eaves and as near the proper spacing as possible. The overall appearance of the finished roof will depend on this. Suggested layout calculation procedures are provided by the tile manufacturers. The layout method for spacing the battens on the roofs shown in Figs. 14–1 and 14–4 may be outlined as follows:

1. Install a kick strip directly above the fascia board. The kick strip is not required if a raised fascia board is used.
2. Measure from the outside edge of the kick strip or raised fascia board 15¼ inches up the roof deck. This point will represent the top edge of the first batten. Snap a chalk line across the roof deck sheathing (closed roof) or rafters (open roof) and nail the batten in place.
3. Measure down from the roof ridge 1½ inches and snap a chalk line. This line will mark the position of the top edge of the upper-most batten. Nail the batten in place.
4. Measure from the top edge of the lowest batten to the top edge of the highest batten. This measurement represents distance A in Figs. 14–1 and 14–4. Divide this measurement into equal spaces not to exceed 14 inches.
5. Consult the layout calculation chart shown in Table 14–1. The calculations in this table are based on the first batten being installed at 15¼ inches from the outside edge of the kick strip or raised fascia board. The table provides information on equal batten spacing and the total number of battens or courses. Example: If distance A is 84 inches, it can be divided into six equal spaces (84 ÷ 14 = 6). Six spaces will require a total of seven battens.

The ridge and hip tiles on many roofs are supported by wood stringers or nailers. The size used will depend on the size of the tile and the slope or pitch. As a general rule, 2 × 6's standing on edge and toenailed to the ridge board are used to support large curved tiles. Flat tiles, on the other hand, can be fully supported by 2 × 2's or 2 × 4's nailed flat against the ridge board.

The cover tiles on some Spanish, mission, and custom tile roofs are supported by wood nailing strips running perpendicular to the roof eaves. The size of the nailing strips (usually 1 × 2 inches or 1 × 4 inches) will depend on the type and size of the tile and the tile manufacturer's application instructions.

Underlayment

Follow the tile manufacturer's instructions for applying the underlayment. If an underlayment is required, sweep the roof deck clean

Table 14–1.   Layout Calculation Chart

Distance A	Equal Spacing	Total Courses	Distance A	Equal Spacing	Total Courses
84"	14"	7	162"	13½"	13
87"	12⁷⁄₁₆"	8	165"	13¾"	13
90"	12⅞"	8	168"	14"*	13
93"	13⁵⁄₁₆"	8	171"	13⅛"	14
96"	13¾"	8	174"	13⅜"	14
99"	14"*	8	177"	13⅝"	14
102"	12¾"	9	180"	13¹³⁄₁₆"	14
105"	13⅛"	9	183"	13⅛"	14
108"	13½"	9	186"	13¼"	15
111"	13⅞"	9	189"	13½"	15
114"	12⅝"	10	192"	13¹¹⁄₁₆"	15
117"	13"	10	195"	13¹⁵⁄₁₆"	15
120"	13⁵⁄₁₆"	10	198"	13³⁄₁₆"	16
123"	13⅝"	10	201"	13⅜"	16
126"	14"	10	204"	13⁹⁄₁₆"	16
129"	12⅞"	11	207"	13¾"	16
132"	13³⁄₁₆"	11	210"	14"*	16
135"	13½"	11	213"	13⁵⁄₁₆"	17
138"	13¹³⁄₁₆"	11	216"	13½"	17
141"	14"*	11	219"	13¹¹⁄₁₆"	17
144"	13¹⁄₁₆"	12	222"	13⅞"	17
147"	13⅜"	12	225"	13¼"	18
150"	13⅝"	12	228"	13⅜"	18
153"	13⅞"	12	231"	13⁹⁄₁₆"	18
156"	13"	13	234"	13¾"	18
159"	13¼"	13	237"	13¹⁵⁄₁₆"	18

*Slight adjustment must be made for these distances.   *(Courtesy Life Tile Corp.)*

with a broom and cover any knotholes in the sheathing with small pieces of tin before laying the roofing felt. Use mastic tape to secure the tin patches to the sheathing.

The underlayment on a closed roof deck will generally consist of overlapping layers of 36-inch wide asphalt-saturated roofing felt. Use either a double layer of 15-pound or a single layer of 30-pound felt. Lap the courses of roofing felt 4 inches along the horizontal edge and 6 inches along the vertical. Secure the roofing felt with nails driven along the horizontal laps and at the ends to hold it in place. On open roof decks, the underlayment may consist of 16-inch wide strips of 30-pound asphalt-saturated roofing felt laid between each course of tile.

Flashing

Metal flashing is preferred for tile roofs. Copper flashing is often recommended, but the other types of less expensive, rust- and corrosion-resistant metals can also be used just as effectively. Flashing application procedures are described in Chapter 7. Flashing specifically required for tile roofs includes the following:

1. Metal drip edges applied along the roof eaves *before* the underlayment is laid.
2. Metal drip edges applied along the roof rakes *after* the underlayment is laid.
3. Minimum 24-inch wide crimped metal flashing applied over 90-pound mineral-surfaced roll roofing in the valleys.

Some examples of different types of tile roof flashing applications are illustrated in Figs. 14–6 and 14–7. Additional flashing details for tile roofs are described in the appropriate sections of this chapter.

Roofing Tile

Modern roofing tile is made from baked clay, concrete, or metal. Clay tile is the most expensive and is used in quality residential construction. Metal tile is generally used on commercial buildings. This chapter covers the application of both clay and concrete tile.

Clay Roofing Tile

Clay roofing tiles are made from a specially prepared clay that is baked in a kiln maintained at the proper temperature. Baking the clay at a properly maintained temperature is important because it affects the quality of the tile. If the baking temperature in the kiln is too low, the tile will be weak and porous. As a result of its higher porosity, the tile will absorb more moisture and eventually deteriorate. Warping and color variations generally result when the kiln temperature is too high. The methods used in manufacturing clay roofing tile are similar to those used in manufacturing brick.

Clay roofing tiles are available in a variety of colors, ranging from various shades of red to dark brown or blue. Exposure to the elements

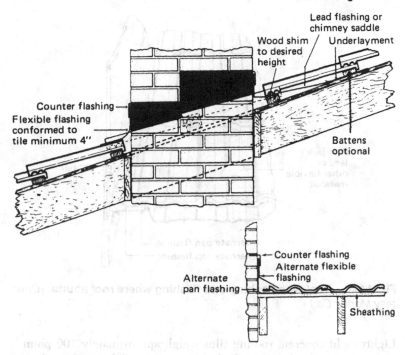

Fig. 14–6. Chimney and vertical wall flashing details. *(Courtesy Monier Co.)*

over an extended period of time will cause clay roofing tiles to fade to lighter shades. This color fading occurs only with clay roofing tiles, not those made of concrete or metal, and the fading occurs in individual tiles on the roof at different rates. Eventually the roof will be covered by tiles of many different shades and tones of the same color, and it is this characteristic of clay roofing tile that provides the roof with such a distinctive and attractive appearance.

Concrete Roofing Tile

Concrete roofing tiles are available in a variety of different designs and colors. They are less expensive than clay roofing tiles and their colors will not fade after extended exposure to weather conditions.

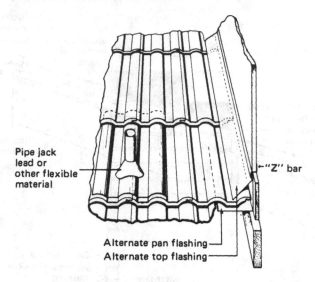

Pipe jack lead or other flexible material

—"Z" bar

Alternate pan flashing
Alternate top flashing

Fig. 14–7. Vent pipe flashing and flashing where roof abutts. *(Courtesy Monier Co.)*

Lightweight concrete roofing tiles weigh approximately 100 pounds per roofing square less than the lightest weight of clay roofing tiles.

Laying Tile

Laying tile can be made easier by snapping vertical and horizontal chalk lines across the roof to serve as a guide for the horizontal courses and vertical rows of tiles (Fig. 14–8). For all types of tiles except interlocking tiles, it is possible to adjust the horizontal guidelines to accommodate tiles that may be smaller than the manufacturer's advertised dimensions (some reduction in size may occur during the baking process for clay tiles) and to insure that full-size, uncut tiles are laid along the roof ridge, hips, and dormers. Adjustments of the horizontal guidelines should always be made in favor of increasing the lap and reducing the exposure.

When loading tile on the roof, space the tiles to insure an equal distribution of weight. This precaution is particularly important for gable roofs. Suggested loading distributions for gable and hip roofs are illustrated in Figs. 14–9 and 14–10.

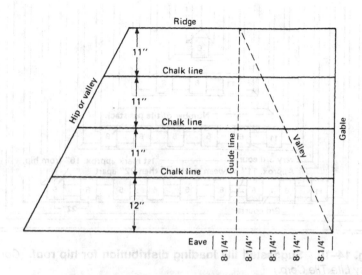

Fig. 14–8. Chalk line locations for tile roof having tile average exposure of 11 inches and average width exposure of 8¼ inches.

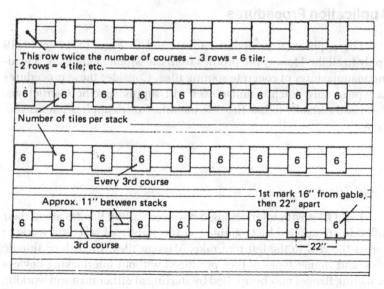

Fig. 14–9. Suggested tile loading distribution for gable roof. *(Courtesy Life Tile Corp.)*

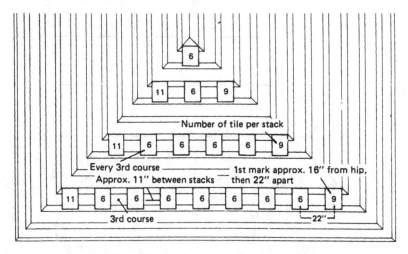

Fig. 14–10. Suggested tile loading distribution for hip roof. *(Courtesy Life Tile Corp.)*

Application Procedures

The application procedures illustrated in Figs. 14–11 to 14–19 apply to all the Monray tile profiles from the Monier Company, a leading manufacturer of concrete roofing tiles. Consider these procedures as typical only. Each manufacturer will supply specific instructions. Read and carefully follow the title manufacturer's application instructions.

Field Tiles

Field tiles with interlocking devices or nailing flanges along their left side are applied by beginning at the right roof rake and working across the roof to the left roof rake. Mission tiles, English or shingle tiles, and similar types of tiles produced without interlocking devices or nailing flanges may be applied by starting at either rake and working across the roof to the opposite rake.

Ridge Details

The top courses of field tiles are brought to within one inch of each other on each side of the ridge and the space is capped with a ridge tile (Fig. 14–11). All ridge tiles are fastened to the roof with nails or wires (Fig. 14–12). If there is no ridge board, and the ridge tiles are fastened with wire, anchor the wire to the nails in the adjacent field tile courses. Lay all ridge tiles in the same direction. Do not use mortar between tile joints. Embed the ridge tiles in a thin layer of mortar in the water course only. Ridge details for roofs with either spaced or solid sheathing are shown in Fig. 14–13.

Gable Juncture Details

Two recommended methods of finishing a ridge where it meets the gable are illustrated in Fig. 14–14. One method calls for mitering the rake and ridge tiles to fit the gable juncture. The other method requires cutting a cover piece from a rake tile to fit the juncture. In either case, all cracks should be filled with mortar to prevent moisture penetration.

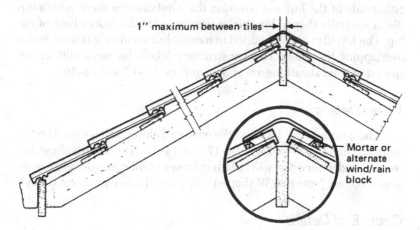

1" maximum between tiles

Mortar or alternate wind/rain block

Fig. 14–11. Roof cross section with ridge details. *(Courtesy Monier Co.)*

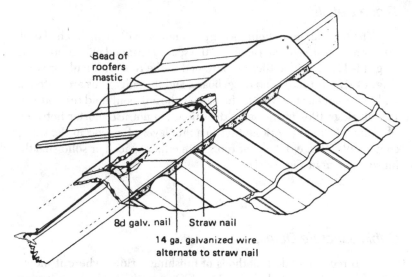

Bead of roofers mastic

8d galv. nail | Straw nail

14 ga. galvanized wire alternate to straw nail

Fig. 14–12.   Ridge tile fastening. *(Courtesy Monier Co.)*

Hip Details

Hips are formed by bringing cut field tiles close to one another on either side of the hip and covering the gap between them with a hip tile. A specially shaped hip starter tile is used at the foot or base of the hip. The hip tiles are embedded in mortar, but no mortar is used in the overlapping tile joints. Hip construction details for roofs with either spaced or solid sheathing are shown in Figs. 14–15 and 14–16.

Valley Details

The construction details of the open and closed valleys used on tile roofs are illustrated in Figs. 14–17 and 14–18. The open valleys are recommended for areas where falling leaves or pine needles are a common problem. Note that W-shaped valley metal is used in open valleys.

Gable End Details

Two frequently used methods of finishing the roof at the gable ends are to install rake tiles or a barge board and flashing (Fig. 14–19). The methods are the same for both spaced and solid roof sheathing.

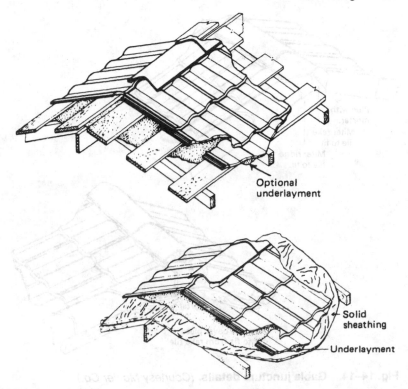

Fig. 14–13. Ridge construction details. *(Courtesy Monier Co.)*

Nails and Fasteners

Large-head noncorrosive copper, aluminum, or galvanized steel nails are used to fasten tiles to a roof deck. For best results, always use the type nail specified by the tile manufacturer. Tile roofing nails should be long enough to penetrate ¾ inch into or through the sheathing or battens.

The nails are driven through one or more machine-punched holes in the tile. They should be driven in far enough to just barely touch the surface of the tile. Driving them further may cause the tile to crack or break.

Use noncorrosive 14-gauge copper wire anchored to copper nails when wire fastening tiles to the roof. Tile tie systems consisting of cop-

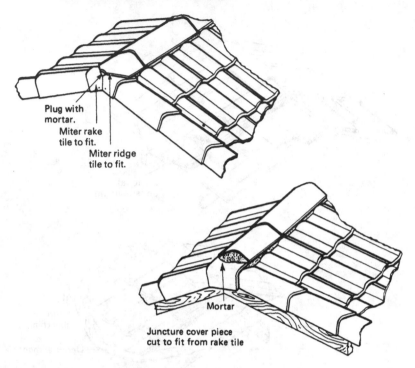

Plug with
mortar.

Miter rake
tile to fit.

Miter ridge
tile to fit.

Mortar

Juncture cover piece
cut to fit from rake tile

Fig. 14–14. Gable juncture details. *(Courtesy Monier Co.)*

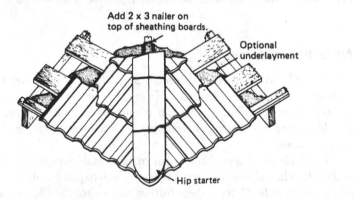

Add 2 x 3 nailer on
top of sheathing boards.

Optional
underlayment

Hip starter

Fig. 14–15. Hip construction on roofs with spaced sheathing.
(Courtesy Monier Co.)

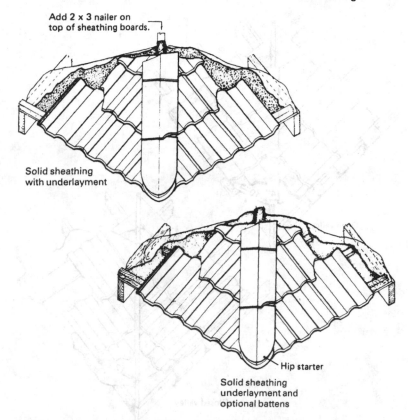

Add 2 x 3 nailer on top of sheathing boards.

Solid sheathing with underlayment

Hip starter

Solid sheathing underlayment and optional battens

Fig. 14–16. Hip construction on roofs with solid sheathing. *(Courtesy Monier Co.)*

per, galvanized wire, or brass strips are also available from some manufacturers for securing tiles.

Cutting and Trimming Tile

Tiles must be cut or trimmed to fit along rakes, hips, ridges, and valleys, or around vent pipes, stacks, and other objects that protrude through the roof surface.

Care must be taken when cutting or trimming tile because it is a hard, brittle material that is easy to crack or chip. Mark the cutting line

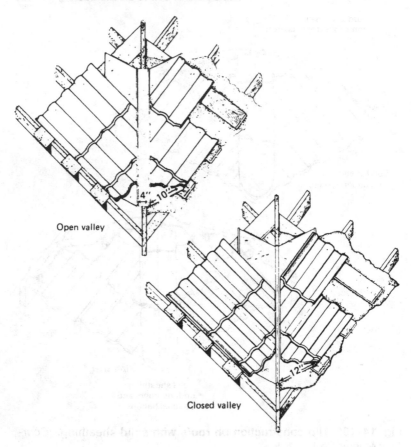

Fig. 14–17. Valley construction on roofs with spaced sheathing. *(Courtesy Monier Co.)*

with a straightedge and place the tile on a flat, wood surface. Cut the tile with a circular saw fitted with a masonry (carborundum) blade.

Mortar

Mortar is sometimes used to provide additional holding power for the tiles, particularly along hips and ridges, and to protect the roof from moisture penetration.

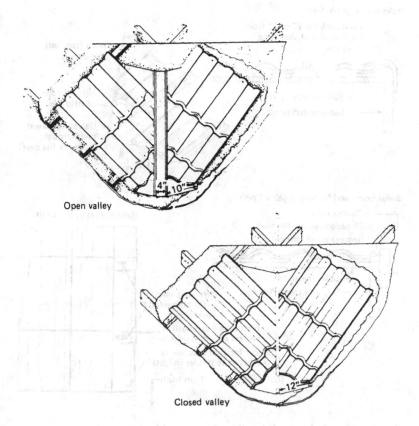

Open valley

Closed valley

Fig. 14–18. Valley construction on roofs with solid sheathing. *(Courtesy Monier Co.)*

Use a cement mortar capable of establishing a strong bond and tight seal under the tile. When mixing mortar use 3 parts sand to 1 part plastic cement. Color pigments are available to match or approximate the color of the tile.

Tile should be immersed in water for at least two minutes before being applied to the roof deck. Dampening the tile avoids the problem of the mortar drying out before it sets and cures.

Rake tiles as gable finish:

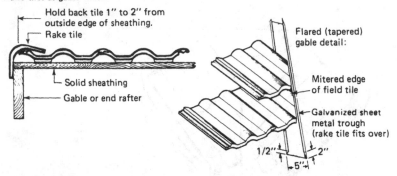

Barge board and flashing as gable finish:

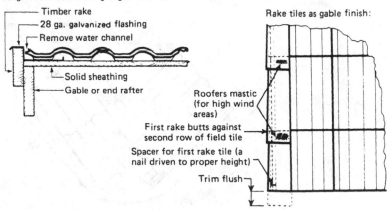

Fig. 14–19. Gable end details. *(Courtesy Monier Co.)*

Types of Tile Roofs

Roofing tiles are produced in the form of field tiles and a variety of different accessory tiles. The field tiles are used to cover the flat expanse of the roof deck and may consist of either individual flat (butted or interlocking) tiles or overlapping pans and covers. Hip and ridge tiles are designed to complement the field tiles for hip and ridge applications. Ridge closures or cat faces are individual units used to close the ridge at gable ends where the rake and ridge units meet. Hip starter units are applied at the bottom of each roof hip. The hip is completed with standard hip tiles and a special terminal tile where the hip meets the ridge. Rake and eave tiles or eave closures are used to close the roof rakes and eaves.

The traditional tile roofs are the flat, Spanish, and mission styles. Spanish tile roofs consist of overlapping S-shaped tiles. Mission or barrel-type roofs are formed from overlapping curved pans and covers. Many custom tile roofs can be created by combining flat plans with cover tiles of different designs. Roman and Greek tile roofs fall within this category. Tile manufacturers also create their own variations of traditional tile roof styles.

Flat Tile Roofs

A flat tile roof bears a strong resemblance to a slate roof. The principal differences are that flat roofing tiles are slightly thicker than slate and have greater uniformity. Examples of field and accessory tiles used on flat tile roofs are shown in Figs. 14–20 and 14–21.

Concrete flat field tiles are manufactured with small nibs or lugs at the head (top) by which they can be hung from horizontal roof battens without nailing (Fig. 14–22). Every third or fourth course of tiles is nailed to provide additional holding power. The lower ends of these tiles are sometimes curved downward to provide a close, tight fit against the underlying course. Flat tiles with nibs on both the bottom and top ends are also available.

Some flat tiles are designed to interlock along the sides (Fig. 14–23). The interlocking of each tile course results in a more watertight roof surface. The water channel on the side of the tile collects rain water and directs it down the roof to the gutters. The tiles shown in Fig. 14–23 overlap to provide a watertight roof surface.

Flat or English shingle tiles are attached to the roof with two large-head roofing nails through machine-punched holes near the top end of each tile (Fig. 14–24). These flat tiles resemble slate in both appearance and method of application.

Flat tiles can be applied to a closed (solid sheathing) roof deck with a pitch under 3-in-12, but the surface must be covered with a

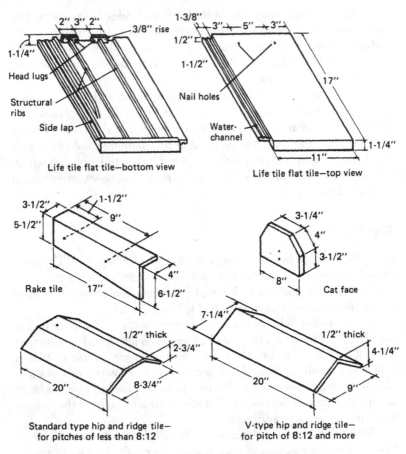

Fig. 14-20. Field and accessory tiles. *(Courtesy Life Tile Corp.)*

watertight membrane before the tiles are laid (Fig. 14-25). The common practice is to cover the roof deck with two layers of 30-pound roofing felt mopped together with hot asphalt. A 12-inch wide strip of 90-pound roofing felt inserted along the eaves of the roof provides additional reinforcement to the angle at the kick strip.

Redwood lath strips 2-inches wide are nailed to the roof perpendicular to the eaves. These strips should be nailed over the rafters and

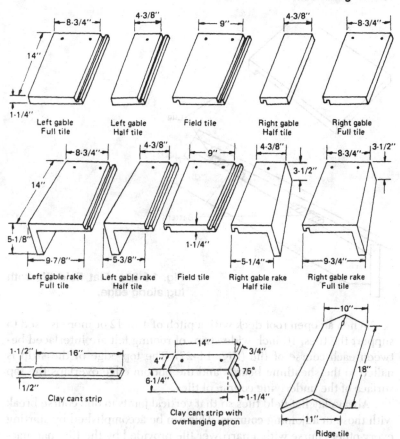

Fig. 14–21. Lincoln interlocking flat field and accessory tiles. *(Courtesy Gladding, McBean & Co.)*

spaced no farther apart than 24 inches on center. They are embedded in the hot asphalt when the roof is mopped.

Top mop the entire roof deck (including the lath strips) with hot asphalt and then nail the 1 × 2-inch battens to the lath strips with 8d nails. When flat tiles are applied to a closed (solid sheathing) roof deck with a pitch of 3-in-12 or more, it is not necessary to hot mop the underlayment (Fig. 14–26).

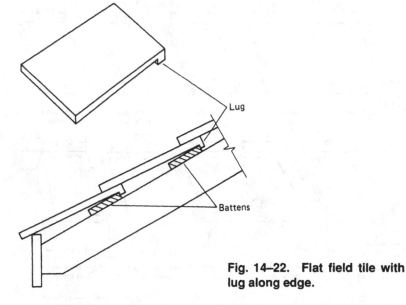

Fig. 14–22. Flat field tile with lug along edge.

When an open roof deck with a pitch of 4-in-12 or more is used to support the tiles, 16-inch wide strips of roofing felt are interlaced between each course of tile (Fig. 14–27). The top edge of the strip is nailed to the sheathing boards and the bottom edge overlaps the top surface of the underlying course of tiles.

Always lay flat field tiles so that vertical joints in one course break with those in adjoining courses. This can be accomplished by starting every other course with a narrower tile provided by the tile manufacturer for this purpose or cutting one from a standard size field tile.

Chimney, pipe, and vent flashing details for flat tile roofs are illustrated in Figs. 14–28 and 14–29. The tile should be cut back far enough

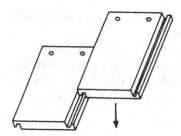

Fig. 14–23. Interlocking flat field tile.

Fig. 14–24. English shingle tile (noninterlocking flat field tile).

to allow a free flow of water. Do not use mortar or obstruct the flow of water between the tile and any projections.

Valley details for a flat tile roof are shown in Figs. 14–30 and 14–31. The valleys should be cut as close to the proper angle as possible. Leave approximately a 4-inch width (2 inches on each side) for an open valley. Butt the tiles for a closed valley.

The valley flashing metal must be cut in a V shape at the roof eave so that side hems are to the outside of the fascia or kick strips (Fig. 14–31). Remove lugs when laying tile over the flashing metal. Do not nail through the valley flashing metal. Mastic may be used to mount small tile pieces. Miscellaneous construction details for flat tile roofs are shown in Figs. 14–32 to 14–35.

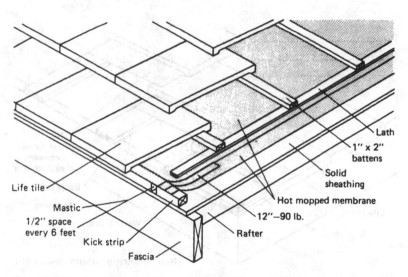

Fig. 14–25. Roof deck with solid sheathing and slope under 3-in-12. *(Courtesy Life Tile Corp.)*

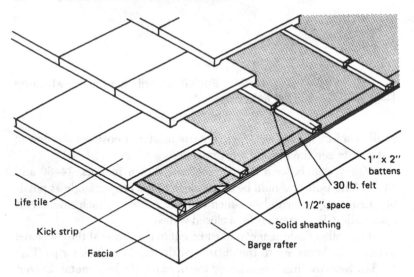

Fig. 14–26. Roof deck with solid sheathing and slope of 3-in-12 or steeper. *(Courtesy Life Tile Corp.)*

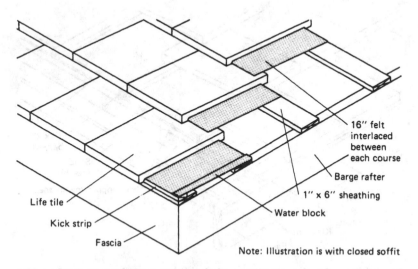

Fig. 14–27. Roof deck with spaced sheathing and slope of 4-in-12 or steeper. *(Courtesy Life Tile Corp.)*

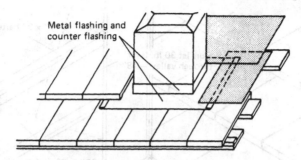

Fig. 14–28.   Chimney flashing details. *(Courtesy Life Tile Corp.)*

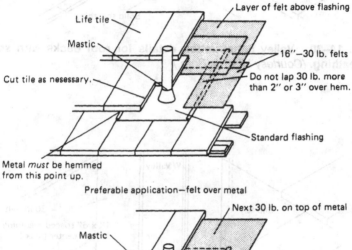

Layer of felt above flashing

Life tile

Mastic

16″–30 lb. felts

Do not lap 30 lb. more
than 2″ or 3″ over hem.

Cut tile as nesessary.

Standard flashing

Metal *must* be hemmed
from this point up.

Preferable application—felt over metal

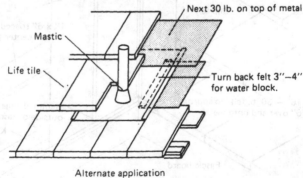

Next 30 lb. on top of metal

Mastic

Life tile

Turn back felt 3″–4″
for water block.

Alternate application

Fig. 14–29.   Recommended pipe and vent flashing methods. *(Courtesy Life Tile Corp.)*

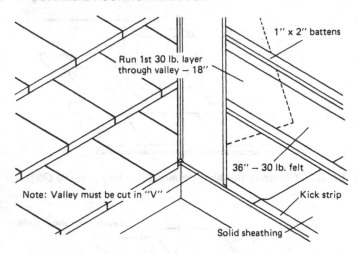

Fig. 14–30. Valley construction details for roof decks with solid sheathing. *(Courtesy Life Tile Corp.)*

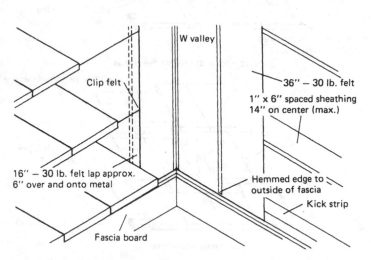

Fig. 14–31. Valley construction details for roof decks with spaced sheathing. *(Courtesy Life Tile Corp.)*

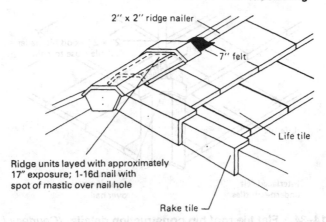

Fig. 14–32. Flat tile roof ridge construction details. *(Courtesy Life Tile Corp.)*

French Tile Roofs

The French tile roof is a variation of the flat tile roof. It is constructed by covering the roof deck with interlocking tiles having fluted surfaces (Fig. 14–36). Grooves or channels on the exposed surface of the tile direct the water runoff away from the joints. Construction details of a typical French tile roof are illustrated in Fig. 14–37. Note the

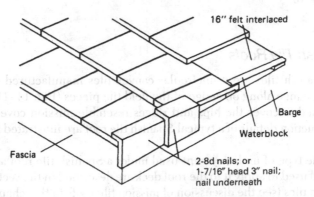

Fig. 14–33. Flat tile roof rake construction details. *(Courtesy Life Tile Corp.)*

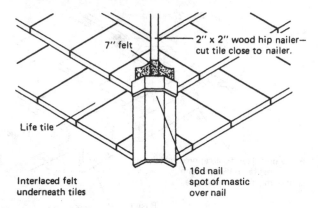

Fig. 14–34. Flat tile roof hip construction details. *(Courtesy Life Tile Corp.)*

use of a cant strip along the edge of the eave. The cant strip replaces the starter strip and provides and necessary elevation for the first tile course (Fig. 14–38).

Hip cover tiles are nailed to a wood strip or stringer which is attached to the hip ridge (Fig. 14–39). The space between the stringer and the rows of tiles is sealed with roofing cement. The main ridge cover tiles differ in design from those used to cover a hip ridge Fig. 14–36. Other construction details of French tile roofs are illustrated in Fig. 14–40.

Spanish Tile Roofs

Spanish tiles are semicircular convex tiles manufactured with a nailing flange along one edge of the field tile pieces (Fig. 14–41). The tiles used to cover the hips and ridges resemble mission cover tiles. Construction details of a typical Spanish tile roof are illustrated in Fig. 14–42.

The type of underlayment used under a Spanish tile roof and the method used to apply it to the roof deck are described in the section on mission tiles (see the discussion of mission tile roofs in this chapter).

The cover portion of each Spanish field tile is laid to overlap the flange of the tile in the adjoining vertical row to form a continuous S pattern across the roof. Each field tile is nailed directly to the roof deck

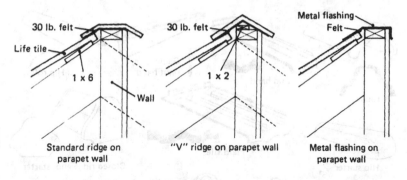

Standard ridge on parapet wall

"V" ridge on parapet wall

Metal flashing on parapet wall

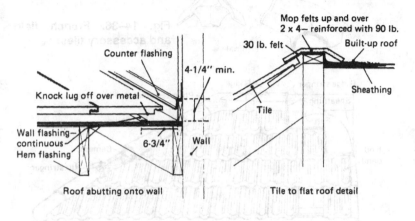

Roof abutting onto wall

Tile to flat roof detail

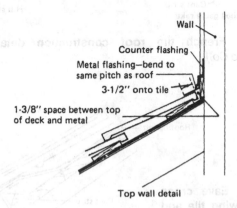

Top wall detail

Fig. 14–35. Miscellaneous flat tile roof construction details. *(Courtesy Life Tile Corp.)*

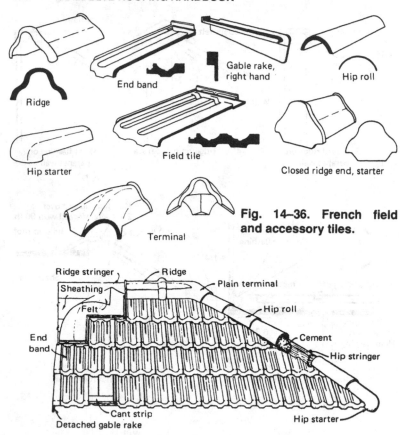

Ridge

End band

Gable rake, right hand

Hip roll

Hip starter

Field tile

Closed ridge end, starter

Terminal

Fig. 14–36. French field and accessory tiles.

Ridge stringer

Sheathing

Felt

Ridge

Plain terminal

Hip roll

End band

Cement

Hip stringer

Cant strip

Detached gable rake

Hip starter

Fig. 14–37. French tile roof construction details. *(Courtesy Ludowicki Tile Co.)*

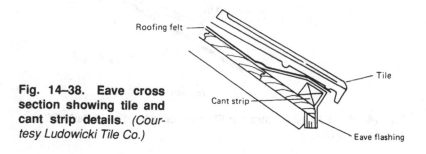

Roofing felt

Tile

Fig. 14–38. Eave cross section showing tile and cant strip details. *(Courtesy Ludowicki Tile Co.)*

Cant strip

Eave flashing

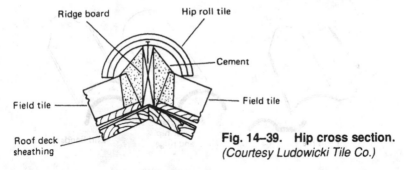

Ridge board Hip roll tile

Cement

Field tile   Field tile

Roof deck
sheathing

Fig. 14–39. Hip cross section.
(Courtesy Ludowicki Tile Co.)

by nails driven through machine-punched holes in the tile flange (Fig. 14–43).

Each vertical row of tiles is closed at the eaves with a specially designed eave closure tile (Fig. 14–44). End bands are used to complete tile courses at the rakes where they are fastened to a wood nailing strip that is parallel to the edge of the roof. When installed, the end bands overlap the gable rake tile on one side and the nailing flange of the last field tile in each horizontal course (Fig. 14–45). Other con-

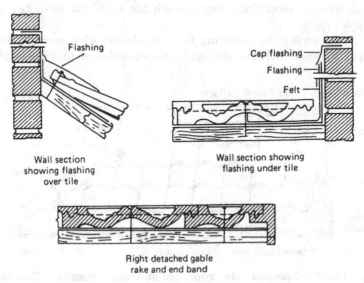

Flashing

Cap flashing

Flashing

Felt

Wall section
showing flashing
over tile

Wall section showing
flashing under tile

Right detached gable
rake and end band

Fig. 14–40. Miscellaneous construction details for French tile roofs. *(Courtesy Ludowicki Tile Co.)*

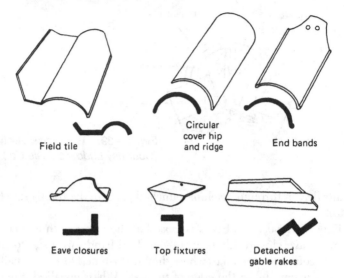

Field tile

Circular
cover hip
and ridge

End bands

Eave closures

Top fixtures

Detached
gable rakes

Fig. 14–41. Spanish field and accessory tiles.

struction details are illustrated in Fig. 14–46. Flashing details common
to all types of roofs (including Spanish tile roofs) are described in
Chapter 7).

Begin roofing by mounting the eave closure tiles over the drip
edge along each roof eave. *Always* lay the eave closure tiles from right

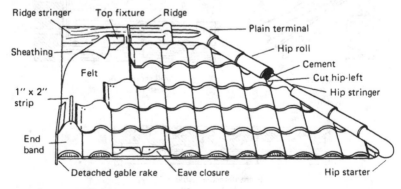

Fig. 14–42. Spanish tile roof construction details. *(Courtesy Ludowicki Tile Co.)*

Fig. 14–43. Nail locations for Spanish tiles.

to left along the roof eave (*never* left to right). This procedure is necessary with Spanish field tiles because the nailing flange is on the left side of each tile. This flange will be covered by the curved or barrel section of the field tile in the adjacent vertical row. Make certain that the spacing of each closure tile produces the desired width exposure in the vertical rows of field tiles.

Lay the first course of field tiles from right to left along the roof eave. The rake tiles can be applied as each course of field tiles is completed or after all the field tiles on the roof are in place. The joint between each field tile and each gable rake tile should be sealed with mortar.

Each successive course of field tiles should overlap the underlying one by 3 inches. Use stringers under the ridge and hip tiles. Fill the space around the stringers with mortar. Mortar should also be used to seal any cracks or breaks where the last course of field tiles meet the ridge and hip tiles.

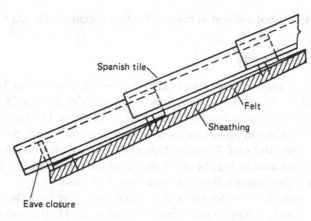

Fig. 14–44. Roof section at eave. *(Courtesy Ludowicki Tile Co.)*

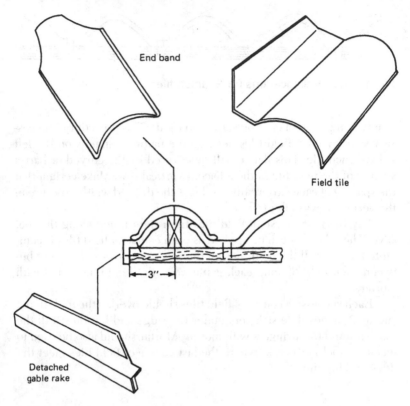

Fig. 14–45.   Roof section at rake. *(Courtesy Ludowicki Tile Co.)*

Mission Tile Roofs

Mission tile roofs are constructed by laying semicircular barrel-shaped tiles in vertical rows across the roof deck (Fig. 14–47). The concave surface (or pan) of the tile in one row is overlapped by the convex (or cover) surface in an adjoining row to form a continuous S-shaped pattern across the roof. Construction details of a typical mission tile roof are illustrated in Fig. 14–48. Because of the type of tile used, this roof is also sometimes called a *straight barrel* tile roof.

The underlayment for roofs with pitches ranging from 3-in-12 to 5-in-12 consists of 30-pound roofing felt laid parallel to the roof eaves. The roofing felt is laid with a 4-inch side lap and 6-inch end lap. It is

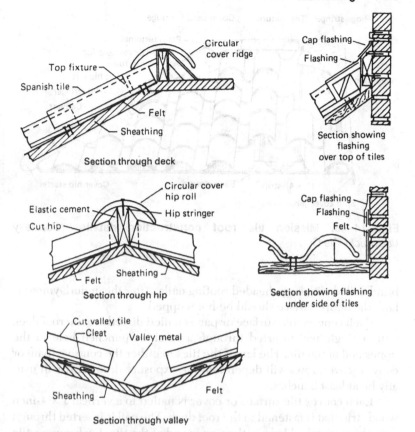

Section through deck

Circular cover ridge

Top fixture

Spanish tile

Felt

Sheathing

Section showing flashing over top of tiles

Cap flashing

Flashing

Section through hip

Circular cover hip roll

Elastic cement

Hip stringer

Cut hip

Felt

Sheathing

Section showing flashing under side of tiles

Cap flashing

Flashing

Felt

Section through valley

Cut valley tile

Cleat

Valley metal

Sheathing

Felt

Fig. 14–46. Miscellaneous construction details for Spanish tile roof. *(Courtesy Ludowicki Tile Co.)*

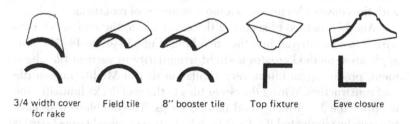

3/4 width cover for rake   Field tile   8″ booster tile   Top fixture   Eave closure

Fig. 14–47. Mission field and accessory tiles. *(Courtesy Ludowicki Tile Co.)*

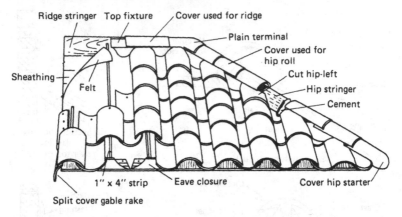

Fig. 14–48. Mission tile roof construction details. *(Courtesy Ludowicki Tile Co.)*

blind nailed with large-headed roofing nails. After the underlayment is laid, the entire surface should be hot mopped.

Each concave tile surface or pan is nailed directly to the roof deck with a single nail inserted through a machine-punched hole in the upper end of the tile. The lap of the tiles in either the concave (pan) or convex (cover) rows will depend on the exposure desired. It will usually be at least 3 inches.

Each convex tile surface or cover is nailed to a vertical 1 × 4-inch wood strip that is fastened to the roof deck. The nail is inserted through a machine-punched hole in the upper end of the tile. Each convex tile must be laid to overlap the edges of the concave tile surfaces or pans in the adjoining parallel rows. Overlapping the edges of the concave tiles with the convex tiles provides a more waterproof roof surface.

Another method of securing the cover tile to the roof check is to wire it to a nail driven into the sheathing (Figs. 14–49 to 14–51). This application method creates a slight irregularity in vertical tile alignment, producing an effect very similar to that of Mediterranean tile roof construction. Wiring the cover tiles to the roof deck eliminates the need for the 1 × 4-inch wood nailing strips. The double row of eave tiles can be eliminated if a 1 × 2-inch kick strip or raised fascia board is used along the roof edge.

A special three-quarter width cover is provided by some tile manufacturers for closing the end of the roof at the rake. The end of each

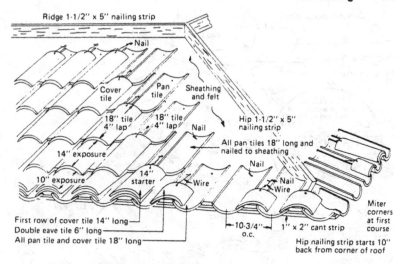

Fig. 14–49. Tile application by the wiring method. *(Courtesy Gladding, McBean & Co.)*

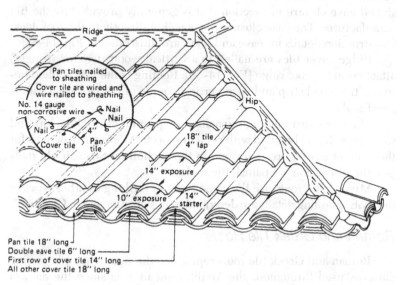

Fig. 14–50. Field tiles laid. *(Courtesy Gladding, McBean & Co.)*

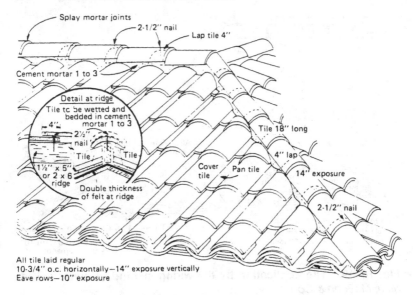

Splay mortar joints

2-1/2" nail

Lap tile 4"

Cement mortar 1 to 3

Detail at ridge

Tile to be wetted and bedded in cement mortar 1 to 3

4"

2½" nail

Tile 18" long

Tile   Tile

Cover tile   Pan tile   4" lap

1½" x 5" or 2 x 6 ridge

Double thickness of felt at ridge

14" exposure

2-1/2" nail

All tile laid regular
10-3/4" o.c. horizontally—14" exposure vertically
Eave rows—10" exposure

Fig. 14–51. Completed roof. *(Courtesy Gladding, McBean & Co.)*

vertical row of convex tiles is closed at the eave with a specially designed eave closure tile section that is generally provided by the tile manufacturer. The eave closure tile is nailed directly to the roof deck. Construction details for eave and rakes are illustrated in Fig. 14–52.

Ridge cover tiles are nailed to a vertical wood strip (or stringer) attached to the roof ridge (Fig. 14–53). Roofing cement is applied between the wood strip and each vertical course of tile to form a waterproof seal.

The space surrounding the nailing strips under the gable, hip, and ridge tiles should be filled with cement mortar. Mixing instructions for the cement mortar are described in the section on Spanish tile roofs (see the discussion of Spanish tile roofs).

Mission tile roofs use the same types of flashing as other types of tile roofs. Flashing details are described in this chapter and in Chapter 7.

Roman and Greek Tile Roofs

Roman and Greek tile roofs reproduce the distinctive tile roofing patterns used throughout the Mediterranean area since the days of early Greece and Rome. Roman tile roofs use flat pans with curved

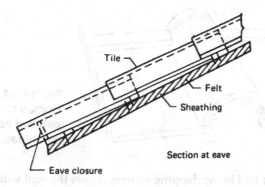

Tile

Felt

Sheathing

Section at eave

Eave closure

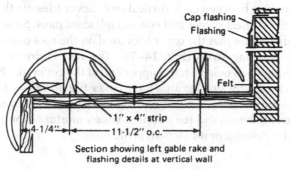

Cap flashing

Flashing

Felt

4-1/4"

1" x 4" strip

11-1/2" o.c.

Section showing left gable rake and
flashing details at vertical wall

Fig. 14–52. Eave and rake construction details. *(Courtesy Ludowicki Tile Co.)*

cover tiles (Fig. 14–54). An angular-shaped cover tile is used with the flat pans on a Greek tile roof (Fig. 14–55).

The construction details of a Roman tile roof on which the cover tiles are fastened to the roof with 14-gauge copper wire and nails are illustrated in Figs. 14–56 to 14–58. The pans (flat tiles with raised

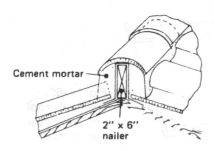

Cement mortar

2" x 6"
nailer

Fig. 14–53. Hip and ridge construction details. *(Courtesy Gladding, McBean & Co.)*

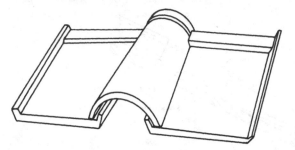

**Fig. 14–54.
Roman tile roof
pan and cover
combination.**

edges) are first laid in overlapping courses across the roof with the required spacing between each vertical row. Cover tiles are then used to overlap the vertical spaces and connect adjacent pans. Some tile manufacturers require that the cover tiles used on the roof plane be nailed to vertical nailing strips (Fig. 14–59). The nail is driven through a machine-punched hole near the upper end of the cover tile. Regardless of which method is used to attach the cover tile on the roof plane, hip and ridge cover tiles are nailed directly to a wood board or stringer.

Flashing details and the points at which mortar is applied are the same as for Spanish or mission-style tile roofs.

Tile Roof Repair and Maintenance

A properly laid tile roof generally requires little maintenance. Occasionally an individual tile will crack, break, or tear loose, but damage is usually the result of a tree limb striking the roof or someone walking on the roof instead of a defect in the tile. Damaged tiles must be removed and replaced with new ones or a leak may develop.

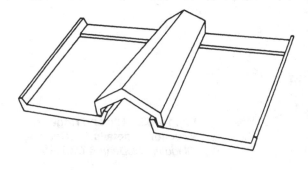

**Fig. 14–55.
Greek tile roof
pan and cover
combination.**

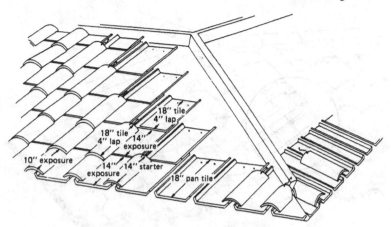

Fig. 14–56. Pan and cover fastening details. *(Courtesy Gladding, McBean & Co.)*

A cracked tile can be repaired with a number of different synthetic sealers. Ask a local building supply dealer for advice. If the tile is too damaged to repair, remove it and replace it with a new one. Remove the damaged tile by first cutting the nails fastening it to the roof with a shingle ripper (also called a *nail ripper*) or a hacksaw blade. Pull the

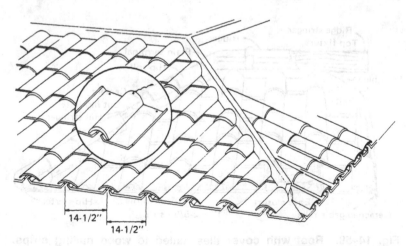

Fig. 14–57. Field tiles laid and roof ready to receive ridge and hip tiles. *(Courtesy Gladding, McBean & Co.)*

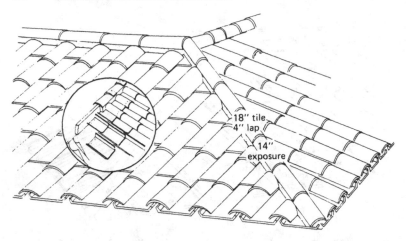

Fig. 14–58. Completed roof. *(Courtesy Gladding, McBean & Co.)*

damaged tile loose and sweep out any dirt or debris. Examine the surface of the roofing felt underlayment under the tile for rips or holes. These should be repaired with a patch and roofing cement before installing the new tile.

Select a new tile of the same size, shape, and color. Slide the replacement tile in place and drill two new nail holes with an electric

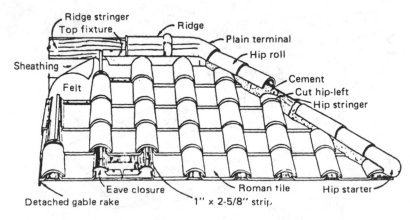

Fig. 14–59. Roof with cover tiles nailed to wood nailing strips. *(Courtesy Ludowicki Tile Co.)*

drill (Fig. 14–60). Drill through both the overlapping tile and the underlying replacement tile at a point approximately 1 inch below the old nail holes. Nail the tiles in place with roofing nails and cover the nail holes with roofing cement.

Cracked or replacement flat-shingle tiles are sometimes reinforced with a piece of copper or aluminum flashing long enough to fit

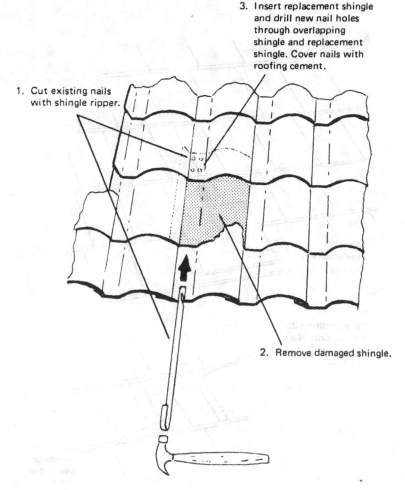

3. Insert replacement shingle and drill new nail holes through overlapping shingle and replacement shingle. Cover nails with roofing cement.

1. Cut existing nails with shingle ripper.

2. Remove damaged shingle.

Fig. 14–60. Replacing damaged tile.

under the entire length of the tile and wide enough to extend under adjoining tiles. The flashing should be about ¼ inch longer than the length of the tile when installed. The extra ¼ inch is bent up to hold the tile in place (Fig. 14–61). Cut off the existing nails with a hacksaw blade before inserting the flashing, drill two new holes, and cover the nail heads with roofing cement.

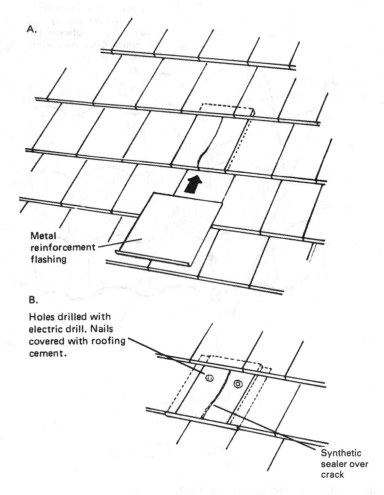

A.

Metal reinforcement flashing

B.

Holes drilled with electric drill. Nails covered with roofing cement.

Synthetic sealer over crack

Fig. 14–61.   Repairing cracked tile.

Reroofing with Tile

Tile has been used in reroofing but it is not a common practice. Reroofing is a practical, low-cost solution to roofing problems when a relatively inexpensive roof needs to be upgraded. As a result, an inexpensive roof covering material is generally recommended for reroofing; more expensive tile is reserved for new construction.

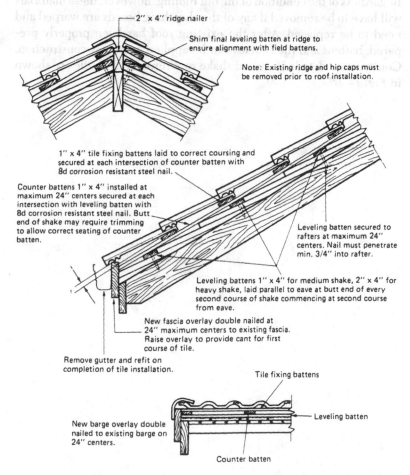

2" x 4" ridge nailer

Shim final leveling batten at ridge to ensure alignment with field battens.

Note: Existing ridge and hip caps must be removed prior to roof installation.

1" x 4" tile fixing battens laid to correct coursing and secured at each intersection of counter batten with 8d corrosion resistant steel nail.

Counter battens 1" x 4" installed at maximum 24" centers secured at each intersection with leveling batten with 8d corrosion resistant steel nail. Butt end of shake may require trimming to allow correct seating of counter batten.

Leveling batten secured to rafters at maximum 24" centers. Nail must penetrate min. 3/4" into rafter.

Leveling battens 1" x 4" for medium shake, 2" x 4" for heavy shake, laid parallel to eave at butt end of every second course of shake commencing at second course from eave.

New fascia overlay double nailed at 24" maximum centers to existing fascia. Raise overlay to provide cant for first course of tile.

Remove gutter and refit on completion of tile installation.

Tile fixing battens

Leveling batten

New barge overlay double nailed to existing barge on 24" centers.

Counter batten

Fig. 14–62. Reroofing over wood shake roof with tile. *(Courtesy Monier Co.)*

If the decision is to reroof with tile, be sure to inspect the roof framing to make certain it is capable of supporting the extra weight of a tile roof. If the roof framing does not appear to be strong enough, reinforce it with additional bracing and rafters.

Prepare the existing roof so that it provides a flat, smooth surface for the tile courses. Repair or replace damaged sections of the old roofing, and make certain that the materials are in relatively good condition and do not show signs of rot or advanced stages of deterioration. Regardless of the condition of the old roofing, however, these materials will have to be removed if any of the sheathing boards are warped and need to be replaced. After the existing roof has been properly prepared, battens and tile courses can be applied as in new construction. Construction details of a wood shake roof reroofed with tile are shown in Fig. 14–62.

CHAPTER 15

Installing Gutters and Downspouts

The roof drainage system is designed to carry water away from the exterior walls of the structure. This is done to protect the edges of the roof and the walls from water damage. If water is allowed to enter the wall cavities, it will soak the insulation and reduce its effectiveness. Water inside the walls will also cause wood framing studs to rot, create stains on inner wall surfaces, and damage the sheathing and siding materials. Carrying the water away from the walls prevents seepage into the basement. A major cause of moisture in the basement is the collection of water along the foundation walls. One way to prevent this problem is to ensure that water runoff from the roof is carried a suitable distance away from the wall. Finally, a strong and efficient roof drainage system will considerably reduce winter snow and ice damage along the roof eaves (Fig. 15–1).

The gutter-and-downspout drainage system is the most commonly used method of draining water runoff from the roofs of residential structures. This type of drainage system is used with sloping roofs. The water is collected in gutters hung from the eave of the roof or attached to the fascia board or rafter ends. The gutters empty the water into vertical downspouts that are used to carry the water down and away from the exterior walls of the structure (Fig. 15–2). On flat roofs the water is often drained from one or more openings located on the roof

Fig. 15–1. Winter snow and ice buildup. *(Courtesy Genova, Inc.)*

and then conducted through an inside wall conduit to an underground drain pipe. A gutter system is not used with this type of roof.

Gutter-and-Downspout Drainage System

A typical gutter-and-downspout drainage system contains the following components and parts:

1. Gutters
2. Downspouts
3. Elbows
4. Inside and outside miters
5. Gutter ends or end caps

Fig. 15–2. Aluminum gutter and downspout. *(Courtesy Alcoa Building Products, Inc.)*

6. Slip joint connectors
7. Gutter screens
8. Hangers, hooks, and brackets
9. Pipe bands, straps, and fasteners

Most of these components and parts are found in the aluminum

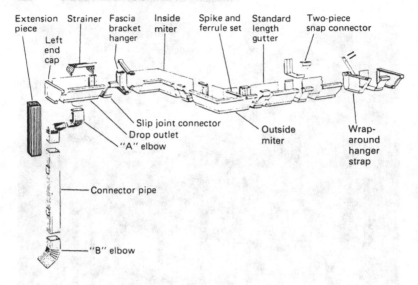

Fig. 15–3. **Typical gutter and downspout system.** *(Courtesy Howmet Aluminum Corp.)*

gutter-and-downspout roof drainage system illustrated in Fig. 15–3. The *gutter* (also called an *eave trough* or *trough*) receives the water runoff from the roof and channels it to a downspout. The *downspout* (sometimes called a *conductor pipe* or *leader*) carries the water from the gutter to the ground. Gutters are available in half-round or formed types (Fig. 15–4). The downspouts may be either round or rectangular

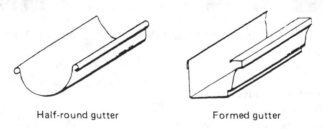

Half-round gutter   Formed gutter

Fig. 15–4. **Typical gutter cross sections.**

in shape. The downspouts are corrugated to provide greater strength against bursting.

A *conductor elbow* should be attached to the bottom of the downspout to direct the flow of water away from the exterior wall. Elbows are also used at each offset gutter and downspout connection. These are the inside and outside miters shown in Fig. 15–3.

Roof drainage systems are made from a variety of materials, including aluminum, galvanized steel, copper and bronze, vinyl, and wood. The most popular type is made of aluminum. Construction details of a typical aluminum gutter and downspout are shown in Fig. 15–2. Galvanized steel gutters and downspouts are also widely used, but they do not provide the same protection against rust and corrosion, and maintenance costs are correspondingly higher.

Vinyl gutters and downspouts are designed for easy do-it-yourself installation (Fig. 15–5). Only a handsaw and a screwdriver are required. Most components and parts can be easily snapped together and installed by one person. The gutters, downspouts, and accessories are

Fig. 15–5. Vinyl gutter and downspout. *(Courtesy Plastmo Vinyl Raingutters)*

made of solid PVC vinyl that will not chip, peel, rust, or rot. The material is resistant to temperature extremes as well as the scratching and denting common to metal gutters and downspouts.

Copper and bronze gutters are more expensive than either aluminum or galvanized steel systems, but they resist rust and corrosion better. Although copper and bronze gutters do not require a protective coating, they may stain the exterior wall surface after an extended period of time. This problem can be avoided by coating them with paint or a suitable varnish or using lead-coated copper.

Wood roof drainage systems are rarely used. They are usually used to drain wood shingle or shake roofs. Construction details of a typical installation are shown in Fig. 15–6. The gutters are mounted on the fascia board that runs along the roof edge (eave). They are screwed to the fascia with 1 × 2-inch furring blocks spaced 24 inches apart (24 inches on center). Galvanized or corrosion-resistant screws are used, and they must be long enough to pass through the furring block and into the fascia board. Wood gutters of the best quality are made from the treated heartwood of cypress, redwood, and western red cedar. Common grades of wood gutter are made of Douglas fir. Square-cut butt joints are used to join wood gutter sections. The joints are fastened

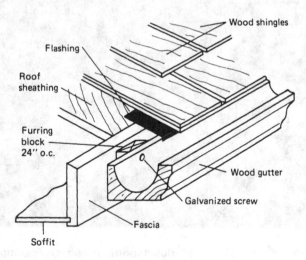

Fig. 15–6. Wood gutter installation.

with dowels or splines, and sealed. A moisture-resistant paint or sealer can be applied to all surfaces to minimize water damage.

Designing a Roof Drainage System

A gutter and downspout system should be designed so that its capacity will meet the maximum anticipated rainfall for the area under normal weather conditions. When a gutter system becomes overloaded, it usually overflows between the downspouts or between a downspout and the end of a section of gutter. When a section of guttering has a downspout at one end, there will be a buildup of water in the central portion. This results from the fact that the gutter farthest from the downspout receives water runoff from the section of the roof directly above it, while the central portion of the guttering receives water from the roof above it as well as the water traveling along the gutter toward the downspout. At the downspout end of the gutter, the water discharges into the downspout and is carried away from the structure.

The maximum carrying capacity of a gutter system is commonly dictated by the size, shape, and cross-sectional area of the gutter. The size of the gutters is roughly determined by the size and spacing of the downspouts (Fig. 15–7). If the downspouts are spaced 35 feet apart or less, the downspout should be the same area as the gutter. If the downspouts are spaced more than 35 feet apart, the width of the gutter should be increased proportionally. Maximum gutter size is limited, however, by such constraints as material strength, weight, volume of material required, and aesthetic appearance. To circumvent these constraints, architects have attempted to increase guttering capacity by mounting the gutters on a slope or by increasing the number of downspouts in the system. Sloping gutters have the disadvantage of looking unsightly when compared to the lines of the structure, and increasing the number of downspouts involves extra expense. A more recent attempt at increasing guttering capacity has been the redesign of the drop outlets. By using drop outlets that increase the flow of water down the downspouts, the capacity of the gutter system can be significantly increased without increasing the cross-sectional area of the guttering or adding more downspouts. All of these factors must be taken

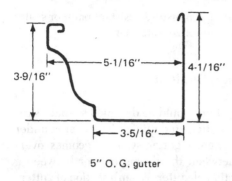

5-1/16"

4-1/16"

3-9/16"

3-5/16"

5" O. G. gutter

Fig. 15–7. Gutter dimensions. *(Courtesy Alcoa Building Products, Inc.)*

into consideration when designing a gutter and downspout system for a structure.

Gutters are sold in standard 10-foot lengths. If the roof length along the eaves measures 32 feet, 40 feet of gutter will have to be ordered with the excess 8 feet cut from one of the 10-foot sections. The order will be based on the total length of the roof that drains into the gutters. For example, a typical gable roof with no minor roofs, which measures 32 feet along the eaves, will require 64 feet of gutters (32 feet for each eave).

The steps to be followed when estimating and ordering the materials for a gutter system may be outlined as follows:

1. Draw a diagram of the existing or planned gutter system.
2. Measure and record on the diagram the length of each gutter run (straight gutter) along the roof eaves.
3. Divide each gutter run by ten to determine the number and location of joints or seams in the run, and the required number of standard 10-foot sections. Note that gutter lengths shorter than 10 feet must be cut from a 10-foot gutter section.
4. Record the number of drop outlets required for the gutter system, the number of downspout pipe straps, and the lengths of the downspouts. Allow two elbows for each drop outlet and one leader and splash block for each downspout.
5. Record the number of inside and outside mitered corners, right- and left-end gutter caps, slip joint connectors (one for each gutter joint), and fasteners required for the gutter system.

Use one slip connector for each cut section of gutter, three hanger straps for each 10-foot length of gutter, and two pipe bands for each 10-foot length of downspout. An elbow directing the water away from the wall should be used at each point a downspout empties onto the ground. The number of inside and outside miter sections will depend on the design of the structure. The same holds true for connections between gutters and downspouts. Most gutter systems require three elbows at each drop outlet (Fig. 15–8).

Gutters should be pitched slightly downward toward the downspout. The water will flow better if this is done properly. The downward pitch of the gutters must be gradual and uniform. Look for low spots in a newly installed gutter system and correct them. Low spots will collect water, and the weight of this water may eventually loosen the joints between gutter sections.

In general, each downspout should not drain more than a 35-foot length gutter. On long roofs, downspouts should be placed at either end of the structure and at the center. Other downspouts should be added if necessary to meet the 35-foot rule.

The number of downspouts and their cross-sectional area will determine the water carrying capacity of a roof drainage system. For example, a 2 × 3-inch downspout will carry water from approximately 700 square feet of roof area, whereas a 3 × 4-inch downspout will drain approximately 1,200 square feet. These figures are based on *average* rainfall figures. In areas of the country where unusually heavy downpours occur, such as in Florida, additional downspouts should be added to the roof drainage system.

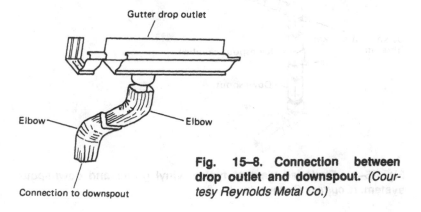

Gutter drop outlet

Elbow

Elbow

Connection to downspout

Fig. 15–8. Connection between drop outlet and downspout. *(Courtesy Reynolds Metal Co.)*

Expansion joints are used in a roof drainage system to relieve the stresses caused by the expansion and contraction of the metal. Unless properly relieved, this movement can loosen the fasteners securing the gutters and downspouts to the structure. In severe cases, it may even distort a metal system. Expansion joints should be used on all hip roof installations, on straight runs over 40 feet in length, and at any point in the roof drainage system subject to unusual stress or restricted movement.

Vinyl Gutters and Downspouts

The principal components and parts of a vinyl gutter and downspout system are illustrated in Fig. 15–9. Both formed and half-round gutters are available. As is the case with metal gutters, formed vinyl gutters are used with square or rectangular downspouts (Fig. 15–9), whereas half-round gutters are used with round downspouts (Fig. 15–5). Vinyl gutters and downspouts are not corrugated because the

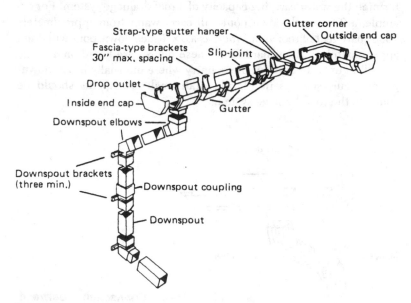

Fig. 15–9. Exploded view of typical vinyl gutter and downspout system. *(Courtesy Genova, Inc.)*

solid PVC plastic is considered strong enough to handle water runoff under most drainage conditions.

Installing Vinyl Gutters and Downspouts

Vinyl gutter and downspout systems can be installed with a minimum number of easily obtainable tools. These tools include a hacksaw with a fine-tooth blade for cutting the gutters and downspouts to the required length, a 10-foot measuring tape, a chalk line, a carpenter's level, a screwdriver, a soft-lead pencil, a drill and bits, a white cloth or paper towel, and a ladder. The gutter system manufacturer will supply the cement and cleaning solvent. If the surfaces are to be painted, they should be cleaned first with the manufacturer's cleaning solvent or acetone.

Vinyl gutters can be mounted on the fascia board with brackets or hung by strap hangers from the roof edge (Figs. 15–10 and 15–11).

Fig. 15–10. Gutter brackets mounted on fascia. *(Courtesy Genova, Inc.)*

Fig. 15–11. Strap hanger installation for vinyl gutters. *(Courtesy Genova, Inc.)*

Hanger straps are used on installations without suitable fascia boards or rafter tails for nailing. They must be bent to match the roof angle and hold the gutter level. Vinyl gutter manufacturers also provide bracket spacers or adapters for mounting the gutters on nonvertical fascias or exposed rafter ends (Fig. 15–12).

If there is an old gutter system in place, remove it and inspect the

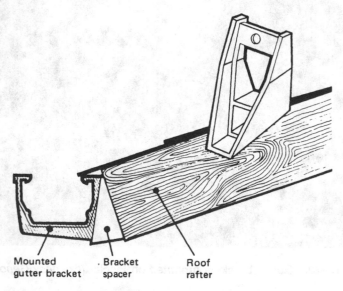

Mounted gutter bracket   . Bracket spacer   Roof rafter

Fig. 15–12. Bracket spacer. *(Courtesy Genova, Inc.)*

fascia boards and roof eaves for damage. Make any necessary repairs, and then clean and paint the surface.

The first step in installing the gutter system is to determine the location of the downspouts. If the gutters are to be mounted on the fascia boards with brackets, snap a level (or sloped) chalk line for each gutter run. The chalk line should be approximately 1½ inches below the edge of the roof to allow proper water spillage into the gutter. If it is sloped, it must incline gently toward the downspout. Strap-type gutter hangers are aligned by means of a string stretched parallel to the roof eave.

Install the drop outlets, gutter elbows, and slip joints for each gutter run (Fig. 15–13). The top edge of each fitting should be flush with the chalk line and screwed to the fascia board. If strap-type hangers are used to support the gutter sections, they should be nailed directly to the roof deck or over the shingle. The drop outlets and gutter elbows are then supported from both sides by gutter brackets that are screwed to the strap hangers. The top outside edge of the fittings should touch the alignment string stretched parallel to the eave.

Fascia-type brackets and strap-type gutter hangers should be

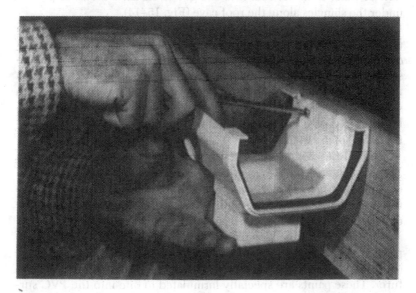

Fig. 15–13. Mounting drop outlet to fascia board with corrosion-resistant screws. *(Courtesy Genova, Inc.)*

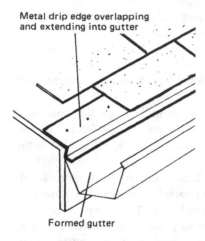

Metal drip edge overlapping
and extending into gutter

Formed gutter

Fig. 15–14. Metal drip edge.
(Courtesy GSW Building Products Division)

evenly spaced 30 inches apart or less between the fittings. The maximum recommended spacing in areas where heavy snows are common is 24 inches. If the roof has insufficient shingle overhang to direct the flow of water from the roof into the gutter, install a metal drip edge under the shingles along the roof eave (Fig. 15–14).

Lay the gutter sections in their brackets and snap them down in place (Fig. 15–15). *Vinyl expands and contracts when temperatures change.* This expansion and contraction must be allowed for when installing a vinyl gutter system or sections may bend and pull loose from the roof. Short gutter sections can be cut to length with a fine-tooth saw. Gutters, downspouts, and fittings are joined with a silicone-lubricated, sliding rubber seal that allows for expansion and contraction at all joints (Fig. 15–16).

Install the downspouts by working down from each drop outlet. Most installations require an offset to extend from the gutter at the roof edge back to the exterior wall of the structure. Brackets are used to fasten the downspouts to the wall. Connect elbows at the bottom of the downspouts to direct the water away from the walls.

Vinyl gutters and downspout parts can be painted or left unpainted. If you wish to paint them, it is best to do so before assembly (Fig. 15–17). Use only paints provided by the vinyl gutter manufacturer. These paints are specially formulated to bite into the PVC surface as though molded in. They will also serve as an excellent primer base for other paints. Do *not* paint the insides of the gutters.

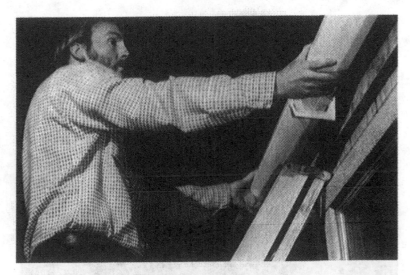

Fig. 15–15. Installing gutter sections in gutter brackets. *(Courtesy Genova, Inc.)*

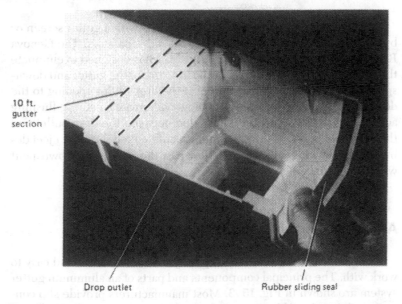

10 ft.
gutter
section

Drop outlet   Rubber sliding seal

Fig. 15–16. Joint between gutter section and drop outlet. *(Courtesy Genova, Inc.)*

Fig. 15–17. Painting vinyl parts. *(Courtesy Genova, Inc.)*

The gutters can be protected from debris with a gutter screen or by using specially designed drop outlets and separators. The Genova Raingo gutter system drop outlet (Fig. 15–18) is designed to eliminate the swirling action of the water at the juncture of the gutter and downspout. The steep slope of the gutter at the drop outlet opening to the downspout smooths out the water flow and creates a strong flushing action. The debris separator is installed between the drop outlet and the downspout (Fig. 15–18). It contains knife-edged slats that eject debris while allowing rain water to flow past the slats to the downspout without clogging.

Aluminum Gutters and Downspouts

Aluminum gutters and downspouts are light, durable, and easy to work with. The principal components and parts of an aluminum gutter system are shown in Fig. 15–3. Most manufacturers provide slip connectors that make soldering unnecessary when joining sections. Gutter

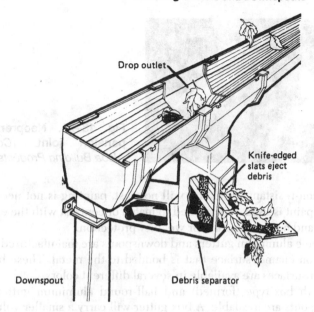

Fig. 15–18. Drop outlet and debris separator. *(Courtesy Genova, Inc.)*

sections are also designed to overlap when joined (Fig. 15–19). Neoprene expansion joints are used to join long runs, on hip roofs, or wherever movement is restricted (Fig. 15–20). The neoprene section expands or contracts as required while at the same time providing a water-tight joint for the gutter sections. Because aluminum is a

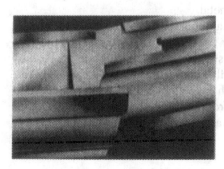

Fig. 15–19. Overlapping gutter sections. *(Courtesy Alcoa Building Products, Inc.)*

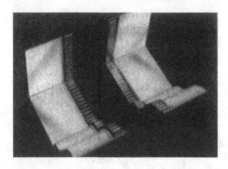

Fig. 15–20. Neoprene expansion joint. *(Courtesy Alcoa Building Products, Inc.)*

corrosion-resistant metal and will not rust, painting is not necessary. When paint is used, it is applied to match or contrast with the exterior siding and not for moisture or weather protection.

Some aluminum gutters and downspouts are manufactured with a baked-on enamel surface that is bonded to the metal. These bonded enamel surfaces are available in several different colors.

Both box-type (formed) and half-round aluminum gutters and downspouts are available. A box gutter will carry a smaller volume of water runoff per foot of roof area than the half-round type, but it is a structurally stronger and more rigid design.

An aluminum mastic is used to seal the joints and provide a watertight connection. To be on the safe side, the brand of aluminum mastic recommended by the gutter and downspout manufacturer should be. used.

Aluminum gutters and downspouts are available in 10-foot lengths with either embossed or plain finishes, and they are generally sold at most local building supply outlets. Because aluminum gutters are lightweight, long runs can be assembled on the ground and placed in position without distorting the metal.

The tools used to install an aluminum gutter and downspout system are essentially the same as those used to install a vinyl system. Make sure the hacksaw blade is suitable for cutting aluminum. Aluminum gutters can be cut more easily by inserting a length of wood in the gutter and then cutting through both the wood and the metal. The wood insert prevents the metal from bending or wobbling as it is being cut. *Never* lean a ladder against an aluminum gutter. The metal is too soft to support the weight.

Installing Aluminum Gutters and Downspouts

Gutters and downspouts must be *tightly* assembled or they may leak and allow water to flow against the exterior wall surfaces. This will usually stain or damage the exterior wall and may cause leakage through the wall into the interior of the structure. This potentially expensive water damage can be avoided by correctly installing the gutters and downspouts and then inspecting them on a regular, semiannual basis.

Before assembling the gutters and downspouts, verify the dimensions by arranging the various components and parts of the drainage system along the sides of the structure. Inspect, clean, and make any necessary repairs before assembling the system.

In modern construction, formed gutters are generally assembled and attached to the roof after it has been shingled. The gutters are held in position by brackets that are screwed directly to the fascia or to a furring strip attached to the fascia. When formed gutters are attached to a furring strip, wrap-around hangers should be used for reinforcement. The hangers should be spaced 48 inches apart. Some of the methods used to mount aluminum gutters are shown in Figs. 15–21 to 15–24.

Downspouts should be fastened to the exterior wall by metal straps spaced approximately four to six feet apart. Use a strap that allows a space between the wall and downspout when it is fastened to the wall. The downspout must be installed in a vertical position. If the downspout is pulled toward the wall and out of its vertical position by

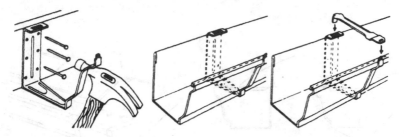

Fig. 15–21. Installing Snap-Lok bracket. *(Courtesy Howmet Aluminum Corp.)*

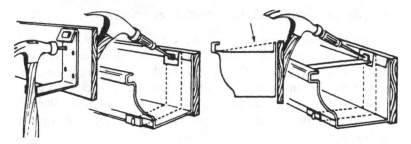

Fig. 15–22. Installing fascia-type gutter bracket. (*Courtesy Howmet Aluminum Corp.*)

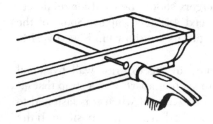

Fig. 15–23. Nailing spike through ferrule to fascia board. (*Courtesy Howmet Aluminum Corp.*)

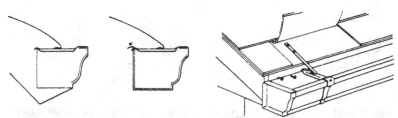

Fig. 15–24. Wraparound hanger strap installed. (*Courtesy Howmet Aluminum Corp.*)

the improper use of a strap, the downspout sections may eventually separate.

Downspout Drainage Connections

A downspout that empties the water directly onto the ground can cause soil erosion that will ruin the appearance of the lawn. In addition to killing any nearby flowers or shrubs, this can lead to the more serious problem of water seepage through the foundation walls.

The water flow from a downspout can be directed away from the structure by connecting the end of the downspout to a drain tile (Fig. 15–25). The downspout is inserted into the bell end of the tile and the joint is sealed with cement. A piece of wire mesh or screen should be placed around the downspout inside the tile before the cement is applied. This will support the cement cap and prevent the cement from seeping down into the tile while it is still wet. The purpose of the cement cap is to plug the opening so that water will not back up and spill out of the tile during heavy water runoff from the roof. Capping the opening also keeps stones, dirt, and other debris out of the drain tile.

An 8-foot run of open-end drain tile installed below grade will carry the runoff a safe distance from the foundation walls. The joints in the open-end drain tiles do not have to be sealed. A large drain tile placed in a vertical position at the lower end of the run can be used to form a dry well. Place the drain tile on its end with the bell opening immediately below the opening of the last tile in the run. Fill the tile with gravel and small stones. Then, cover the opening of the tile with a cast-concrete cap and fill in the hole with dirt up to grade level.

A splash block provides another means of directing water runoff away from the exterior walls, although it is not as effective as using drain tiles. A typical installation is shown in Fig. 15–26. The 3-foot splash block is made of concrete with a channel that inclines downward and away from the downspout.

Repair and Maintenance

Make it a habit to regularly check all gutters and downspouts at least twice a year. The first inspection should take place in late fall

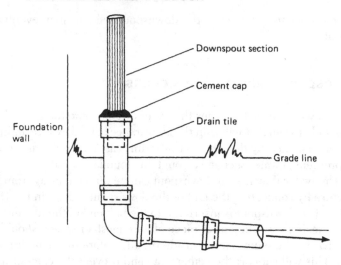

Fig. 15–25. Downspout and drain connection.

after the leaves have fallen. The second inspection should be made in the early spring to check for damage caused by winter weather conditions. Other inspections are recommended after particularly heavy snowfalls or rain storms.

Clean out the gutters and downspouts in the fall after all the leaves have fallen from the trees. Accumulations of leaves in the gutters may plug the openings to downspouts and prevent adequate drainage.

Flush the gutters and downspouts with a garden hose and check them for leaks (Fig. 15–27). Leaks should be repaired immediately because winter weather conditions will only make them worse.

Replace damaged gutters or downspouts immediately. Check downspout straps for looseness or damage. Retighten the loose straps and replace the damaged ones. The same should be done with gutter hanger straps, brackets, and fasteners.

Check for sagging gutters. Sags in a gutter run must be eliminated or water will collect at these points and not flow toward the downspouts.

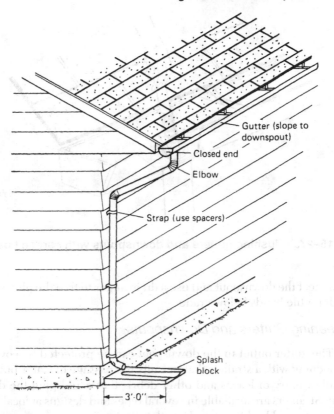

Fig. 15–26. Downspout and splash block.

Clearing Plugged Downspouts

Leaves and other debris will plug the opening to the downspout unless a gutter screen is used. In cold weather, water will accumulate above the plugged area and freeze. The ice then expands and may cause the downspout to break open along its joint.

A plugged downspout can be cleared by the same kind of auger used to unplug stopped up drains. Once a passage has been opened, the downspout can be flushed with a garden hose. If the downspout is connected to a below-grade run of drain tile, it may be necessary to

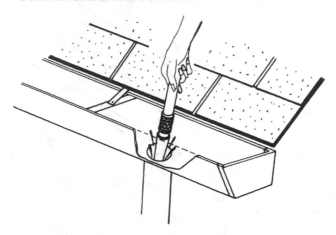

Fig. 15–27. Flushing gutters and downspouts with garden hose.

disconnect the downspout and use a drain auger to free the plug where the drain tile bends below grade.

Screening Gutters and Downspouts

The gutter outlet to the downspout can be protected by covering the opening with a strainer that permits the passage of water but does not allow entry of leaves and other debris that could plug the downspout. Strainers are available in several types and designs at local hardware stores and building supply outlets.

Screening the gutters is another effective method of preventing the accumulation of leaves and other debris. One end of the screen should be inserted under the shingles and nailed to the roof deck (Fig. 15–28).

Protection Against Water Damage

Aluminum and copper gutters will not rust. Galvanized sheet metal gutters are also normally rust-resistant as long as the protective galvanized coating remains intact. Check the coating on the gutters and downspouts for scratches or other damage. Damage to the galvanized coating can be caused by careless handling during installation, by the movement of a loose slip connector, gutter hanger, or down-

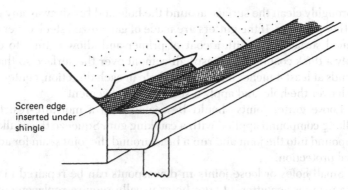

Screen edge
inserted under
shingle

Fig. 15–28. Gutter screen.

spout, or by the pulling apart of soldered joints. Clean the damaged area, dry it thoroughly, and cover it with a protective, corrosion-resistant coating such as roofing cement. These are asphalt-based coatings that are generally available through a local building supply outlet.

The outside surfaces of galvanized sheet metal gutters can be protected from rust and corrosion by coating them with a good corrosion-resistant paint. Always clean the surfaces thoroughly before applying the paint.

Wood gutters can be waterproofed by coating the inside surfaces with a penetrating wood preservative. Apply the coating and allow it time to dry thoroughly before installing the gutters.

Repairing Leaks

Leaks can develop from holes in gutters, broken gutter seams, or loose joints in gutter runs. Loose or broken gutter seams or gutter holes ¼ inch in diameter or smaller can be sealed with plastic asphalt cement, butyl gutter and lap seal, or asphalt-saturated roofing tape. To insure good adhesion, use steel wool or a stiff wire brush to thoroughly clean the area over which the sealant will be applied. Remove loose dirt and other particles from the cleaning operation and apply the sealant. Extend the sealant coating out at least 4 inches in each direction from the hole.

Gutter holes larger than ¼ inch in diameter must be patched and sealed. Patches may be cut from aluminum, galvanized steel, or canvas.

Thoroughly clean the surface around the hole and brush away any dirt or other particles. If the gutters are made of galvanized steel, cover the cleaned surface area with a coat of primer and allow it time to dry. Apply a thick coat of plastic asphalt cement over the surface so that it extends at least 4 inches out from the hole in each direction, center the patch over the hole, and apply a second coat of cement.

Loose gutter joints should be sealed with a moisture-resistant caulking compound applied with a caulking gun. Squeeze the caulking compound into the joint and run a bead around the joint seam for additional protection.

Small holes or loose joints in downspouts can be repaired in the same manner as gutters. Large holes usually require replacement of the downspout because water pressure makes the retention of a patch very difficult.

CHAPTER 16

Roofing for Historic Buildings*

The purpose of this chapter is to help you identify the features or elements of a roof that contribute to the historic character of a building and show you how to preserve that character.

Historic Character

There are different ways of understanding old buildings. They can be seen as examples of specific building types, which are usually related to a building's function, such as schools, courthouses, or churches. Buildings can be studied as examples of using specific materials such as concrete, wood, steel, or limestone. They can also be considered as examples of an historical period, which is often related to a specific architectural style, such as Gothic Revival farmhouses, one-story bungalows, or Art Deco apartment buildings.

There are many other facets of an historic building besides its

---

*(By Sarah M. Sweetser. Material in this chapter was previously published by U.S. Department of the Interior, National Park Service, Preservative Assistance Division, Technical Preservation Services in the Preservation Briefs series.)

functional type, materials, construction, or style that contribute to its historic qualities or significance. Some of these qualities are feelings conveyed by the sense of time and place or in buildings associated with events or people. A complete understanding of any property may require documentary research about its style, construction, function, furnishings or contents; knowledge about the original builder, owners, and later occupants; and knowledge about the evolutionary history of the building. Even though buildings may be of historic, rather than architectural significance, it is their tangible elements that embody its significance for association with specific events or persons and it is those *tangible elements* both on the exterior and interior that should be preserved.

Following is a method you can use to identify those visual and tangible aspects of the historic building. While this may aid in the planning process for carrying out any ongoing or new use or restoration of the building, this approach is not a substitute for developing an understanding about the significance of an historic building and the district in which it is located.

If the various materials, features, and spaces that give a building its visual character are not recognized and preserved, then essential aspects of its character may be damaged in the process of change.

A building's character can be irreversibly damaged or changed in many ways, for example, by removal of a distinctive Spanish tile roof and replacement with asphalt shingles.

Identify a Building's Visual Character

Here is a two-step approach that can be used by anyone to identify those materials, features, and spaces that contribute to the visual character of a building. This approach involves first examining the building from afar to understand its overall setting and architectural context; then moving up very close to appreciate its materials and the craftsmanship and surface finishes evident in these materials.

Step 1: Overall Aspects—Identifying the overall visual character of a building is nothing more than looking at its distinguishing physical aspects without focusing on its details. The major contributors to a building's overall character are embodied in the general aspects of its *setting;* the *shape* of the building; its *roof* and roof features, such as chimneys or cupolas; the various *projections* on the building, such as porches or bay windows; the *recesses* or voids in a building; such as open galleries, arcades, or recessed balconies; the *openings* for windows and doorways; and finally the various exterior *materials* that contribute to the building's character. Step 1 involves looking at the building from a distance to understand the character of its site and setting, and it involves walking around the building where that is possible (Figs. 16–1 and 16–2). Some buildings will have one or more sides that are more important than the others because they are more highly visible. This does not mean that the rear of the building is of no value whatever but it simply means that it is less important to the overall character. On the other hand, the rear may have an interesting back porch or offer a private garden space or some other aspect that may contribute to the visual character. Such a general approach to looking at the building and site will provide a better understanding of its overall character without having to resort to an infinitely long checklist of its possible features and details. Regardless of whether a building is complicated or relatively plain, it is these broad categories that contribute to an understanding of the overall character rather than the specifics of architectural features such as shingle type and size.

Step 2: Close Range — Step 2 involves looking at the building at close range or arm's length, where it is possible to see all the surface qualities of the materials, such as their *color* and *texture,* or surface evidence of craftsmanship or age (Figs. 16–3 to 16–6). In some instances, the visual character is the result of the juxtaposition of materials that are contrastingly different in their color and texture. The surface qualities of the materials may be important because they impart the very sense of craftsmanship and age that distinguishes historic buildings from other buildings. Furthermore, many of these closeup qualities can be easily damaged or obscured by work that affects those surfaces.

There is an almost infinite variety of surface materials, textures, and finishes that are part of a building's character which are fragile and easily lost.

Using this two-step approach, it is possible to identify all those elements and features that help define the visual character of the building. In most cases, there are a number of aspects about the exterior that are important to the character of an historic building.

Fig. 16–1. The overall shape and specific details of this roof are essential elements of the building's historic character, and should not be modified, despite the use of alternative surface materials. *(Gamwell House, Bellingham, Washington)*

Fig. 16–2. The front wall of this modest commercial building has a simple three-part shape that is the controlling aspect of its overall visual character. The upward projecting parapet and the decorative stonework completely hide the roof. The hidden flat roof contributes little to the historic character. *(Photo by Emogene A. Bevitt)*

Planning

A historic building is a product of the cultural heritage of its region, the technology of its period, the skill of its builders, and the materials used for its construction. To assist in planning, the following process has been developed by the National Park Service and is applicable to all historic buildings. This planning process is a sequential approach to the preservation of historic wood frame buildings. It begins with the premise that historic materials should be retained wherever possible. When retention, including retention with some repair, is not

Fig. 16–3. Repairs on this pantile roof were made with new tiles held in place with metal hangers. *(Main Building, Ellis Island, New York)*

possible, then *replacement of the irreparable historic material can be considered.* The purpose of this approach is to determine the appropriate level of treatment for the preservation of historic wood frame buildings including their roofs. The planning process has the following four steps:

1. Identify and preserve those materials and features that are important in defining the *building's historic character.* This may include features such as chimneys, roof railings, decorative flashing and shingles.

2. Undertake routine maintenance on historic materials and features. Routine maintenance generally involves the least amount of work needed to preserve the materials and features of the building. For example, maintenance of a wood shingle roof would include cleaning pine needles off valleys and out of gutters.

3. Repair historic materials and features. For a historic material such as wood roofing, repair would generally involve patching and piecing-in with new material according to recognized preservation methods.

4. Replace severely damaged or deteriorated historic materials and features in kind. Replacing sound or repairable historic material is

Fig. 16–4. The Victorians loved to use different colored slates to create decorative patterns on their roofs, an effect which cannot be easily duplicated by substitute materials. Before any repair work on a roof such as this, the slate sizes, colors, and position of the patterning should be carefully recorded to assure proper replacement. *(Ebenezer Maxwell Mansion, Philadelphia, Pennsylvania, photo courtesy of William D. Hershey)*

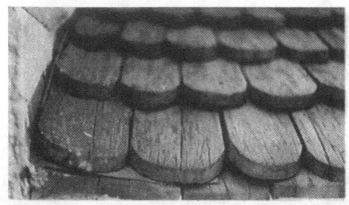

Fig. 16–5. Replacement of particular historic details is important to the individual historic character of a roof, such as the treatment at the eaves of this rounded butt wood shingle roof. Also note that the surface of the roof was carefully sloped to drain water away from the side of the dormer. In the restoration, this function was augmented with the addition of carefully concealed modern metal flashing. *(Mount Vernon, Virginia)*

Fig. 16–6. Galvanized sheet-metal shingles imitating the appearance of pantiles remained popular from the second half of the 19th century into the 20th century. *(Episcopal Church, now the Jerome Historical Society Building, Jerome, Arizona, 1927)*

never recommended; however, if the historic material cannot be repaired because of the extent of deterioration or damage, then it will be necessary to *replace* an entire character-defining feature such as the building's roofing. The preferred treatment is always replacement in kind, that is, with the same material. Because this approach is not always feasible, consider the use of a compatible substitute material. A substitute material should only be considered, however, if the form, detailing, and overall appearance of the substitute material conveys the visual appearance of the historic material, and the application of the substitute material does not damage, destroy or obscure historic features.

Significance of the Roof

A weather-tight roof is basic in the preservation of a structure, regardless of its age, size, or design. In the system that allows a building to work as a shelter, the roof sheds the rain, shades from the sun, and buffers the weather.

During some periods in the history of architecture, the roof imparts much of the architectural character. It defines the style and contributes to the building's aesthetics. The hipped roofs of Georgian architecture, the turrets of Queen Anne, the Mansard roofs, and the graceful slopes of the Shingle Style and Bungalow designs are examples of the use of roofing as a major design feature.

But no matter how decorative the patterning or how compelling the form, the roof is a highly vulnerable element of a shelter that will inevitably fail. A poor roof will permit the accelerated deterioration of historic building materials—masonry, wood, plaster, paint—and will cause general disintegration of the basic structure. Furthermore, there is an urgency involved in repairing a leaky roof since such repair costs will quickly become prohibitive. Although such action is desirable as soon as a failure is discovered, temporary patching methods should be carefully chosen to prevent inadvertent damage to sound or historic roofing materials and related features. Before any repair work is performed, the historic value of the materials used on the roof should be understood. Then a complete internal and external inspection of the roof should be planned to determine all the causes of failure and to identify the alternatives for repair or replacement of the roofing.

Historic Roofing Materials in America

Clay Tile

European settlers used clay tile for roofing as early as the mid-17th century; many pantiles (S-curved tiles), as well as flat roofing tiles, were used in Jamestown, Virginia. In some cities such as New York and Boston, clay was popularly used as a precaution against such fire as those that engulfed London in 1666 and scorched Boston in 1679 (Fig. 16–3).

Tile roofs found in the mid-18th century Moravian settlements in Pennsylvania closely resembled those found in Germany. Typically, the tiles were 14-15 inches long, 6-7 inches wide with a curved butt. A lug on the back allowed the tiles to hang on the lathing without nails or pegs. The tile surface was usually scored with finger marks to promote drainage. In the Southwest, the tile roofs of the Spanish missionaries (mission tiles) were first manufactured (ca. 1780) at the Mission San Antonio de Padua in California. These semicircular tiles were made by molding clay over sections of logs, and they were generally 22 inches long and tapered in width.

The plain or flat rectangular tiles most commonly used from the 17th through the beginning of the 19th century measured about 10 inches by 6 inches by ½ inch, and had two holes at one end for a nail or peg fastener. Sometimes mortar was applied between the courses to secure the tiles in a heavy wind.

In the mid-19th century, tile roofs were often replaced by sheet-metal roofs, which were lighter and easier to install and maintain. However, by the turn of the century, the Romanesque Revival and Mission style buildings created a new demand and popularity for this picturesque roofing material.

Slate

Another practice settlers brought to the New World was slate roofing. Evidence of roofing slates have been found also among the ruins of mid-17th-century Jamestown. But because of the cost and the time required to obtain the material, which was mostly imported from Wales, the use of slate was initially limited. Even in Philadelphia (the second largest city in the English-speaking world at the time of the Revolution) slates were so rare that "The Slate Roof House" distinctly referred to William Penn's home built late in the 1600s. Sources of native slate were known to exist along the eastern seaboard from Maine to Virginia, but difficulties in inland transportation limited its availability to the cities, and contributed to its expense. Welsh slate continued to be imported until the development of canals and railroads in the mid-19th century made American slate more accessible and economical.

Slate was popular for its durability, fireproof qualities, and aesthetic potential. Because slate was available in different colors (red, green, purple, and blue-gray), it was an effective material for decorative patterns on many 19th-century roofs (Gothic and Mansard styles) (Fig. 16–4). Slate continued to be used well into the 20th century, notably on many Tudor revival style buildings of the 1920s.

Shingles

Wood shingles were popular throughout the country in all periods of building history. The size and shape of the shingles as well as the detailing of the shingle roof differed according to regional craft practices (Fig. 16–5). People within particular regions developed preferences for the local species of wood that most suited their purposes. In New England and the Delaware Valley, white pine was frequently used; in the South, cypress and oak; in the far west, red cedar or redwood. Sometimes a protective coating such as a mixture of brick dust and fish oil, or a paint made of red iron oxide and linseed oil was applied to increase the durability of the shingle.

Commonly in urban areas, wooden roofs were replaced with more fire resistant materials, but in rural areas this was not a major concern. On many Victorian country houses, the practice of wood shingling survived the technological advances of metal roofing in the 19th century, and near the turn of the century enjoyed a full revival in its namesake, the Shingle Style. Colonial revival and the Bungalow styles in the 20th century assured wood shingles a place as one of the most fashionable, domestic roofing materials.

Metal

Metal roofing in America is principally a 19th-century phenomenon. Before then the only metals commonly used were lead and copper. For example, a lead roof covered "Rosewell," one of the grandest mansions in 18th-century Virginia. But more often, lead was used for protective flashing. Lead, as well as copper, covered roof surfaces where wood, tile, or slate shingles were inappropriate because of the roof's pitch or shape.

Copper with standing seams covered some of the more notable early American roofs including that of Christ Church (1727–1744) in Philadelphia. Flat-seamed copper was used on many domes and cupolas. The copper sheets were imported from England until the end of the 18th century when facilities for rolling sheet metal were developed in America.

Sheet iron was first known to have been manufactured here by the Revolutionary War financier, Robert Morris, who had a rolling mill near Trenton, New Jersey. At his mill Morris produced the roof of his own Philadelphia mansion, which he started in 1794. The architect Benjamin H. Latrobe used sheet iron to replace the roof on Princeton's "Nassau Hall," which had been gutted by fire in 1802.

The method for corrugating iron was originally patented in England in 1829. Corrugating stiffened the sheets, and allowed greater span over a lighter framework, as well as reduced installation time and labor. In 1834 the American architect William Strickland proposed corrugated iron to cover his design for the marketplace in Philadelphia.

Galvanizing with zinc to protect the base metal from rust was developed in France in 1837. By the 1850s the material was used on post offices and customhouses, as well as on train sheds and factories. In 1857 one of the first metal roofs in the South was installed on the U.S. Mint in New Orleans. The Mint was thereby "fireproofed" with a 20-gauge galvanized, corrugated iron roof on iron trusses (Fig. 16–6).

Tin-plate iron, commonly called "tin roofing," was used extensively in Canada in the 18th century, but it was not as common in the United States until later. Thomas Jefferson was an early advocate of tin roofing, and he installed a standing-seam tin roof on "Monticello" (ca. 1770–1802). The Arch Street Meetinghouse (1804) in Philadelphia had tin shingles laid in a herringbone pattern on a "piazza" roof.

However, once rolling mills were established in this country, the low cost, light weight, and low maintenance of tin plate made it the most common roofing material. Embossed tin shingles, whose surfaces created interesting patterns, were popular throughout the country in the late 19th century (Fig. 16–8). Tin roofs were kept well-painted, usually red; or, as the architect A. J. Davis suggested, in a color to imitate the green patina of copper.

Fig. 16–7. Repeated repair with asphalt, which cracks as it hardens, has created a blistered surface on this sheet-metal roof and built-in gutter, which will retain water. Repairs could be made by carefully heating and scraping the surface clean, repairing the holes in the metal with a flexible mastic compound or a metal patch, and coating the surface with a fiber paint. *(Roane County Courthouse, Kingston, Tennessee, photo courtesy of Building Conservation Technology, Inc.)*

Fig. 16–8. Tin shingles, commonly embossed to imitate wood or tile, or with a decorative design, were popular as an inexpensive, textured roofing material. These shingles (8⅜ inches by 12½ inches on the exposed surface) were designed with interlocking edges, but they have been repaired by surface nailing, which may cause future leakage. *(Ballard House, Yorktown, Virginia, photo by Gordie Whittington, National Park Service)*

Fig. 16–9. A Chicago firm's catalog dated 1896 illustrates a method of unrolling, turning the edges, and finishing the standing seam on a metal roof.

Terne plate differed from tin plate in that the iron was dipped in an alloy of lead and tin, giving it a duller finish. Historic, as well as modern, documentation often confuses the two, so much that it is difficult to determine how often actual "terne" was used.

Zinc came into use in the 1820s, at the same time tin plate was becoming popular. Although a less expensive substitute for lead, its advantages were controversial, and it was never widely used in this country.

Other Materials

Asphalt shingles and roll roofing were used in the 1890s. Many roofs of asbestos, aluminum, stainless steel, galvanized steel, and lead-coated copper may soon have historic values as well. Awareness of these and other traditions of roofing materials and their detailing will contribute to more sensitive preservation treatments.

Locating the Problem

Failures of Surface Materials

When trouble occurs, it is important to contact a professional, either an architect, a reputable roofing contractor, or a craftsman familiar with the inherent characteristics of the particular historic roofing system involved. These professionals may be able to advise on immediate patching procedures and help plan more permanent repairs. A thorough examination of the roof should start with an appraisal of the existing condition and quality of the roofing material itself. Particular attention should be given to any southern slope because year-round exposure to direct sun may cause it to break down first.

Wood—Some historic roofing materials have limited life expectancies because of normal organic decay and wear. For example, the flat surfaces of wood shingles erode from exposure to rain and ultraviolet rays from the sun. Some species are more hardy than others, and heartwood, for example, is stronger and more durable than sapwood.

Ideally, shingles are split with the grain perpendicular to the surface. This is because if shingles are sawn across the grain, moisture may enter the grain and cause the wood to deteriorate. Prolonged moisture on or in the wood allows moss or fungi to grow, which will further hold the moisture and cause rot.

Metal—Of the inorganic roofing materials used on historic buildings, the most common are perhaps the sheet metals: lead, copper, zinc, tin plate, terne plate, and galvanized iron. In varying degrees each of these sheet metals is likely to deteriorate from chemical action by pitting or streaking. This can be caused by airborne pollutants; acid rainwater; acids from lichen or moss; alkalis found in lime mortars or portland cement, which might be on adjoining features and washes down on the roof surface; or tannic acids from adjacent wood sheathings or shingles made of red cedar or oak.

Corrosion from *galvanic action* occurs when dissimilar metals, such as copper and iron, are used in direct contact. Corrosion may also occur even though the metals are physically separated; one of the metals will react chemically against the other in the presence of an electrolyte such as rainwater. In roofing, this situation might occur when either a copper roof is decorated with iron cresting, or when steel nails are used in copper sheets. In some instances the corrosion can be prevented by inserting a plastic insulator between the dissimilar materials. Ideally, the fasteners should be a metal sympathetic to those involved.

Iron rusts unless it is well-painted or plated. Historically this problem was avoided by use of tin plating or galvanizing. But this method is durable only as long as the coating remains intact. Once the plating is worn or damaged, the exposed iron will rust. Therefore, any iron-based roofing material needs to be undercoated, and its surface needs to be kept well-painted to prevent corrosion.

One cause of sheet metal deterioration is fatigue. Depending upon the size and the gauge of the metal sheets, wear and metal failure can occur at the joints or at any protrusions in the sheathing as a result from the metal's alternating movement to thermal changes. Lead will tear because of *creep*, or the gravitational stress that causes the material to move down the roof slope.

Slate—Perhaps the most durable roofing materials are slate and tile. Seemingly indestructible, both vary in quality. Some slates are hard and tough without being brittle. Soft slates are more subject to erosion and to attack by airborne and rainwater chemicals, which cause the slates to wear at nail holes, to delaminate, or to break (Fig. 16–10). In winter, slate is very susceptible to breakage by ice, or ice dams.

Fig. 16–10. This detail shows slate delamination caused by a combination of weathering and pollution. In addition, the slates have eroded around the repair nails, incorrectly placed in the exposed surface of the slates. *(Lower Pontalba Building, New Orleans, photo courtesy of Building Conservation Technology, Inc.)*

Tile—Tiles will weather well, but tend to crack or break if hit, as by tree branches, or if they are walked on improperly. Like slates, tiles cannot support much weight. Low quality tiles that have been insufficiently fired during manufacture will craze and spall under the effects of freeze and thaw cycles on their porous surfaces.

Failures of Support Systems

Once the condition of the roofing material has been determined, the related features and support systems should be examined on the exterior and on the interior of the roof. The gutters and downspouts need periodic cleaning and maintenance since a variety of debris fill them, causing water to back up and seep under roofing units. Water will eventually cause fasteners, sheathing, and roofing structure to deteriorate. During winter, the daily freeze-thaw cycles can cause ice floes to develop under the roof surface. The pressure from these ice floes will dislodge the roofing material, especially slates, shingles, or tiles. Moreover, the buildup of ice dams above the gutters can trap enough moisture to rot the sheathing or the structural members.

Many large public buildings have built-in gutters set within the perimeter of the roof. The downspouts for these gutters may run within the walls of the building, or drainage may be through the roof surface or through a parapet to exterior downspouts. These systems can be effective if properly maintained; however, if the roof slope is inadequate for good runoff, or if the traps are allowed to clog, rainwater will form pools on the roof surface. Interior downspouts can collect debris and thus back up, perhaps leaking water into the surrounding walls. Exterior downspouts may fill with water, which in cold weather may freeze and crack the pipes. Conduits from the built-in gutter to the exterior downspout may also leak water into the surrounding roof structure or walls.

Failure of the flashing system is usually a major cause of roof deterioration. Flashing should be carefully inspected for failure caused by either poor workmanship, thermal stress, or metal deterioration (both of flashing material itself and of the fasteners). With many roofing materials, the replacement of flashing in an existing roof is a major operation, which may require taking up large sections of the roof surface. Therefore, the installation of top quality flashing material on a new or replaced roof should be a primary consideration. *Remember, some roofing and flashing materials are not compatible.*

Roof fasteners and clips should also be made of a material compatible with all other materials used, or coated to prevent corrosion. For example, the tannic acid in oak will corrode iron nails. Some roofs such as slate and sheet metals may fail if nailed too rigidly.

If the roof structure appears sound and nothing indicates recent movement, the area to be examined most closely is the roof substrate — the sheathing or the battens. The danger spots would be near the roof plates, under any exterior patches, at the intersections of the roof planes, or at vertical surfaces such as dormers. Water penetration, indicating a breach in the roofing surface or flashing, should be readily apparent, usually as a damp spot or stain. Probing with an ice pick may reveal any rot which may indicate previously undetected damage to the roofing membrane. Insect infestation evident by small exit holes and frass (a sawdust-like debris) should also be noted. Condensation on the underside of the roofing is undesirable and indicates improper ventilation. Moisture will have an adverse effect on any roofing material; a good roof stays dry inside and out.

Repair or Replace

Understanding potential weaknesses of roofing material also requires knowledge of repair difficulties. Individual slates can be replaced normally without major disruption to the rest of the roof, but replacing flashing on a slate roof can require substantial removal of surrounding slates. If it is the substrate or a support material that has deteriorated, many surface materials such as slate or tile can be reused if handled carefully during the repair. Such problems should be evaluated at the outset of any project to determine if the roof can be effectively patched, or if it should be completely replaced.

Will the repairs be effective? Maintenance costs tend to multiply once trouble starts. As the cost of labor escalates, repeated repairs could soon equal the cost of a new roof.

The more durable the surface is initially, the easier it will be to maintain. Some roofing materials such as slate are expensive to install, but if top quality slate and flashing are used, it will last 40-60 years with minimal maintenance. Although the installation cost of the roof will be high, low maintenance needs will make the lifetime cost of the roof less expensive.

Historical Research

In a restoration project, research of documents and physical investigation of the building usually will establish the roof's history. Documentary research should include any original plans or building specifications, early insurance surveys, newspaper descriptions, or the personal papers and files of people who owned or were involved in the history of the building. Old photographs of the building might provide evidence of missing details.

Along with a thorough understanding of any written history of the building, a physical investigation of the roofing and its structure may reveal information about the roof's construction history. Starting with an overall impression of the structure, are there any changes in the roof slope, its configuration, or roofing materials? Perhaps there are obvious patches or changes in patterning of exterior brickwork where a gable roof was changed to a gambrel, or where a whole upper story was added. Perhaps there are obvious stylistic changes in the roof line, dormers, or ornamentation. These observations could help one understand any important alteration, and could help establish the direction of further investigation.

Because most roofs are physically out of the range of careful scrutiny, the "principle of least effort" has probably limited the extent and quality of previous patching or replacing, and usually considerable evidence of an earlier roof surface remains. Sometimes the older roof will be found as an underlayment of the current exposed roof. Original roofing may still be intact in awkward places under later features on a roof. Often if there is any unfinished attic space, remnants of roofing may have been dropped and left when the roof was being built or repaired. If the configuration of the roof has been changed, some of the original material might still be in place under the existing roof. Sometimes whole sections of the roof and roof framing will have been left intact under the higher roof. The profile and/or flashing of the earlier roof may be apparent on the interior of the walls at the level of the alteration. If the sheathing or lathing appears to have survived changes in the roofing surface, they may contain evidence of the roofing systems. These may appear either as dirt marks, which provide "shadows" of a roofing material, or as nails broken or driven down into the wood, rather than pulled out during previous alterations or repairs. Wooden headers in the roof framing may indicate that earlier chimneys or skylights have been removed. Any metal ornamentation that might have existed may be indicated by anchors or unusual markings along the ridge or at other edges of the roof. This primary evidence is essential for a full understanding of the roof's history.

Caution should be taken in dating early "fabric" on the evidence of a single item, as recycling of materials is not a mid-20th-century innovation. Carpenters have been reusing materials, sheathing, and framing members in the interest of economy for centuries. Therefore, any analysis of the materials found, such as nails or sawmarks on the wood, requires an accurate knowledge of the history of local building practices before any final conclusion can be accurately reached. It is helpful to establish a sequence of construction history for the roof and roofing materials; any historic fabric or pertinent evidence in the roof should be photographed, measured, and recorded for future reference. During the repair work, useful evidence might unexpectedly appear. It is essential that records be kept of any type of work on a historic building, before, during, and after the project. Photographs are generally the easiest and fastest method, and should include overall views and details at the gutters, flashing, dormers, chimneys, valleys, ridges, and eaves. All photographs should be immediately labeled to insure accurate identification at a later date. Any patterning or design on the roofing deserves particular attention (Fig. 16–11). For example, slate roofs are often decorative and have subtle changes in size, color, and texture, such as a gradually decreasing coursing length from the eave to the peak. If not carefully noted before a project begins, there may be problems in replacing the surface. The standard reference for this phase of the work is *Recording Historic Buildings*, compiled by Harley J. McKee for the Historic American Buildings Survey, National Park Service, Washington, D.C., 1970.

Replacing the Historic Roofing Material

Professional advice will be needed to assess the various aspects of replacing a historic roof. With some exceptions, most historic roofing materials are available today. If not, an architect or preservation group who has previously worked with the same type material may be able to recommend suppliers. Special roofing materials, such as tile or embossed metal shingles, can be produced by manufacturers of related products that are commonly used elsewhere, either on the exterior or interior of a structure. With some creative thinking and research, the historic materials usually can be found.

Craft Practices

Determining the craft practices used in the installation of a historic roof is another major concern in roof restoration. Early builders took great pride in their work, and experience has shown that the rustic or irregular designs commercially labeled "Early American" are a

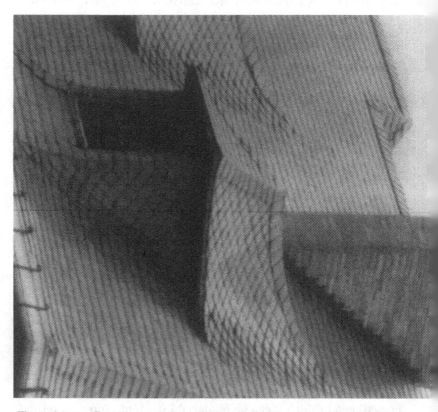

Fig. 16–11. Because of the roof's visibility, the slate detailing around the dormers is important to the character of this structure. Note how the slates swirl from a horizontal pattern on the main roof to a diamond pattern on the dormer roofs and side walls. *(18th and Q Streets, NW, Washington, DC)*

20th-century-invention. For example, historically, wood shingles underwent several distinct operations in their manufacture including splitting by hand, and smoothing the surface with a draw knife. In modern nomenclature, the same item would be a taper-split shingle which has been dressed. Unfortunately, the rustic appearance of today's commercially available handsplit and re-sawn shingle bears no resemblance to the handmade roofing materials used on most early American buildings.

Early craftsmen worked with a great deal of common sense; they understood their materials. For example they knew that wood shingles should be relatively narrow; shingles much wider than about 6 inches would split when walked on, or they may curl or crack from varying temperature and moisture. It is important to understand these aspects of craftsmanship, remembering that people wanted their roofs to be weather-tight and to last a long time. The recent use of "mother-goose" shingles on historic structures is a gross underestimation of the early craftsman's skills.

Supervision

Finding a modern craftsman to reproduce historic details may take some effort. It may even involve some special instruction to raise his understanding of certain historic craft practices. At the same time, it may be pointless (and expensive) to follow historic craft practices in any construction that will not be visible on the finished product. But if the roofing details are readily visible, their appearance should be based on architectural evidence or on historic prototypes. For instance, the spacing of the seams on a standing-seam metal roof will affect the building's overall scale and should therefore match the original dimensions of the seams.

Many older roofing practices are no longer performed because of modern improvements. Research and review of specific detailing in the roof with the contractor before beginning the project is highly recommended. For example, one early craft practice was to finish the ridge of a wood shingle roof with a roof "comb" — that is, the top course of one slope of the roof was extended uniformly beyond the peak to shield the ridge, and to provide some weather protection for the raw horizontal edges of the shingles on the other slope. If the comb is known to have been the correct detail, it should be used. Though this method leaves the top course vulnerable to the weather, a disguised strip of flashing will strengthen this weak point.

Detail drawings or a sample mock-up will help ensure that the contractor or craftsman understands the scope and special requirements of the project. It should never be assumed that the modern carpenter, slater, sheet metal worker, or roofer will know all the historic details. Supervision is as important as any other stage of the process.

Alternative Materials

The use of the historic roofing material on a structure may be restricted by building codes or by the availability of the materials, in which case an appropriate alternative will have to be found.

Some municipal building codes allow variances for roofing materials in historic districts. In other instances, individual variances may be obtained. Most modern heating and cooking is fueled by gas, electricity, or oil—none of which emit the hot embers that historically have been the cause of roof fires. Where wood burning fireplaces or stoves are used, spark arrestor screens at the top of the chimneys help to prevent flaming material from escaping, thus reducing the number of fires that start at the roof. In most states, insurance rates have been equalized to reflect revised considerations for the risks involved with various roofing materials.

In a rehabilitation project, there may be valid reasons for replacing the roof with a material other than the original. The historic roofing may no longer be available, or the cost of obtaining specially fabricated material may be prohibitive. But the decision to use an alternative material should be weighed carefully against the primary concern to keep the historic character of the building. If the roof is flat and is not visible from any elevation of the building, and if there are advantages to substituting a modern built-up composition roof for what might have been a flat metal roof, then it may make better economic and construction sense to use a modern roofing method. But if the roof is readily visible, the alternative material should match as closely as possible the scale, texture, and coloration of the historic roofing material.

Asphalt shingles or ceramic tiles are common substitute materials intended to duplicate the appearance of wood shingles, slates, or tiles. Fire-retardant, treated wood shingles are currently available. The treated wood tends, however, to be brittle, and may require extra care (and expense) to install. In some instances, shingles laid with an interlay of fire-retardant building paper may be an acceptable alternative.

Lead-coated copper, terne-coated steel, and aluminum/zinc-coated steel can successfully replace tin, terne plate, zinc, or lead. Copper-coated steel is a less expensive (and less durable) substitute for sheet copper.

The search for alternative roofing materials is not new. As early as the 18th century, fear of fire caused many wood shingle or board roofs to be replaced by sheet metal or clay tile. Some historic roofs were failures from the start, based on over-ambitious and naive use of materials as they were first developed. Research on a structure may reveal that an inadequately designed or a highly combustible roof was replaced early in its history, and therefore restoration of a later roof material would have a valid precedent. In some cities, the substitution of sheet metal on early row houses occurred as soon as the rolled material became available.

Cost and ease of maintenance may dictate the substitution of a material wholly different in appearance from the original. The practical problem (wind, weather, and roof pitch) should be weighed against the historical consideration of scale, texture, and color. Sometimes the effect of the alternative material will be minimal. But on roofs with a high degree of visibility and patterning or texture, the substitution may seriously alter the architectural character of the building.

Temporary Stabilization

It may be necessary to carry out an immediate and temporary stabilization to prevent further deterioration until research can determine how the roof should be restored or rehabilitated, or until funding can be provided to do a proper job. A simple covering of exterior plywood or roll roofing might provide adequate protection, but any temporary covering should be applied with caution. One should be careful not to overload the roof structure, or to damage or destroy historic evidence or fabric that might be incorporated into a new roof at a later date. In this sense, repairs with caulking or bituminous patching compounds should be recognized as potentially harmful, since they are difficult to remove, and at their best, are very temporary.

Precautions

The architect or contractor should warn the owner of any precautions to be taken against the specific hazards in installing the roofing material. Soldering of sheet metals, for instance, can be a fire hazard, either from the open flame or from overheating and undetected smoldering of the wooden substrate materials.

Thought should be given to the design and placement of any modern roof appurtenances such as plumbing stacks, air vents, or TV antennas. Consideration should begin with the placement of modern plumbing on the interior of the building, otherwise a series of vent stacks may pierce the roof membrane at various spots creating maintenance problems as well as aesthetic ones. Air handling units placed in the attic space will require vents which, in turn, require sensitive design. Incorporating these in unused chimneys has been very successful in the past.

Whenever gutters and downspouts are needed that were not on the building historically, the additions should be made as unobtrusively as possible, perhaps by painting them out with a color compatible with the nearby wall or trim.

Maintenance

Although a new roof can be an object of beauty, it will not be protective for long without proper maintenance. At least twice a year, the roof should be inspected against a checklist. All changes should be recorded and reported. Guidelines should be established for any foot traffic that may be required for the maintenance of the roof. Many roofing materials should not be walked on at all. For some—slate, asbestos, and clay tile—a self-supporting ladder might be hung over the ridge of the roof, or planks might be spanned across the roof surface. Such items should be specifically designed and kept in a storage space accessible to the roof. If exterior work ever requires hanging scaffolding, use caution to insure that the anchors do not penetrate, break, or wear the roofing surface, gutters, or flashing.

Any roofing system should be recognized as a membrane that is designed to be self-sustaining, but that can be easily damaged by intrusions such as pedestrian traffic or fallen tree branches. Certain items should be checked at specific times. For example, gutters tend to accumulate leaves and debris during the spring and fall and after heavy rain. Hidden gutter screening both at downspouts and over the full length of the gutter could help keep them clean. The surface material

Fig. 16–12. Decorative features such as cupolas require extra maintenance. The flashing is carefully detailed to promote run-off, and the wooden ribbing must be kept well-painted. This roof surface, which was originally tin plate, has been replaced with lead-coated copper for maintenance purposes. *(Lyndhurst, Tarrytown, New York, photo courtesy of the National Trust for Historic Preservation)*

would require checking after a storm as well. Periodic checking of the underside of the roof from the attic after a storm or winter freezing may give early warning of any leaks. Generally, damage from water or ice is less likely on a roof that has good flashing on the outside and is well ventilated and insulated on the inside. Specific instructions for the maintenance of the different roof materials should be available from the architect or contractor.

Summary

The essential ingredients for replacing and maintaining a historic roof are:

- Understanding the historic character of the building and being sympathetic to it.
- Careful examination and recording of the existing roof and any evidence of earlier roofs.
- Consideration of the historic craftsmanship and detailing and implementing them in a renewal wherever visible.
- Supervision of the roofers or maintenance personnel to assure preservation of historic fabric and proper understanding of the scope and detailing of the project.
- Consideration of alternative materials where the original cannot be used.
- Cyclical maintenance program to assure that the staff understands how to take care of the roofs and of the particular trouble spots to safeguard.

With these points in mind, it will be possible to preserve the architectural character and maintain the physical integrity of the roofing on a historic building.

CHAPTER 17

Historic Wooden Shingle Roofs*

Wooden shingle roofs are important elements of many historic buildings. The special visual qualities imparted by both the *historic shingles* and the *installation patterns* should be preserved when a wooden shingle roof is replaced. This requires an understanding of the size, shape, and detailing of the historic shingle and the method of fabrication and installation. These factors combined to create roofs expressive of particular architectural styles, which were often influenced by regional craft practices. The use of wooden shingles from the early settlement days to the present illustrates an extraordinary range of styles (Figs. 17–1 to 17–4).

Wooden shingle roofs need periodic replacement. They can last from 15 to over 60 years, but the shingles should be replaced before there is deterioration of other wooden components of the building. Appropriate replacement shingles are available, but careful research, design, specifications, and the selection of a skilled roofer are necessary

*(By Sharon C. Park, AIA. This chapter was previously published by U.S. Department of the Interior, National Park Service Preservation Assistance Division, Technical Preservation Services, in the Preservation Briefs Series.)

Fig. 17–1. The Rolfe-Warren House, a tidewater Virginia property, was restored to its 18th-century appearance in 1933. The handsplit and dressed wooden shingles are typical of the tidewater area with special features such as curved butts, projecting ridge comb and closed swept valleys at the dormer roof connections. Circa 1970. *(Photo: Association for the Preservation of Virginia Antiquities.)*

to assure a job that will both preserve the appearance of the historic building and extend the useful life of the replacement roof.

Unfortunately, the wrong shingles are often selected or are installed in a manner incompatible with the appearance of the historic roof. There are a number of reasons why the wrong shingles are selected for replacement roofs. They include the failure to identify the appearance of the original shingles; unfamiliarity with available products; an inadequate budget; or a *confusion in terminology*. In any discussion about historic roofing materials and practices, it is important to understand the historic definitions of terms like "shingles," as well as the modern definitions or use of those terms by craftsmen and the industry. Historically, from the first buildings in America, these wooden roofing products were called *shingles*, regardless of whether they were

Fig. 17–2. Handsplit and dressed shingles were also used on less elaborate buildings as seen in the restoration of the circa 1840 kitchen at the Winedale Inn, Texas. The uneven surfaces of the handsplit shingles were generally dressed or smoothed with a draw-knife to keep the rainwater from collecting in the wood grain and to ensure that the shingles lay flat on the sub-roof. *(Photo: Thomas Taylor.)*

the earliest handsplit or the later machine-sawn type. The term *shake* is a relatively recent one, and today is used by the industry to distinguish the sawn products from the split products, but through most of our building history there has been no such distinction.

Considering the confusion among architects and others regarding these terms as they relate to the appearance of early roofs, it should be stated that there is a considerable body of documentary information about historic roofing practices and materials in this country, and that many actual specimens of historic shingles from various periods and places have been collected and preserved so that their historic appear-

Fig. 17–3. Readily available and inexpensive sawn shingles were used not only for roofs, but also for gables and wall surfaces. The circa 1891 Chambers House, Eugene, Oregon used straight sawn butts for the majority of the roof and hexagonal butts for the lower portion of the corner tower. Decorative shingles in the gable ends and an attractive wooden roof cresting feature were also used. *(Photo: Lane County Historical Society.)*

ances are well established. Essentially, the rustic-looking shake that we see used so much today has little in common with the shingles that were used on most of our early buildings in America.

Throughout this chapter, the term *shingle* will be used to refer to historic wooden roofs in general, whether split or sawn, and the term *shake* will be used only when it refers to a commercially-available product. The variety and complexity of terminology used for currently available products will be seen in the accompanying chart entitled "Shingles and Shakes."

This chapter discusses what to look for in historic wooden shingle

Fig. 17–4. With the popularity of the revival of historic styles in the late 19th and early 20th centuries, a new technique was developed to imitate English thatch roofs. For the Tudor Revival thatch cottages, steaming and curving of sawn shingles provided on undulating pattern to this picturesque roof shape. *(Photo: Courtesy of C.H. Roofing.)*

roofs and when to replace them. It discusses ways to select or modify modern products to duplicate the appearance of a historic roof, offers guidance on proper installation, and provides information on coatings and maintenance procedures to help preserve the new roof.

Wooden Shingle Roofs in America

Because trees were plentiful from the earliest settlement days, the use of wood for all aspects of construction is not surprising. Wooden shingles were lightweight, made with simple tools, and easily installed. Wooden shingle roofs were prevalent in the Colonies, while in Europe at the same time, thatch, slate, and tile were the prevalent roofing materials. Distinctive roofing patterns exist in various regions of the country that were settled by the English, Dutch, Germans, and Scandinavians. These patterns and features include the size, shape, and exposure length of shingles, special treatments such as swept valleys, combed ridges, and decorative butt end or long side-lapped beveled handsplit shingles. Such features impart a special character to each building, and prior to any restoration or rehabilitation project the physical and photographic evidence should be carefully researched in order to document the historic building as much as possible. Care should be taken not to assume that aged or deteriorated shingles in photographs represent the historic appearance.

Shingle Fabrication

Historically, wooden shingles were usually thin (⅜ inch to ¾ inch), relatively narrow (3 inches to 8 inches), of varying length (14 inches to 36 inches), and almost always smooth. The traditional method for making wooden shingles in the 17th and 18th centuries was to handsplit them from log sections known as bolts (Fig. 17–5A). These bolts were quartered or split into wedges. A mallet and froe (or ax) were used to split or rive out thin planks of wood along the grain. If a tapered shingle was desired, the bolt was flipped after each successive strike with the froe and mallet. The wood species varied according to available local woods, but only the heartwood, or inner section, of the log was usually used. The softer sapwood generally was not used because it deteriorated quickly. Because handsplit shingles were somewhat irregular along the split surface, it was necessary to dress or plane the shingles on a shavinghorse with a drawknife or drawshave (Fig. 17–5B) to make them fit evenly on the roof. This reworking was necessary to provide a tight-fitting roof over typically open shingle lath or sheathing boards. Dressing, or smoothing of shingles, was almost universal, no matter what wood was used or in what part of the country the building was located, except in those cases where a temporary or very utilitarian roof was needed.

Shingle fabrication was revolutionized in the early 19th century by steam-powered saw mills (Fig. 17–6). Shingle mills made possible the production of uniform shingles in mass quantities. The sawn shingle of uniform taper and smooth surface eliminated the need to hand dress. The supply of wooden shingles was therefore no longer limited by local factors. These changes coincided with (and in turn increased) the popularity of architectural styles such as Carpenter Gothic and Queen Anne that used shingle to great effect.

Handsplit shingles continued to be used in many places well after the introduction of machine sawn shingles. There were, of course, other popular roofing materials, and some regions rich in slate had fewer examples of wooden shingle roofs. Some western boom towns used sheet metal because it was light and easily shipped. Slate, terneplate, and clay tile were used on ornate buildings and in cities that limited the use of flammable wooden shingles. Wooden shingles, however, were never abandoned. Even in the 20th century, architectural styles such as the Colonial Revival and Tudor Revival used wooden shingles.

Modern wooden shingles, both sawn and split, continue to be made, but it is important to understand how these new products differ from the historic ones and to know how they can be modified for use on historic buildings. Modern commercially available shakes are generally thicker than the historic handsplit counterpart and are usually left "undressed" with a rough, corrugated surface. The rough surface shake, furthermore, is often promoted as suitable for historic preservation projects because of its rustic appearance. It is an erroneous assumption that the more irregular the shingle, the more authentic or historic it will appear.

Historic Detailing and Installation Techniques

While the size, shape and finish of the shingle determine the roof's texture and scale, the installation patterns and details give the roof its unique character. Many details reflect the craft practices of the builders and the architectural style prevalent at the time of construction. Other details had specific purposes for reducing moisture penetration to the structure. In addition to the most visible aspects of a shingle roof, the details at the rake boards, eaves, ridges, hips, dormers, cupolas, gables, and chimneys should not be overlooked.

The way the shingles were laid was often based on functional and practical needs. Because a roof is the most vulnerable element of a building, many of the roofing details that have become distinctive features were first developed simply to keep water out. Roof combs on the

Fig. 17–5A. Custom handsplit shingles are still made the traditional way with a mallet and froe or ax. For these cypress shingles, a "bolt" section of log the length of the shingle has been sawn and is ready to be split into wedge-shaped segments. Handsplit shingles are fabricated with the ax or froe cutting the wood along the grain and separating, or riving, the shingle away from the remaining wedge. Note the long wooden shingles covering the work shed. *(Photos: Al Honeycutt, North Carolina Division of Archives and History.)*

Fig. 17–5B. The rough handsplit surfaces are dressed on a shavinghorse using a drawknife. *(Photos: Al Honeycutt, North Carolina Division of Archives and History.)*

windward side of a roof protect the ridge line. Wedges, or cant strips, at dormer cheeks roll the water away from the vertical wall. Swept valleys and fanned hips keep the grain of the wood in the shingle parallel to the angle of the building joint to aid water run-off. The slight projec-

Fig. 17–6A. Modern machine-made shingles are sawn. Eastern White Pine quarter split shingle block on equalizer saw is trimmed to parallel the ends. The thickness and taper can be precisely controlled. *(Photo: Steve Ruscio, The Shingle Mill.)*

tion of the shingles at the eaves directs the water run-off either into a gutter or off the roof away from the exterior wall. These details varied from region to region and from style to style. They can be duplicated even with the added protection of modern flashing.

In order to have a weather-tight roof, it was important to have adequate coverage, proper spacing of shingles, and straight grain shingles. Many roofs were laid on open shingle lath or open sheathing boards (Fig. 17–7). Roofers typically laid three layers of shingles with approximately ⅓ of each shingle exposed to the weather. Spaces between shingles (⅛ inches to ½ inch depending on wood type) allowed the shingles to expand when wet. It was important to stagger each overlapping shingle by a minimum of 1½ inches to avoid a direct path for moisture to penetrate a joint. Doubling or tripling the starter course at the eave gave added protection to this exposed surface. In order for the roof to lay as flat as possible, the thickness, taper and surface of the

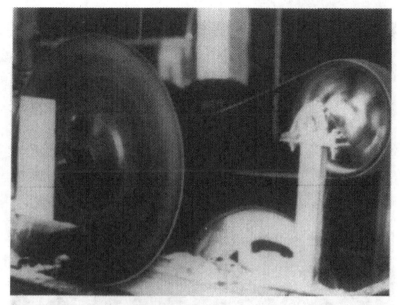

Fig. 17–6B. The restored 19th-century shingle mill saw cuts tapered flitches or shingles. The thickness and taper can be precisely controlled. *(Photo: Steve Ruscio, The Shingle Mill.)*

shingles was relatively uniform; any unevenness on handsplit shingles had already been smoothed away with a drawknife. To keep shingles from curling or cupping, the shingle width was generally limited to less than 10 inches.

Not all shingles were laid in evenly spaced, overlapping, horizontal rows. In various regions of the country, there were distinct installation patterns; for example, the biaxially-tapered long shingles occasionally found in areas settled by the Germans (Fig. 17–8). These long shingles were overlapped on the side as well as on top. This formed a ventilation channel under the shingles that aided drying. Because ventilation of the shingles can prolong their life, roofers paid attention to these details (Fig. 17–9).

Early roofers believed that applied coatings would protect the wood and prolong the life of the roof. In many cases they did; but in many cases, the shingles were left to weather naturally and they, too, had a long life. Eighteenth-century coatings included a pine pitch coat-

Fig. 17–7. The reshingling of the circa 1856 Stovewood House in Decorah, Iowa, revealed the original open sheathing boards and pole rafters. Sawn cedar shingles were used as a replacement for the historic cedar shingles seen still in place at the ridge. A new starter course is being laid at the eaves. *(Photo: Norwegian-American Museum, Decorah, Iowa.)*.

ing not unlike turpentine, and boiled linseed oil or fish oil mixed with oxides, red lead, brick dust, or other minerals to produce colors such as yellow, Venetian red, Spanish brown, and slate grey. In the 19th century, in addition to the earlier colors, shingles were stained or painted to complement the building colors: Indian red, chocolate brown, or brown-green. During the Greek Revival and later in the 20th century with other revival styles, green was also used. Untreated shingles age to a silver-grey or soft brown depending on the wood species.

The craft traditions of the builders often played an important role in the final appearance of the building. The Historic Details and Installation Patterns Chart (Fig. 17–10) identifies many of the features found on historic wooden roofs. These elements, different on each building, should be preserved in a re-roofing project.

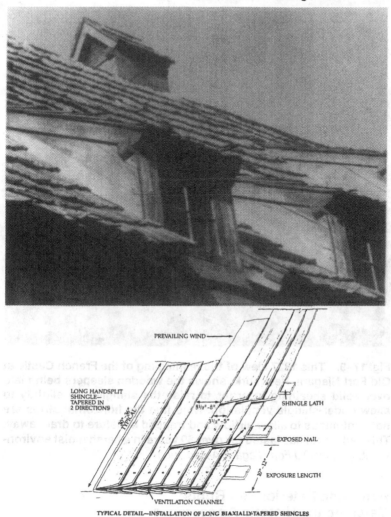

TYPICAL DETAIL—INSTALLATION OF LONG BIAXIALLY-TAPERED SHINGLES

Fig. 17–8. The long biaxially tapered handsplit shingles on the Ephrata Cloisters in Pennsylvania were overlapped both vertically and horizontally. The insert sketch shows channels under the shingles that provided ventilation and drainage of any trapped moisture. The aged appearance of these handsplit and dressed shingles belies their original smoothness. Replacement shingles should match the original, not the aged appearance. *(Photo: National Park Service; Sketch: Reed Engle.)*

Fig. 17–9. This 1927 view of the reshingling of the French Castle at Old Fort Niagara, New York, shows the wooden sleepers being laid over solid sheathing in order to raise the shingles up slightly to allow under-shingle ventilation. Note that the horizontal strips are not continuous to allow airflow and trapped moisture to drain away. This cedar roof has lasted for over 60 years in a harsh moist environment. *(Photo: Old Fort Niagara Assoc. Inc.)*

Replacing Deteriorated Roofs: Matching the Historic Appearance

Historic wooden roofs using straight edgegrain heartwood shingles have been known to last over 60 years. Fifteen to thirty years, however, is a more realistic lifespan for most premium modern wooden shingle roofs. Contributing factors to deterioration include the thinness of the shingle, the durability of the wood species used, the exposure to the sun, the slope of the roof, the presence of lichens or moss growing on the shingle, poor ventilation levels under the shingle or in

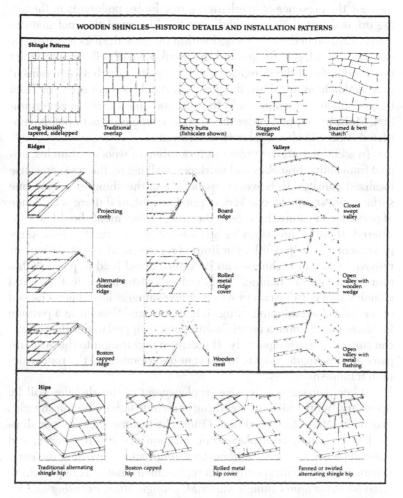

WOODEN SHINGLES—HISTORIC DETAILS AND INSTALLATION PATTERNS

Shingle Patterns

Long biaxially-tapered, sidelapped

Traditional overlap

Fancy butts (fishscales shown)

Staggered overlap

Steamed & bent "thatch"

Ridges

Projecting comb

Alternating closed ridge

Boston capped ridge

Board ridge

Rolled metal ridge cover

Wooden crest

Valleys

Closed swept valley

Open valley with wooden wedge

Open valley with metal flashing

Hips

Traditional alternating shingle hip

Boston capped hip

Rolled metal hip cover

Fanned or swirled alternating shingle hip

Fig. 17–10. The Historic Details and Installation Patterns Chart illustrates a number of special features found on wooden roofs. Documented examples of these features, different for every building and often reflecting regional variations, should be accurately reproduced when a replacement roof is installed. *(Chart: Sharon C. Park; delineation by Kaye Ellen Simonson.)*

the roof, the presence of overhanging tree limbs, pollutants in the air, the original installation method, and the history of the roof maintenance. Erosion of the softer wood within the growth rings is caused by rainwater, wind, grit, fungus and the breakdown of cells by ultraviolet rays in sunlight. If the shingles cannot adequately dry between rains, if moss and lichens are allowed to grow, or if debris is not removed from the roof, moisture will be held in the wood and accelerate deterioration. Moisture trapped under the shingle, condensation, or poorly ventilated attics will also accelerate deterioration.

In addition to the eventual deterioration of wooden shingles, impact from falling branches and workmen walking on the roof can cause localized damage. If, however, over 20% of the shingles on any one surface appear eroded, cracked, cupped or split, or if there is evidence of pervasive moisture damage in the attic, replacement should be considered. If only a few shingles are missing or damaged, selective replacement may be possible. For limited replacement, the old shingle is removed and a new shingle can be inserted and held in place with a thin metal tab, or "babbie." This reduces disturbance to the sound shingles above. In instances where a few shingles have been cracked or the joint of overlapping shingles is aligned and thus forms a passage for water penetration, a metal flashing piece slipped under the shingle can stop moisture temporarily. If moisture is getting into the attic, repairs must be made quickly to prevent deterioration of the roof structural framing members.

When damage is extensive, replacement of the shingles will be necessary, but the historic sheathing or shingle lath under the shingles may be in satisfactory condition. Often, the historic sheathing or shingle laths, by their size, placement, location of early nail holes, and water stain marks, can give important information regarding the early shingles used. Before specifying a replacement roof, it is important to *establish the original shingle material, configuration, detailing and installation* (Fig. 17–11). If the historic shingles are still in place, it is best to remove several to determine the size, shape, exposure length, and special features from the unweathered portions. If there are already replacement shingles on the roof, it may be necessary to verify through photographic or other research whether the shingles currently on the roof were an accurate replacement of the historic shingles.

The following information is needed in order to develop accurate specifications for a replacement shingle:

Fig. 17–11. The replacement sawn red cedar shingles matched the deteriorated shingles exactly for this barn re-roofing. The old shingles, seen to the far left, were removed as the new shingles were installed. Even the horizontal coursing matched because the exposure length for both old and new shingles was the same. *(Photo: Williamsport Preservation Training Center.)*

Original wood type (White Oak, Cypress, Eastern White Pine, Western Red Cedar, etc.)

Size of shingle (length, width, butt thickness, taper)

Exposure length and nailing pattern (amount of exposure, placement and type of nails)

Type of fabrication (sawn, handsplit, dressed, beveled, etc.)

Distinctive details (hips, ridges, valleys, dormers, etc.)

Decorative elements (trimmed butts, variety of pattern, applied color coatings, exposed nails)

Type of substrate (open shingle lath or sheathing, closed sheathing, insulated attics, sleepers, etc.)

Replacement roofs must comply with local codes which may require, for example, the use of shingles treated with chemicals or pressure-impregnated salts to retard fire. These requirements can usually be met without long-term visual effects on the appearance of the replacement roof.

The accurate duplication of a wooden shingle roof will help ensure the preservation of the building's architectural integrity. Unfortunately, the choice of an inappropriate shingle or poor installation can severely detract from the building's historic appearance (Fig. 17–12). There are a number of commercially-available wooden roofing products as well as custom roofers who can supply specially-made shingles for historic preservation projects (see Shingle and Shake Chart, Fig. 17–13). Unless restoration or reconstruction is being undertaken, shingles that match the visual appearance of the historic roof without replicating every aspect of the original shingles will normally suffice. For example, if the historic wood species is no longer readily available, western red cedar or eastern white cedar may be acceptable. Or, if the shingles are located high on a roof, sawn shingles or commercially available shakes with the rustic faces planed off may adequately reproduce the appearance of an historic handsplit and dressed shingle.

There will always be certain features, however, that are so critical to the building's character that they should be accurately reproduced. Following is guidance on matching the most important visual elements.

Highest Priority in Replacement Shingles:

- best quality wood with a similar surface texture
- matching size and shape: thickness, width, length
- matching installation pattern: exposure length, overlap, hips, ridges, valleys, etc.
- matching decorative features: fancy butts, colors, exposed nails

Areas of Acceptable Differences:

- species of wood
- method of fabrication of shingle, if visual appearance matches
- use of fire-retardants, or preservative treatments, if visual impact is minimal

Fig. 17–12. Inappropriately selected and installed wooden shingles can drastically alter the historic character of a building. This tavern historically was roofed with handsplit and dressed shingles of a relatively smooth appearance. In this case, a commercially-available shake was used to effect a "rustic" appearance. *(Photo: National Park Service.)*

- use of modern flashing, if sensitively installed
- use of small sleepers for ventilation, if the visual impact is minimal and rake boards are sensitively treated
- method of nailing, if the visual pattern matches

Treatments and Materials to Avoid:

- highly textured wood surfaces and irregular butt ends, unless documented
- standardized details (prefab hips, ridges, panels, etc.) unless documented
- too wide shingles or those with flat grain (which may curl), unless documented

AVAILABLE WOODEN SHINGLES AND SHAKES FOR RE-ROOFING				
TYPE		SIZE	DESCRIPTION	NOTES
Custom split & dressed	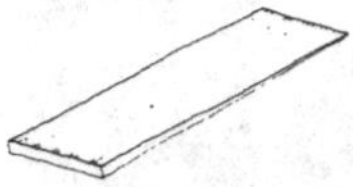	Made to match historic shingles	Handsplit the traditional way with froe & mallet. Tapered. Surfaces dressed for smoothness	Appropriate if: • Worked to match uniformly dressed original shingles
Tapersplit*	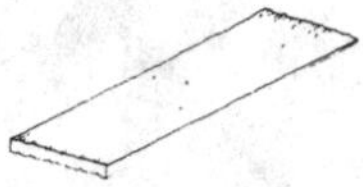	Typically: L = 15", 18", 24" W = 4"-14" Butts vary 1/2"-3/4"	Commercially available. Handsplit the traditional way with froe & mallet. Tapered. Bundles contain varying widths & butt thicknesses. Surfaces may be irregular along grain.	Appropriate if: • irregular surfaces are dressed • butt thicknesses ordered uniform • wide shingles are split
Straightsplit	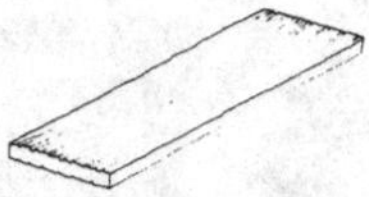	Typically: L = 15", 18", 24" W = 4"-14" Butts vary mediums = 3/8-3/4" heavies = 3/4-1¼"	Commercially available. Hand or machine split without taper. Bundles contain varying butt thicknesses; often very wide shingles. Surface may be irregular along the grain. Thick shingles not historic.	Not appropriate for most preservation projects • Limited use of thin, even straightsplits on some cabins, barns, etc.
Handsplit* resawn	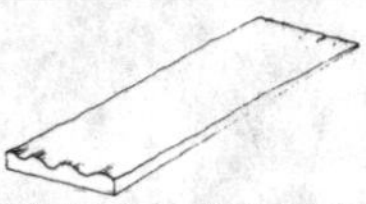	Typically: L = 15", 18", 24" W = 4"-14" Butts vary mediums = 3/8-3/4" heavies = 3/4-1¼"	Commercially available. Machine split and sawn on the backs to taper. Split faces often irregular, even corrugated in appearance. Butt thickness vary and may be too wide.	Not appropriate for preservation projects
Tapersawn*	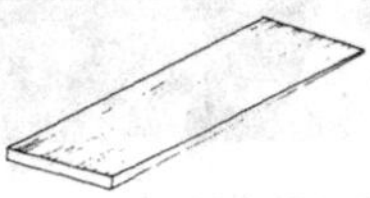	Typically: L = 15", 18", 24" W = 4"-14" Butts vary 1/2"-3/4"	Commercially available. Made from split products with sawn surfaces. Tapered. Butt thicknesses vary and shingles may be too wide. Saw marks may be pronounced.	Appropriate if: • butt thicknesses ordered uniform • wide shingles are split • pronounced saw marks sanded
Sawn-straight butt		Typically: L = 16"-.40 (<3/8") 18"-.45 24"-.50 (1/2") W = Varies by order	Custom or commercially available. Tapered. Sawn by circular saw.	Appropriate to reproduce historic sawn shingles
Sawn-fancy butt		Typically: L = 16"-.40 (<3/8") 18"-.45 24"-.50 (1/2") W = Varies by order	Custom or commercially available. Tapered. Sawn by circular saw. A variety of fancy butts available	Appropriate to reproduce historic fancy butts
Steam-bent		Varies by order to match. "Thatch" roofs	Custom or commercially available. Tapered. Thin sawn shingles are steamed and bent into rounded forms.	Appropriate to reproduce "thatch" shingles

Fig. 17–13. This chart identifies a variety of shingles and shakes used for reroofing buildings. The * identifies product names used by the Cedar Shake and Shingle Bureau, although shingles and shakes of the types described are available in other woods. Manufacturers define "shakes" as split products while "shingles" refer to sawn products. Shingle, however, is the historic term used to describe wooden roofing products, regardless of how they were made. Whether shingles or shakes are specified for reroofing, they should match the size and appearance of the historic shingles. *(Chart: Sharon C. Park; delineation by Kaye Ellen Simonson.)*

What is Currently Available

Types of Wood—Western red cedar, eastern white cedar, and white oak are most readily available today. For custom orders, cypress, red oak, and a number of other historically used woods may still be available. Some experiments using non-traditional woods (such as yellow pine and hemlock) treated with preservative chemicals are being tested for the new construction market, but are generally too thick, curl too easily, or have too pronounced a grain for use on historic buildings.

Method of Manufacture—Commercially available modern shingles and shakes are for the most part machine-made. While commercially available shakes are promoted by the industry as handsplit, most are split by machine (this reduces the high cost of hand labor). True handsplit shingles, made the traditional way with a froe and mallet, are substantially more expensive, but are more authentic in appearance than the rough, highly textured machine-split shakes. An experienced shingler can control the thickness of the handsplit shingle and keep the shingle surface grain relatively even. To have an even roof installation, it is important to have handsplit shingles of uniform taper and to have less than 1/8th variation across the surface of the shingle. For that reason, it is important to dress the shingles or to specify uniform butt thickness, taper, and surfaces. Commercially available shakes are shipped with a range of butt sizes within a bundle (e.g., ½ inch, ⅝ inch, ¾ inch as a mix) unless otherwise specified. Commercially available shakes with the irregular surfaces sawn off are also available. In many cases, except for the residual circular saw marks, these products appear not unlike a dressed handsplit shingle.

Sawn shingles are still made much the same way as they were historically—using a circular saw. The circular saw marks are usually evident on the surface of most sawn shingles. There are a number of grooved, striated, or steamed shingles of the type used in the 20th century to effect a rustic or thatched appearance. Custom sawn shingles with fancy butts or of a specified thickness are still available through mill shops. In fact, shingles can be fabricated to the weathered thickness in order to be integrated into an existing historic roof. If sawn shingles are being used as a substitute for dressed handsplit shingles, it may be desirable to belt sand or plane the surface of the sawn shingles to reduce the prominence of the circular saw marks.

As seen from the Shingle and Shake chart, few of the commercially available shakes can be used without some modification or careful specification. Some, such as heavy shakes with a corrugated face, should be avoided altogether. While length, width, and butt configuration can be specified, it is more difficult to ensure that the thickness and the texture will be correct. For that reason, whatever shingle or shake is desired, it is important to view samples, preferably an entire bundle, before specifying or ordering. If shingles are to be trimmed at the site for special conditions, such as fanned hips or swept valleys, additional shingles should be ordered.

Coatings and Treatments—Shingles are treated to obtain a fire-retardant rating; to add a fungicide preservative (generally toxic); to revitalize the wood with a penetrating stain (oil as well as water-based); and to give color.

While shingles can be left untreated, local codes may require that only fire-retardant shingles be used. In those circumstances, there are several methods of obtaining rated shingles (generally class "B" or "C"). The most effective and longest-lasting treatment is to have treated salts pressure-impregnated into the wood cells after the shingles have been cut. Another method (which must be periodically renewed) is to apply chemicals to the surface of the shingles. If treated shingles need trimming at the site, it is important to check with the manufacturer to ensure that the fire-retardant qualities will not be lost. Pressure-impregnated shingles, however, may usually be trimmed without loss of fire-retardant properties.

The life of a shingle roof can be drastically shortened if moss, lichens, fungi, or bacterial spores grow on the wood. Fungicides (such as chromated copper arsenate, CCA) have been found to be effective in inhibiting such fungal growth, but most are toxic. Red cedar has a natural fungicide in the wood cells and unless the shingles are used in unusually warm, moist environments, or where certain strains of spores are found, an applied fungicide is usually not needed. For most woods, the Forest Products Laboratory of the U.S. Department of Agriculture has found that fungicides do extend the life of the shingles by inhibiting growth on or in the wood. There are a variety available. Care should be taken in applying these chemicals and meeting local code requirements for proper handling.

Penetrating stains and water repellent sealers are sometimes recommended to revitalize wood shingles subject to damage by ultraviolet rays. Some treatments are oil-borne, some are water-borne, and some are combined with a fungicide or a water repellent. If any of these treatments is to be used, they should be identified as part of the specifications. Manufacturers should be consulted regarding the toxicity or other potential complications arising from the use of a product or of several in combination. It is also important not to coat the shingles with vapor-impermeable solutions that will trap moisture within the shingle and cause rotting from beneath.

Specifications for the Replacement Roof

Specifications and roofing details should be developed for each project. Standard specifications may be used as a basic format, but they should be modified to reflect the conditions of each job. Custom shingles can still be ordered that accurately replicate a historic roof, and if the roof is simple, an experienced shingler could install it without complicated instructions. Most rehabilitation projects will involve competitive bidding, and each contractor should be given very specific information as to what type of shingles are required and what the installation details should be. For that reason, both written specifications and detailed drawings should be part of the construction documents.

For particularly complex jobs, it may be appropriate to indicate that only roofing contractors with experience in historic preservation projects be considered (Fig. 17–14). By pre-qualifying the bidders, there is greater assurance that a proper job will be done. For smaller jobs, it is always recommended that the owner or architect find a roofing contractor who has recently completed a similar project and that the roofers are similarly experienced.

Specifications identify exactly what is to be received from the supplier, including the wooden shingles, nails, flashing, and applied coatings (Fig. 17–15). The specifications also include instructions on removing the old roofing (sometimes two or more earlier roofs), and on preparing the surface for the new shingles, such as repairing damage to the lath or sheathing boards. If there are to be modifications to a standard product, such as cutting beveled butts, planing off residual surface circular saw marks, or controlling the mixture of acceptable widths (3 inches to 8 inches), these too should be specified. Every instruction for modifying the shingles themselves should be written into the specifications or they may be overlooked.

Fig. 17–14A. The later non-historic shingles were removed from Appomattox Manor (circa 1840 with later additions) and roofing paper was installed for temporary protection during the re-shingling. *(Photos: John Ingle)*

Fig. 17–14B. These weathered historic 19th-century handsplit and dressed shingles were found in place under a later altered roof. Note the straight butt eave shingles under the curved butts of the historic dormer shingles.

Fig. 17–14C. The replacement shingles (see Fig. 17–15) matched the historic shingles and were of such high quality that little hand dressing was needed at the site. The building paper, a temporary protection, was removed as the shingles were installed on the sheathing boards.

The specifications and drawn details should describe special features important to the roof. Swept valleys, combed ridges, or wedged dormer cheek run-offs should each be detailed not only with the patterning of the shingles, but also with the placement of flashing or other unseen reinforcements. There are some modern products that appear to be useful. For example, papercoated and reinforced metal-laminated flashing is easy to use and, in combination with other flashing, gives added protection over eaves and other vulnerable areas; adhesives give a stronger attachment at projecting roofing combs that could blow away in heavy wind storms. Light-colored sealants may be less obvious than dark mastic often used in conjunction with flashing or repairs. These modern treatments should not be overlooked if they can prolong the life of the roof without changing its appearance.

Fig. 17–14D. The fanned hips (seen here), swept valleys, and projecting ridge combs were installed as part of the re-roofing project. Special features, when documented, should be reproduced when reshingling historic roofs.

Case Study

This historic roofing project (Fig. 17–14) called for the replacement of the roof on Appomattox Manor, Hopewell, Virginia. The original 19th-century handsplit and dressed wooden shingles 18 inches long, 3 inches–4 inches wide, and ⅝ inches thick were found in place. The butts were curved and evidence of a red stain remained. The specifications (Fig. 17–15) and details were researched so that the appearance of the historic shingles and installation patterns could be matched in the re-shingling project.

Roofing Practices to Avoid

Certain common roofing practices for modern installations should be avoided in re-roofing a historic building unless specifically ap-

Fig. 17–14E. In order to achieve a Class B fire-rating, the shingles were dipped in fire-retardant chemicals and allowed to dry prior to installation. Iron oxide was added to this chemical dip to stain the shingles to match the historic red color. These coatings will need periodic reapplication.

proved in advance by the architect. These practices interfere with the proper drying of the shingles or result in a sloppy installation that will accelerate deterioration (Fig. 17–16). They include improper coverage and spacing of shingles, use of staples to hold shingles, inadequate ventilation, particularly for heavily insulated attics, use of heavy building felts as an underlayment, improper application of surface coatings causing stress in the wood surfaces, and use of inferior flashing that will fail while the shingles are still in good condition.

Avoid skimpy shingle coverage and heavy building papers. It has become a common modern practice to lay impregnated roofing felts under new wooden shingle roofs. The practice is especially prevalent in roofs that do not achieve a full triple layering of shingles. Historically, approximately one third of each shingle was exposed, thus making a three-ply or three-layered roof. This assured adequate coverage. Due to the expense of wooden shingles today, some roofers expose more of the shingle if the pitch of the roof allows, and compensate for

---

Type of wood to be used: Western Red Cedar.
Grade of wood and manufacturing process: Number One, Tapersplit Shakes, 100% clear, 100% edgegrain, 100% heartwood, no excessive grain sweeps, curvatures not to exceed ½" from level plain in length of shake; off grade (7% tolerance) material must *not* be used.
Size of the shingle: 18" long, ⅝" butt tapered to ¼" head, 3"–4" wide, sawn curved butts, 5-½" exposure.
Surface finish and any applied coatings: relatively smooth natural grain, no more than ⅛" variation in surface texture, butt thickness to be uniform throughout bundles. Site dipped with fire-rated chemicals tinted with red iron oxide for opaque color.
Type of nails and flashing: double hot dipped galvanized nails sized to penetrate sheathing totally; metal flashing to be 20 oz. lead-coated copper, or terne-coated stainless steel; additional flashing reinforcement to be aluminum foil type with fiber backing to use at hips, ridges, eaves, and valleys.
Types of sheathing: uninsulated attic, any deteriorated ¾" sheathing boards, spaced ½"-¾", to be replaced in kind.

Fig. 17–15. Excerpts from Specifications.

less than three layers of shingles by using building felts interwoven at the top of each row of shingles. This absorptive material can hold moisture on the underside of the shingles and accelerate deterioration. If a shingle roof has proper coverage and proper flashing, such felts are unnecessary as a general rule. However, the selective use of such felts or other reinforcements at ridges, hips, and valleys does appear to be beneficial.

Beware of heavily insulated attic rafters. Historically, the longest lasting shingle roofs were generally the ones with the best roof ventilation. Roofs with shingling set directly on solid sheathing and where there is insulation packed tightly between the wooden rafters without adequate ventilation run the risk of condensation-related moisture damage to wooden roofing components. This is particularly true for air-conditioned structures. For that reason, if insulation must be used, it is

Fig. 17–16. These commercially available roofing products with rustic split faces are not appropriate for historic preservation projects. In addition to the inaccurate appearance, the irregular surfaces and often wide spaces between shingles will allow wind-driven moisture to penetrate up and under them. The excessively wide boards will tend to cup, curl, and crack. Moss, lichens, and debris will have a tendency to collect on these irregular surfaces, further deteriorating the roofing. *(Photo: Sharon C. Park.)*

best to provide ventilation channels between the rafters and the roof decking, to avoid heavy felt building papers, to consider the use of vapor barriers, and perhaps to raise the shingles slightly by using "sleepers" over the roof deck. This practice was popular in the 1920s in what the industry called a "Hollywood" installation, and examples of roofs lasting 60 years are partly due to this under-shingle ventilation (Fig. 17–9).

Avoid staples and inferior flashing. The common practice of using pneumatic staple guns to affix shingles can result in shooting staples through the shingles, in crushing the wood fibers, or in cracking the shingle. Instead, corrosion-resistant nails, generally with barked or de-formed shanks long enough to extend about ¾ inch into the roof deck-ing, should be specified. Many good roofers have found that the pneu-matic nail guns, fitted with the proper nails and set at the correct pressure with the nails just at the shingle surface, have worked well and reduced the stress on shingles from missed hammer blows. If red cedar is used, copper nails should not be specified because a chemical reaction between the wood and the copper will reduce the life of the roof. Hot-dipped zinc-coated, aluminum, or stainless steel nails should be used. In addition, copper flashing and gutters generally should not be used with red cedar shingles as staining will occur, although there are some historic examples where very heavy gauge copper was used which outlasted the roof shingles. Heavier weight flashing (20 oz.) holds up better than lighter flashing, which may deteriorate faster than the shingles. Some metals may react with salts or chemicals used to treat the shingles. This should be kept in mind when selecting treat-ments. Terne-coated stainless steel and lead-coated copper are gener-ally the top of the line if copper is not appropriate.

Avoid patching deteriorated roof lath or sheathing with plywood or composite materials. Full size lumber may have to be custom-ordered to match the size and configuration of the original sheathing in order to provide an even surface for the new shingles. It is best to avoid plywood or other modern composition boards that may deteriorate or delaminate in the future if there is undetected moisture or leakage. If large quantities of shingle lath or sheathing must be removed and re-placed, the work should be done in sections to avoid possible shifting or collapse of the roof structure.

Avoid spray painting raw shingles on a roof after installation. Rapidly-drying solvent in the paint will tend to warp the exposed sur-face of the shingles. Instead, it is best to dip new shingles prior to in-stallation to keep both sides in the same tension. Once the entire shin-gle has been treated, however, later coats can be limited to the exposed surface.

Maintenance

The purpose of regular or routine maintenance is to extend the life of the roof. The roof must be kept clean and inspected for damage both to the shingles and to the flashing, sheathing, and gutters. If the roof is to be walked on, rubber-soled shoes should be worn. If there is a simple ridge, a ladder can be hooked over the roof ridge to support and distribute the weight of the inspector.

Keeping the roof free of debris is important. This may involve only sweeping off pine needles, leaves and branches as needed. It may involve trimming overhanging branches. Other aspects of maintenance, such as removal of moss and lichen build-up, are more difficult. While they may impart a certain charm to roofs, these moisture-trapping organisms will rot the shingles and shorten the life of the roof. Buildups may need scraping and the residue removed with diluted bleaching solutions (chlorine), although caution should be used for surrounding materials and plants. Some roofers recommend power washing the roofs periodically to remove the dead wood cells and accumulated debris. While this makes the roof look relatively new, it can put a lot of water under shingles, and the high pressure may crack or otherwise damage them. The added water may also leach out applied coatings.

If the roof has been treated with a fungicide, stain, or revitalizing oil, it will need to be re-coated every few years (usually every 4–5). The manufacturer should be consulted as to the effective life of the coating. With the expense associated with installation of wood shingles, it is best to extend the life of the roof as long as possible. One practical method is to order enough shingles in the beginning to use for periodic repairs.

Periodic maintenance inspections of the roof may reveal loose or damaged shingles that can be selectively replaced before serious moisture damage occurs (Fig. 17–17). Keeping the wooden shingles in good condition and repairing the roof, flashing and guttering, as needed, can add years of life to the roof.

Conclusion

A combination of careful research to determine the historic appearance of the roof, good specifications, and installation details designed to match the historic roof, and long-term maintenance, will make it possible to have not only a historically authentic roof, but a cost-effective one. It is important that professionals be part of the team from the beginning. A preservation architect should specify materials

Fig. 17–17. Routine maintenance is necessary to extend the life of the roof. On this roof, the shingles have not seriously eroded, but the presence of lichens and moss is becoming evident and there are a few cracked and missing shingles. The moss spores should be removed, missing shingles replaced, and small pieces of metal flashing slipped under cracked shingles to keep moisture from penetrating. *(Photo: Williamsport Preservation Training Center)*

and construction techniques that will best preserve the roof's historic appearance. The shingle supplier must ensure that the best product is delivered and must stand behind the guarantee if the shipment is not correct. The roofer must be knowledgeable about traditional craft practices. Once the new shingle roof is in place, it must be properly maintained to give years of service.

References

Associations

American Plywood Association

> PO Box 11700
> Tacoma, WA 98411
> 206-565-6600
>
> Publishes the *Comprehensive Guide to Panel Construction Systems for both Residential and Commerical/industrial Buildings.*

Cedar Shake and Shingle Bureau

> 515 116th Avenue, NE, Suite 275
> Bellevue, WA 98004
> 206-453-1321
>
> Develops and distributes design, technical, and application information that applies to cedar roofing, including the *Design and Application Manual for New Roof Construction.*

National Roofing Contractors Association

> O'Hare International Center
> 10255 W. Higgins Road, Suite 600
> Rosemont, Il 60018-5607

708-299-9070, 800-872-7663
800-323-9545 (marketing for publications)

Association for contractors applying asphalt, cool tar pitch, elasto-plastic, slate, tile, metal and wood roofs. Monthly magazine *Professional Roofing*, wide selection of technical publications and installation manuals, publications on safety, insurance, legal issues, industry surveys and business forecasts, videos, software, sample contracts and homeowner information. Some publications available in Spanish.

Roof Consultants Institute

7424 Chapel Hill Road
Raleigh, NC 27607
919-859-0742

International association of roof consultants, 400 members in US and Canada. Includes architects, engineers, and manufacturers. See publications section.

Roofing Industry Educational Institute

Bldg. H. Ste. 110
14 Inverness Drive, East
Englewood, CO 80112
303-790-7200

Offers maintenance manual for commercial roofing. Main slant is education, offers seminars, certification program, information newsletter $12/year. Main slant is commercial roofing, but offers seminars on Steep Roofs also.

Scaffold Industry Association

14039 Sherman Way
Van Nuys, CA 91405-2599
818-782-2012

Publishes *Scaffold Industry Association Newsletter,* as well as informational manuals.

Manufacturers

Roofmaster Products Co.

750 Monterey Pass Road
Monterey Park, CA 91754-3668
213-261-5122

Manufacturers and distributors of roofing equipment, tools and accessories; including safety equipment, ladders, hand tools.

Evergreen Slate Company, Inc.

68 Potter Avenue
Granville, NY 12832
518-642-2530

Slate shingles.

FibreChem Corporation

11000-1 South Commerce Boulevard
P.O. Box 411368
Charlotte, NC 28241
800-346-6147

Fiber-cement shingles, non-asbestos, simulated slate roofing shingles.

Follansbee Steel

PO Box 610
State Street
Follansbee, WV 26037
800-624-6906

Product information and installation instructions for terne roofs. Also makes flashings.

Genova Products, Inc.

7034 E. Court Street
Davison, MI 48423
313-744-4500

Vinyl gutters, roof vents, roof flashing.

Historic Oak Roofing

PO Box 3121
Johnson City, TN 37602
615-282-5429
800-321-3781

Split, smooth, and handsplit oak shingles and shakes. Uses long-lasting white oak exclusively.

Vermont Structural Slate Company, Inc.

3 Prospect Street, Box 98
Fair Haven, VT 05743
802-265-4933

Slate shingles.

Publications

RSI (Roofing, Siding, Insulation)

Edgell Communications
One East First Street
Duluth, MN 55802
218 723-9362

This monthly magazine covers the latest technical and business topics for residential and commercial roofing.

Professional Roofing

O'Hare International Center
10255 W. Higgins Road, Suite 600
Rosemont, Il 60018-5607
708-299-9070, 800-872-7663
800-323-9545 (marketing for publications)

Magazine published monthly by National Roofing Contractors Association. Technical articles, questions and answers, new products.

A Roof Cutter's Secret

by Will Hollady, 180 pages, softcover from

Builders' Resources
RR 2, Box 146
Richmond, VT 05477

Advanced roof framing guide written for experienced framers.

Journal of Light Construction

PO Box 686
Holmes, PA 10943
800-345-8112

This monthly journal carries occasional articles on roofing and has an annual roofing issue.

Fine Homebuilding

Taunton Press
Subscription Dept.
635 Main Street Box 5506
Newtown, CT 06470-5506
800-888-8286

This monthly magazine has occasional articles on roofing topics.

OSHA Construction Industry Standards, Part 1926

Occupational Safety and Healthy Administration

Available from:
U.S. Government Printing Office
Superintendent of Documents
U.S.G.P.O., Washington, DC 20402
202-783-3238

Practical Restoration Reports

RR1 Box 2947
Dept. G
Sanford, ME 04073

This series of technical reports on the restoration and preservation of historic buildings includes topics on gutters and roofing. Write to ask for a list of current titles.

Interface

7424 Chapel Hill Road
Raleigh, NC 27607
919-859-0742

Newsletter published bimonthly by Roof Consultants Institute. Technical articles, industry updates.

Research Organizations

USDA Forest Products Laboratory

1 Gifford Pinchot Drive
Madison, WI 53705
608-231-9200

The Forest Products Laboratory researches practically every aspect of wood products and their uses. Milling, fabrication, finishing and painting, and fire retarding treatments are just a few of the topics covered. The published results of their work is available to the public. Call or write with questions. Typically, a report or publication will be sent. Unusual questions are welcome but may require time to research.

Small Homes Council—Building Research Council

One East St. Mary's Road
Champaign, IL 61820
800-336-0616

For almost 40 years the SHC-BRC has been simplifying construction techniques and adapting large-scale building techniques to the needs of the small builder, designer, and remodeler. Current work focuses on moisture problems in attics, energy-efficient construction and roofing. The council has published books for carpenters and builders on such topics as insulation and restoration. Write or call for a free list of publications.

Index